Tectonic Geomorphology of Mountains

Tectonic Geomorphology of Mountains:

A New Approach to Paleoseismology

William B. Bull

Blackwell Publishing

BLACKWELL PUBLISHING
350 Main Street, Malden, MA 02148-5020, USA
9600 Garsington Road, Oxford OX4 2DQ, UK
550 Swanston Street, Carlton, Victoria 3053, Australia

First published 2007 by Blackwell Publishing Ltd

1   2007

Library of Congress Cataloging-in-Publication Data

Bull, William B., 1930–
    Tectonic geomorphology of mountains : a new approach to paleoseismology / William B. Bull.
        p. cm.
    Includes bibliographical references and index.
    ISBN-13: 978-1-4051-5479-6 (hardback : alk. paper)
    ISBN-10: 1-4051-5479-9 (hardback : alk. paper) 1. Morphotectonics. 2. Paleoseismology. I. Title.

    QE511.44.B85 2007
    551.43'2–dc22

                                        2006100890

A catalogue record for this title is available from the British Library.

Set in 10.74/11pt AGaramond
by SPi Publisher Services, Pondicherry, India
Printed and bound in Singapore
by C.O.S Printers Pte Ltd

For further information on
Blackwell Publishing, visit our website:
www.blackwellpublishing.com

Contents

Preface

Uplift by mountain-building forces changes fluvial landscapes. Pulsatory tectonic activity on a range-bounding fault increases relief, changes rates of geomorphic processes, and modifies the shapes of hills and streams. Landscape responses to uplift occupy a critical time frame for studies of past earthquakes between the brevity of instrumental seismic data and long-term geologic crustal shifts. The appealing challenge for us is to determine how and when nearby and distant parts of the landscape change in consecutive reaches upstream from a tectonically active range front. Each climatic and lithologic setting has a characteristic style and rate of erosion, which adds spice to the scientific challenge. Landscape analyses include the geomorphic consequences of seismic shaking and surface rupture and their associated hazards to human-kind. Tectonic geomorphology is essential for complete paleoseismology investigations. Locations, sizes, times, and patterns of seismic shaking by prehistorical earthquakes can be described and surface rupture and seismic-shaking hazards evaluated.

This book explores tectonic geomorphology of mountain fronts on many temporal and spatial scales to encourage expansion of paleoseismology inquiries from the present emphasis on stratigraphic investigations in trench exposures. Evaluating earthquake hazards is in part a study of mountain-front segments. Cumulative displacements over late Quaternary time spans create landscape assemblages with distinctive signatures that are functions of uplift rate, rock mass strength, and the geomorphic processes of erosion and deposition. Such interactions define classes of relative uplift. Tectonic activity class maps define tectonically inactive regions as well as fronts of slow to rapidly rising mountains. Fault scarps focus our attention on recent surface ruptures and propagation of active faults. Dating and describing the characteristics of single prehistoric surface-ruptures is important. But now we can link sequences of events and depict sequences of prehistorical earthquakes along complex plate boundary fault zones. Examples here include the Alpine fault in New Zealand and the northern Basin and Range Province in the United States.

This book applies a variety of geomorphic concepts to tectonics and paleoseismology. Don't expect landscape summaries for all major mountain ranges. Repetitive descriptions would dilute explanation and application of basic principles. Do expect essential concepts that should help you better understand the landscape evolution of your favorite mountains. Mountain front tectonic geomorphology studies can determine:
1) Which faults are active [Holocene ruptures],
2) Fault slip rates for short time spans [offset landforms] and long time spans [landscape evolution],
3) Time of most recent surface rupture and degree of irregularity of earthquake recurrence interval, and
4) Intensity and extent of seismic shaking.
The amount of related literature cited borders on being unwieldy because of topic diversity of and the rapidly increasing interest of earth scientists in these subjects. I had to pick and choose so as to not overwhelm the content with citations of relevant literature. My citations are merely a gateway to related literature.

Dating times of prehistoric earthquakes and estimating rates of tectonic and geomorphic processes continue to be of paramount importance. Study methods are changing, and precision and accuracy are improving. Diffusion-equation modeling of fault scarps and stratigraphic radiocarbon dates on pre- or post earthquake material collected from trenches have long been bastions for approximate age estimates. Sykes and Nishenko made a plea in 1984 for better ways of dating frequent earthquakes along plate boundary fault zones whose earthquake recurrence intervals may be shorter than the intervals defined by groups of overlapping radiocarbon age estimates. The rapid development of terrestrial cosmogenic nuclides broadens dating perspectives by estimating ages beyond the reach of radiocarbon analyses and by making surface-exposure dating a cornerstone for studies of geomorphic processes. Tree-ring analyses and lichenometry have potential for dating prehistorical earthquakes with a precision of ± 5 years.

Both methods are used here in a study of Alpine fault, New Zealand, earthquake history.

The subjects of the six chapters are wideranging. Acknowledging the scrunch and stretch horizontal components of bedrock uplift is assessed from a geomorphic standpoint in Chapter 1. Diverse, essential conceptual models and methods for fluvial tectonic geomorphology are presented in Chapter 2. Contrasting tectonic landforms and landscape evolution associated with thrust and normal faults are the focus of Chapter 3. Uplift, stream-channel downcutting, and piedmont aggradation are interrelated base-level processes that are used to define relative classes of mountain-front tectonic activity in Chapter 4. The fault scarps of Chapter 5 are incipient mountain fronts with surface-rupture recurrence intervals ranging from 200 years to 200,000 years. Chapter 6 considers how mountains crumble from seismic shaking. It uses coseismic rockfalls and tree-ring analyses for precise, accurate dating of earthquakes of the past 1,000 years and for mapping the intensity of seismic shaking of these prehistorical events.

Readers should know basic geologic principles as these essays are written for earth scientists and students of geomorphic processes, landscape evolution, and earthquake studies. This book is appropriate for upper division and graduate-level courses in active tectonics, geologic hazards, tectonic geomorphology, physical geography and geomorphology, engineering geology, and paleoseismology.

This project began in 1975 when Luna Leopold encouraged me to embark on selected in-depth geomorphic syntheses using book manuscripts as a career development tool. Global climate change and tectonic deformation are major factors influencing the behavior of fluvial geomorphic systems. Book goals determined my study emphases in a series of projects. "Geomorphic Responses to Climatic Change" (Bull, 1991) revealed pervasive impacts on geomorphic processes of arid and humid regions. This second book examines tectonic geomorphology of mountain ranges in a paleoseismology context.

Of course the varied content of this book is indeed a team effort by the earth-science community. Students in the Geosciences Department at the University of Arizona played essential roles in every chapter. Peter Knuepfer, Larry Mayer, Les McFadden, Dorothy Merritts, and Janet Slate were among the many who tested the conceptual models of Chapter 2 with field-based studies. The first true positive test of the fault segmentation model (Schwartz and Coppersmith, 1984) in Chapter 3 is the work of Kirk Vincent. Les McFadden and Chris Menges broke new ground with me for the Chapter 4 elucidation of tectonic activity classes of mountain fronts of the Mojave Desert and Transverse Ranges of southern California. Susanna Calvo, Oliver Chadwick, Karen Demsey, Julia Fonseca, Susan Hecker, Phil Pearthree, and Kirk Vincent helped define the essential aspects for studies of normal-fault scarps of the Basin and Range Province in a vast region stretching from Idaho into Mexico. Andrew Wells kindly provided fascinating details about the sensitivity of New Zealand coastal and fluvial landscapes to seismic shaking. The integration of geomorphic and structural features shown in the Figure 1.12 map is the work of Jarg Pettinga. Kurt Frankel and Mike Oskin shared results and concepts of work in progress and Figures 5.35–5.40.

The book project expanded in scope during a decade when a new lichenometry method was developed to date and describe how seismic shaking influences rockfalls and other landslides. Lichenometry projects included expeditions into the Southern Alps and Sierra Nevada with Fanchen Kong, Tom Moutoux, and Bill Phillips. Their careful fieldwork and willingness to express divergent opinions were essential ingredients for this paleoseismology breakthrough. I appreciate the assistance of John King in sampling and crossdating the annual growth rings of trees in Yosemite, and of Jim Brune's help in measuring lichen sizes near the Honey Lake fault zone. Jonathan Palmer introduced me to Oroko Swamp in New Zealand, which turned out to be a key dendroseismology site.

Images are essential for landscape analysis and portrayal. Tom Farr of the Jet Propulsion Laboratory of the California Institute of Technology always seemed to have time to help find the essential NASA and JPL images used here. The banner photo for Chapter 2 and Figure 4.14 are the artistry of Peter Kresan. I thank

Frank Pazagglia for Figure 2.4, Malcolm Clark for the Chapter 4 banner photo, Tom Rockwell for the Figure 5.28 image, Greg Berghoff for Figure 5.34, Scott Miller for Figure 6.2 and Eric Frost for Figure 6.9A.

Formal reviews of the entire book manuscript by Lewis Owen and Philip Owens provided numerous suggestions that greatly improved book organization and content. I am especially indebted to Wendy Langford for her meticulous proofreading and to Rosie Hayden for editorial suggestions. Their thoroughness improved format and uniformity of expression. It was a pleasure to work with the efficient production staff at Blackwell Publishing including Ian Francis, Rosie Hayden, and Delia Sandford.

Essential financial and logistical support for this work was supplied by the U.S. National Science Foundation, National Earthquake Hazards Reduction Program of the U.S. Geological Survey, National Geographic Society, University of Canterbury in New Zealand, Hebrew University of Jerusalem, Royal Swedish Academy of Sciences, and Cambridge University in the United Kingdom.

Chapter 1

Scrunch and Stretch Bedrock Uplift

Earthquakes! Active Tectonics! Evolution of Mountainous Landscapes! Landscapes have a fascinating story to tell us. Tectonic geomorphology intrigues laypersons needing practical information as well as scientists curious about Earth's history.

How fast are the mountains rising? When will the next large earthquake occur? Will the seismic shaking disrupt the infrastructures that we depend on? How do the landscapes surrounding us record mountain-building forces within the Earth's crust, and how does long-term erosion influence crustal processes? Humans are intrigued by tectonic geomorphology on scales that include origins of continents, grandeur of their favorite mountain range, and the active fault near their homes.

Let us expand on the purpose and scope summarized in the Preface by elaborating on the structure of this book. I introduce, describe and use geomorphic concepts to solve problems in tectonics and paleoseismology. The intended geographical focus is global application of examples from southwestern North America and New Zealand. A fluvial emphasis excludes glaciers, sand seas, and active volcanoes. I present data and analyses from diverse tectonic, climatic, and lithologic settings so you can resolve similar problems in other geographical settings.

This book emphasizes responses of fluvial systems to uplift, or more specifically the adjustments of geomorphic processes to base-level fall. Uplift terminology usage continues to change since the hallmark paper by Molnar and England (1990). Geomorphologists may use uplift terms in a different context than structural geologists. So Chapter 1 is a brief review of terminology and types of base-level change induced by tectonic deformation in extensional and contractional settings. Such crustal stretching and scrunching is nicely recorded by landforms ranging in size from mountain ranges to fault scarps.

A variety of useful geomorphic concepts are assembled in Chapter 2 instead of being scattered. Get familiar with these principles. This broad base of essential concepts lets you evaluate and explore new and diverse approaches in tectonic geomorphology. These include a sensitive erosional–depositional threshold, time lags of response to *perturbations* (changes in variables of a system), types of equilibrium (graded) conditions in stream

Photograph of 59,000 and 96,000 marine terraces (Ota et al., 1996) and 330,000 year old mountains (Bull, 1984, 1985) rising out of the sea at Kaikoura, New Zealand systems, local and ultimate base levels, and the process of tectonically induced downcutting to the base level of erosion. These guidelines are a foundation for understanding interrelations between tectonics and topography in the next three chapters.

Chapter 3 compares the landscape evolution and useful tectonic landforms for mountain ranges being raised by slip on active thrust and normal faults. These fluvial systems are affected differently by the two styles of tectonic base-level fall. Strike-slip faulting tends to tear drainage basins apart: a much different subject that is not emphasized here. Some tectonic landforms, like triangular facets, are rather similar in different tectonic settings. But piedmont landforms are much different in thrust- and normal-fault landscapes. Comparable contrasts should be expected elsewhere, such as the countries bordering the Mediterranean Sea, and Mongolia.

The next three chapters discuss tectonic geomorphology for three distinct time spans (Fig. 1.1) of about 2,000,000, 12,000, and 1,000 years. The tectonic-geomorphology theme continues to be applications for paleoseismology. The landscape tectonic

activity classes of Chapter 4 are based on universal geomorphic responses to different rates of base-level fall during the Quaternary time span. The resulting diagnostic landscape assemblages are defined and mapped for diverse tectonic and structural settings in California. This model could have been created, and applied, just as easily for suites of mountain fronts in Japan, China, Mongolia, and Russia.

Fault scarps are the focus of Chapter 5, with an emphasis on the Holocene time span. Choosing to discuss recent surface ruptures in southwestern North America was done in part to hold variations of several controlling factors to a limited range. These include climate and alluvium mass strength. Such studies of incipient mountain fronts can be made just as easily in the Tibetan Plateau, the Middle East, and Africa.

New approaches are overdue to decipher the sequences of frequent earthquakes that characterize plate-boundary fault zones. Chapter 6 develops a new geomorphic way to precisely date earthquakes in New Zealand and to describe their seismic shaking. It then tests the model in California. This geomor-

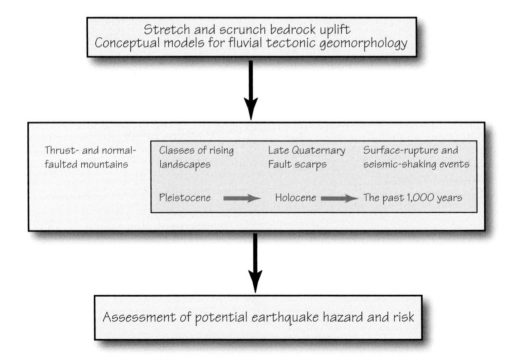

Figure 1.1 Major topics of this book and their application to paleoseismology.

phic approach to paleoseismology provides essential information about the frequency and magnitude of recurrent tectonic perturbations such as surface ruptures and seismic shaking. Other plate-boundary settings, such as the Andes of South America, Anatolian fault zone of Turkey, and the Himalayas may be even better suited for this way to study earthquakes than my main study areas.

This book uses two primary, diverse study regions to develop concepts in tectonic geomorphology for fluvial systems in a global sense. Principal sites in New Zealand are shown in Figure 1.2 and southwestern North America sites in Figure 1.3 together with the links to their chapter section numbers.

1.1 Introduction

Continental landscapes of planet earth are formed in large part by interactions of tectonic and fluvial processes, which are modulated by the pervasive influence of late Quaternary climate changes. *Tectonics* is the study of crustal deformation: the evolution of

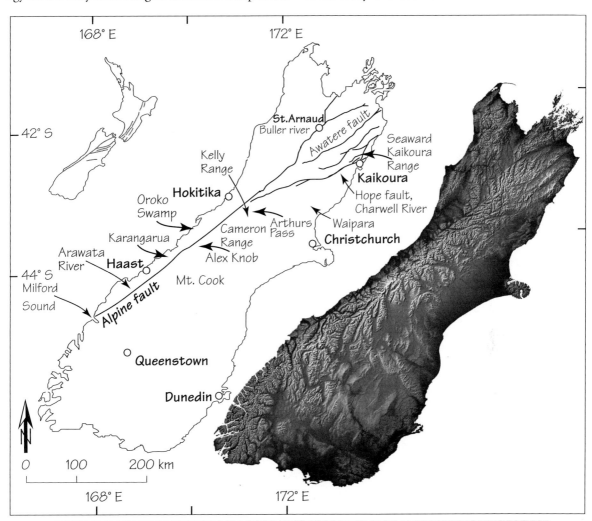

Figure 1.2 Locations of Southern Alps study sites in the South Island of New Zealand discussed in Sections 1.2, 2.4, 2.5, 2.6, and 6.2.1. This is a grayscale version of Shuttle Radar Topography Mission image PIA06662 furnished courtesy of NASA and JPL.

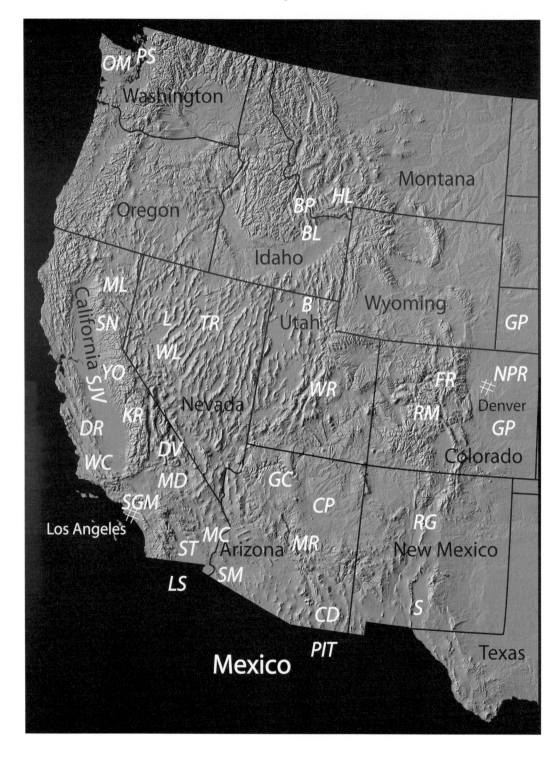

geologic structures ranging from broad transition zones between crustal plates to small faults and folds. *Geomorphology* is the study of landscapes and the processes that shape them. The influences of vertical and horizontal earth deformation on fluvial, coastal, and glacial processes and the resulting landscapes comprise the domain of *tectonic geomorphology*. The main emphasis here is on fluvial system responses to tectonic deformation.

The challenge for all of us is to more fully recognize and use tectonic signals in the landscapes around us. The consequences of earth deformation by specific geologic structures profoundly affect geomorphic processes and landscape evolution. Conversely, evolution of landscape assemblages can be used to decipher the kinematics of faults and folds.

Changes in style, rate, and locations of faulting and folding change the landscape too. An example is the Hope fault of New Zealand where Eusden et al. (2000) describe a 13 km long and 1.3 km wide transpressional duplex structure (adjacent areas of rise and fall) that has migrated northeast along a range-bounding oblique-slip fault that is as active as the San Andreas fault of California, USA. This leading portion of the duplex structure is rising on thrust faults. In the trailing southwest portion, formerly active duplex structures are now collapsing, undergoing a reversal of slip style to become normal faults. Rising geomorphic base levels become falling base levels with dramatic consequences for hills and streams of upstream watersheds. Another example is drainage nets that change as tips of faults propagate (Jackson

et al., 1996). Structural geologists need to recognize how tectonic deformation affects erosion, deposition, and landforms.

Tectonic geomorphology aids tectonic inquiries on many temporal and spatial scales. Some of us seek to understand how horizontal, as well as vertical, earth deformation affects the shapes of hills and streams in a quest to better understand long-term partitioning of strain along plate boundary-fault systems (Lettis and Hanson, 1991). Others study landslides in order to determine earthquake recurrence intervals and to make maps depicting patterns of seismic shaking caused by prehistorical earthquakes (Chapter 6).

Tectonic geomorphology, seismology, and paleoseismology are cornerstone disciplines for studies of *active tectonics* (neotectonics). *Seismology* – historical instrumental studies of earthquakes – contributes much to our understanding of crustal structure and tectonics by 1) defining earthquake hypocenters (location and depth of initial rupture along a fault plane), 2) describing earthquake focal mechanisms (strike-slip, normal, and reverse styles of displacement), 3) evaluating the frequency, magnitude, and spatial distributions of present-day earthquakes, and 4) modeling how yesterday's earthquake changes the distribution of crustal stresses that will cause future earthquakes. *Paleoseismology* – the study of prehistorical earthquakes – utilizes many earth science disciplines including dendrochronology, geochronology, geodesy, geomorphology, seismology, soils genesis, stratigraphy, and structural geology. Tectonic geomorphology is indispensable for complete paleoseis-

Figure 1.3 Locations of study sites in the western United States and northern Mexico and their book section numbers [5.5]. B, Pleistocene Lake Bonneville [5.2.3]; BL, Big Lost River [5.3]; BP, Borah Peak and the Lost River Range [3.3.4]; CD, Curry Draw [2.2.3]; CP, Colorado Plateau; DR, Diablo Range [4.2.3.2]; DV, Death Valley, Panamint Range, and Saline Valley; FR, Front Range [1.3]; GC, Grand Canyon [2.5.2]; GP, Great Plains [1.3]; HL, Hebgen Lake [5.6.2]; KR, Kings River [6.2.2.2]; L, Pleistocene Lake Lahontan [5.2.3]; LS, Laguna Salada [2.1]; MC, McCoy Mountains [3.2.2]; MD, Mojave Desert [4.2.3.2]; ML, Mount Lassen and the southern end of the subduction related Cascade volcanoes [4.1]; MR, Mogollon Rim [4.2.2]; NPR, North Platte River [1.3]; OM, Olympic Mountains [1.2.2, 5.5]; PIT, Pitaycachi fault [2.2.5, 5.4.2]; PR, Panamint Range, Death Valley, and Saline Valley [4.1, 4.2.2, 4.2.3]; PS, Puget Sound [5.5]; RG, Rio Grande River and extensional rift valley [1.3]; RM, Rocky Mountains [1.3]; S, Socorro [6.6.1]; SGM, San Gabriel Mountains [3.2, 3.3.1, 4.2.3.2]; SJV, San Joaquin Valley [4.1, 4.2.3.2]; SM, Sheep Mountain [4.2.2]; SN, Sierra Nevada microplate [4.1, 6.2.2]; ST, Salton Trough [4.2.3.2]; TR, Tobin Range, Pleasant Valley, Dixie Valley, and the Stillwater Range [3.2.1, 4.2.3.2, 5.1, 5.6.2]; WC, Wallace Creek [2.5.2]; WL, Walker Lake [5.2.3]; WR, Wasatch Range [3.3.3]; YO, Yosemite National Park [6.2.2.1, 6.2.2.3]. Digital topography courtesy of Richard J. Pike, US Geological Survey.

mology investigations. For example, stream-channel downcutting and diffusion-equation modeling of scarp erosion to complement stratigraphic information gleaned from trenches across the fault scarp.

Quaternary temporal terms (Table 1.1) have been assigned conventional ages*. The 12-ka age assignment for the beginning of the Holocene is arbitrary and is preceded by the transition between full-glacial and interglacial climatic conditions. Unless specifically noted, radiocarbon ages are conventional (using the old 5,568 year half-life allows comparison with dates in the older literature) and have been corrected for isotope fractionation. The term "calendric radiocarbon age" means that the correct 5,730 year half-life is used and that variations in atmospheric ^{14}C have been accounted for, using the techniques of Stuiver et al. (1998). Calibration of radiocarbon ages (Bard et al., 1990) shows that the peak of full-glacial conditions may be as old as 22 ka instead of the conventional radiocarbon age estimate of 18 ka. The 125 and 790-ka ages are radiometric and paleomagnetic ages that have been fine-tuned using the astronomical clock (Johnson, 1982; Edwards et al., 1987a, b). The 1,650-ka age is near the top of the Olduvai reversed polarity event (Berggren et al., 1995).

Landscape evolution studies accommodate many time spans. Topics such as the consequences of rapid mountain-range erosion on crustal processes involve time spans of more than 1 My. Examinations

Age	Ka
Holocene	
Late	0-4
Middle	4-8
Early	8-12
Pleistocene	
Latest	12-22
Late	12-125
Middle	125-790
Early	790-1650

Table 1.1 Assigned ages of Quaternary temporal terms, in thousands of years before present (ka).

*1 ky = 1000 years; 1 ka = 1 ky before present.
 1 My = 1 million years; 1 Ma = 1 My before present.

of how Quaternary climate changes modulate fluvial system behavior generally emphasize the most recent 50 ky. Understanding the behavior of fault zones concentrates on events of the past 10 ky.

The first concepts discussion about examines several processes that raise and lower the land surfaces of Chapter 1 study sites. Streams respond to uplift by eroding mountain ranges into drainage basins. So Chapter 2 then examines how far streams can cut down into bedrock – their base level limit. We also explore the behavior of fluvial systems to lithologic and climatic controls in different tectonic settings in the context of response times, the threshold of critical power, and tectonically induced downcutting. These concepts will give you a foundation for perceiving tectonic nuances of mountain fronts and hillslopes.

1.2 Pure Uplift, Stretch and Scrunch Bedrock Uplift

1.2.1 Isostatic and Tectonic Uplift

My approach to tectonic geomorphology examines some of the myriad ways that uplift may influence fluvial landscapes. Many new methods and models alter our perceptions of tectonics and topography as we seek to better understand everything from landscapes and prehistorical earthquakes to crustal dynamics. So we begin this chapter by examining the intriguing and occasionally puzzling meanings of the term "uplift". My emphasis is on how subsurface processes affect altitudes of all points in a landscape. Read England and Molnar's 1990 article and you will come away with a fascinating perspective about several components that influence uplift of points on the surface of a large mountain range. The key to using their breakthrough is to recognize the factors influencing uplift of bedrock, not only at the land surface but also at many positions in the Earth's crust. I introduce additional parameters that also influence rock uplift. Both tectonic and geomorphic processes influence bedrock uplift (Fig. 1.4).

S.I. Hayakawa's semantics philosophy (1949) certainly rings true here; "The word is not the object, the map is not the territory". Not only will each of us have different (and changing) impressions of uplift terminology, but also my attempts to neatly organize key variables are hindered by substantial overlap between categories. Fault displacements do more than raise and lower bedrock (*pure uplift*), because

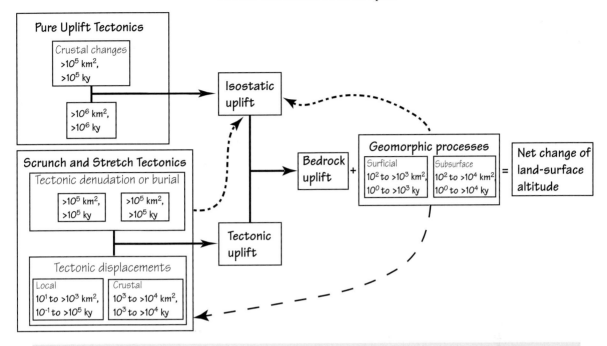

Figure 1.4 Links between tectonic, isostatic, and nontectonic variables affecting landscape altitudes and bedrock uplift. Feedback mechanisms to isostatic and tectonic uplift are shown with dashed lines.

earth deformation usually entails tectonic shortening (*scrunch*) and extension (*stretch*) processes. So, rocks move both vertically and horizontally (Willett, Slingerland, and Hovius, 2001). Do not expect crisp black-and-white definitions, because a model of fuzzy overlap is closer to the truth in the world of scrunch-and-stretch tectonics.

Erosion of mountains is like rain falling on a marine iceberg; the height of both results from buoyant support. Rainfall can never melt enough ice to lower the surface of an iceberg to the water line. This is because ice melted above the waterline is largely replaced by "uplift" of submerged ice. Sea level is a handy reference datum for uplift of ice or mountain ranges. Uplifted materials may be above or below that worldwide waterline. *Altitude* is the specific term for height above present sea level, whereas the engineering term "*elevation*" can have several geologic connotations, including uplift. Isostatic uplift occurs because ice is only 90 % as dense as seawater. If 100 tons is melted from the exposed surface of an iceberg, it is compensated by 90 tons of ice raised by isostatic uplift. This is pure uplift because it is not complicated by shearing or tensional failure of ice.

Similarly, isostatic uplift of mountain ranges continues despite eons of surficial erosion because continental crust "floats" on the denser rocks of the Earth's mantle. Continental crust with a density of about 2,700 kg/m³ is in effect floating on mantle with a density of about 3,300 kg/m³– a density contrast of roughly 82% (90% contrast for oceanic crust with a density of 3,000 kg/m³). The iceberg analogy is appropriate because materials deep in the earth behave as viscous fluids over geologic time spans (Jackson, 2002). Fluvial and glacial denudation of 1,000 m only seems to significantly lower a mountain range because it is largely compensated by 820 m of concurrent isostatic rebound.

Neither ice nor rock landscapes remain the same, unless erosional lowering is the same for all points in a landscape. Relief and altitudes of peaks increase if melt of ice, or erosion of rocks, is mainly along valley floors. Removal of mass above our sea-level datum causes pure isostatic uplift of all parts of the landscape. The average altitude of both the iceberg and the mountain range decreases with time because buoyancy-driven isostasy can never fully compensate for the mass lost by erosion.

A substantial proportion of mountain-range uplift is the result of these crustal isostatic adjustments (Molnar and England, 1990; Gilchrist et al, 1994; Montgomery, 1994; Montgomery and Greenberg, 2000). Isostatic uplift is both regional and continuous (Gilchrist and Summerfield, 1991), and generally does not cause pulses of renewed mountain building. This is done by scrunch and stretch tectonics.

A major difference between icebergs and mountain ranges is that mountains do not float in a Newtonian fluid such as water, which has no shear strength. Continental rock masses float on hot lithospheric materials whose rigidity provides some support. Rocks at shallower depths are stronger (cooler) and respond to changes in load by flexing in an elastic manner. Small, local changes in rock mass will not cause the lithosphere to flex because it has enough strength to support minor changes in load. But beveling of a 10,000 km^2 mountain range will indeed influence crustal dynamics. Prolonged erosion has resulted in substantial cumulative isostatic rebound of the Appalachian Mountains of the eastern United States for more than 100 My.

Tectonic geomorphologists would prefer to discern how different uplift rates influence landforms and geomorphic processes, but reality is not that simple. Mountain-building forces may continue long after tectonic quiescence seems to have begun, as revealed by strath terraces (a tectonic landform discussed in Sections 2.4.1 and 2.6) in pretty dormant places like Australia (Bierman and Turner, 1995). Space and time frameworks of references vary greatly for the Figure 1.4 surface-uplift variables. Generally, they are large and long for pure uplift, tectonic denudation, or burial, and small and short for tectonic displacements and geomorphic processes.

The predicament is that uplift has two components – tectonic and isostatic. Tectonic mountain-building forces may cease but the resulting isostatic adjustments will continue as long as streams transfer mass from mountains to sea. The best we can do at present is to observe landscape responses to the algebraic sum of tectonic and isostatic uplift.

Bedrock uplift = Tectonic uplift + Isostatic uplift   (1.1)

This seems simple, until we attempt to quantify the Figure 1.4 variables that influence tectonic uplift and isostatic uplift.

The term *bedrock* is used here in a tectonic instead of a lithologic context. Bedrock is any earth material that is being raised, with no regard as to the degree of lithification or age. We should note the "fuzziness" of this definition.

Three exceptions are acknowledged; these occur when the nontectonic surficial process of deposition raises a landscape. The most obvious and dramatic is volcanic eruption, which raises landscape altitudes by depositing lava and tephra. Of course volcanic eruptions may also be associated with tectonic shortening and extension.

Tectonic geomorphologists are interested in how climate change affects the behavior of streams in humid and arid regions. Mountain valleys and piedmonts undergo aggradation events as a result of major climate changes (Bull, 1991) that change the discharge of water and sediment. We do not class such stream alluvium as bedrock because its deposition is the result of a nontectonic process that raises valley-floor altitudes. Alluvium laid down before the particular time span that we are interested in would be treated like other earth materials, as bedrock. Studies of Pleistocene uplift would treat Miocene fluvial sand and gravel as bedrock. Thirdly, nontectonic deposition includes eolian processes such as the creation of sand dunes. Least obvious, but far more widespread, is deposition of loessial dust. In New Zealand windblown dust is derived largely from riverbeds after floods and the loessial blanket that covers much of the stable parts of the landscape may contain layers of volcanic ash, such as the 26.5 ka Kawakawa tephra (Roering et al., 2002, 2004). Hillslopes where this ash has been buried by 0.5 to 5 m of loess are landscapes where deposition has slowly raised the altitudes of points on the land surface during the 26 ky time span at average rates of <0.02 to >0.1 m/ky. Deposition – by volcanic ejecta, inability of a stream to convey all bedload supplied from hillslopes, and dust fall – is just one of several nontectonic geomorphic processes that change altitudes of points in a landscape (Fig. 1.4).

I prefer to emphasize bedload transport rates in this book because bedload governs stream-channel responses to bedrock uplift. Rivers transport mainly suspended load to the oceans and deposit silty sand and clay on floodplains. Dissolved load is bedrock conveyed in solution. Both require little stream power, but the unit stream power required to mobilize and transport bedload reduces the energy available for tectonically induced downcutting of stream channels. Saltating cobbles and boulders are tools for abrasion of bedrock. With suspended load being flushed downstream it is bedload that is deposited as fill stream terraces and many alluvial fans. Such landforms are used to analyze responses of fluvial systems to bedrock uplift and to changes in late Quaternary climate too.

Three classes of deposition in Figure 1.5 illustrate the care needed in defining sand and gravel as bedrock. The active range-bounding fault controls the behavior of this fluvial system. Stream channel processes normally switch abruptly from net erosion to net deposition after crossing the fault zone.

Sandy alluvial-fan deposits of Miocene age have been elevated and now underlie watersheds in this hypothetical mountainous landscape. Most of us would agree that the uplifted Miocene fan deposits, although unconsolidated, should be classed as bedrock in a geomorphic sense. They are mountainous terrain into which drainage basins are carved.

The gravelly fill terraces in the valley upstream from the mountain front are the result of climate-change perturbations. Without perturbations the watersheds of tectonically active mountain ranges in the Mojave Desert would have undergone uninter-rupted long-term degradation of their valley floors. But late Quaternary climatic fluctuations significantly affected sediment yield and stream discharge. Climate-change perturbations in arid and humid watersheds can temporarily reverse the tendency for stream-channel downcutting, even in rapidly rising mountain ranges. Climate-change perturbations are dominant because they quickly affect geomorphic processes throughout a drainage basin, whereas uplift on a fault zone is local and the resulting increase in relief progresses upstream relatively slowly.

Climate-change induced aggradation events in the Mojave Desert raised valley floors <5 to >50 m. The range is largely due to lithologic controls on weathering and erosion. Aggradation was the result of insufficient stream power to convey bedload supplied from hillslopes whose vegetation changed drastically when the climate changed. Major aggradation events at about 125 ka and 10 ka were times of widespread stripping of hillslope sediment reservoirs that were no longer protected by dense growth of plants. A climatic perturbation at about 60 ka also coincides with a global sea-level highstand and caused an aggra-

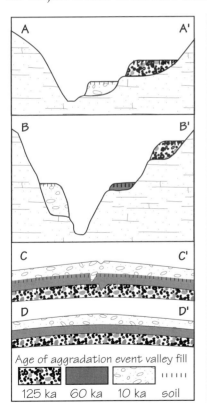

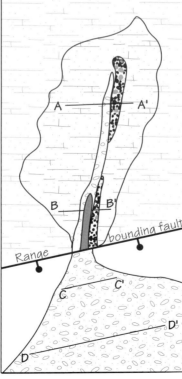

Figure 1.5 Summary of late Quaternary deposition for a typical Mojave Desert, California fluvial system where times and locations of aggradation are controlled by climatic perturbations that overwhelmed the effects of uplift along active fault zones. Hachures show soil profiles that postdate the ends of aggradation events and record brief intervals of nondeposition on the fanhead.

dation event of smaller magnitude. So the sedimentology and thickness of each late Quaternary fill-terrace, and the concurrent increments of alluvial-fan deposition, were different (Table 1.2).

Vertical separations between the beveled bedrock beneath several valley fills record stream-channel downcutting induced by uplift along the range-bounding fault zone in the intervals between climate-change induced aggradation events. The potent 10 ka aggradation event might have buried the equally strong 125 ka aggradation event at cross section A–A if there had been no tectonically-induced lowering of the valley floor. More tectonically induced degradation has occurred at cross section B–B' than at A–A' because it is closer to the active fault zone. Depositional elevation of the stream terrace tread is a clear-cut example of nontectonic elevation of landscape elements. Such deposits should not be classed as bedrock.

How should we regard the area of active alluvial-fan deposition downstream from the range-bounding fault? Surface ruptures on the normal fault create the space for continuing accumulation of basin fill. Such fans are tectonic landforms because nearly constant deposition would not have occurred without continuing uplift. Differential uplift along the fault has been sufficiently rapid to maintain late Quaternary aggradation adjacent to the mountain front (Section 4.2.2).

Major Late Quaternary climatic changes caused the rate of fan aggradation to vary and influenced the locations of fan deposition. Minor, brief climatic fluctuations are superimposed on the long-term climatic controls. They caused brief episodes of stream-channel downcutting in the mountains and temporary entrenchment of the fan apex. Brief local cessation of depositional processes allowed incipient soil-profile development on the fan surfaces adjacent at cross section C–C'. Each aggradation event was strong enough to backfill the fanhead trench, thus allowing fan deposition to continue to radiate out from an apex at the mountain front.

It is debatable as to whether such fan deposits should be regarded as bedrock. Perhaps they should be classed as bedrock because the locations of fan deposition are tectonically controlled. Deposition of a thick fan would not occur here in the absence of active faulting. Alternatively, one might argue that rates of sedimentation vary with late Quaternary climates. Deposition merely tends to partially offset tectonic lowering of basin altitudes in an extensional terrain. Such fans should not be classed as bedrock.

Lithospheric rigidity interjects the important element of scale into our perception of what constitutes uplift. Tectonic-uplift variables behave differently at the local scale of a single hillside or small watershed as compared to large chunks of the Earth's crust. For each point in a landscape, tectonic deformation caused by different styles of faulting and folding is superimposed on regional uplift (or subsidence) caused by broad warping of the lithosphere. This is a matter of different wavelengths for different earth-deformation processes.

Alluvial geomorphic surface	Aggradation age, ka	Basis for age estimate
Q4		Active washes, riparian trees, no rock varnish on cobbles
Q3b	~8	^{14}C dating of plant fossils, lake stratigraphy, rock varnish
Q3a	~12	^{14}C dating of plant fossils, lake stratigraphy, rock varnish
Q2c	~60	$^{230}Th/^{234}U$ ages of pedogenic carbonate, uranium-trend date, calibrated fault slip age estimate, cosmogenic ^{10}Be age estimate
Q2b	~125	$^{230}Th/^{234}U$ ages of pedogenic carbonate
Q2a	240–730	K/A dating of tuff, basalt flow, normal paleomagnetic polarities
Q1	>1,200	K/A dating of basaltic sources dissected into ridges and ravines

Table 1.2 Pulses of climate-change induced alluviation in the Mojave Desert of California. Summarized from Tables 2.13 and 2.15 of Bull (1991).

Tectonic uplift = Local uplift + Crustal uplift   (1.2)

The background regional crustal warping may be slow or fast, but it affects erosion rates of local landforms as well as those of entire mountain ranges. Local faulting creates topographic anomalies such as rising mountain fronts that attract tectonic geomorphologists (Chapter 3). We analyze landforms to separate local tectonic deformation from background regional uplift. But separating tectonic from isostatic uplift can be difficult at the watershed spatial scale because not all earth deformation is purely vertical.

Scrunch and stretch tectonics plays an important role in deformation of Earth's crust. For example, plate-boundary subduction is a tectonic process, but how much of the resulting bedrock uplift is the result of isostatic uplift caused by thickening of the crust? How much is the result of scrunch induced by concurrent folding and thrust faulting? Conversely, in extensional terrains how much of a decrease in altitude is offset by isostatic adjustment resulting from concurrent erosion of mountain ranges? How much of lowering induced by stretch tectonics is offset by aggradation (Fig. 1.5) in basins that receive the deposits? Let's begin with brief summaries of the contents of the "Pure Uplift" and "Geomorphic Processes" boxes of Figure 1.4 to gain background before delving into "Stretch and Scrunch" box.

Many factors affect magnitudes and response times for isostatic uplift. Important slow changes in the crust include accretion, or thinning, of light, buoyant crustal materials. Temperature increase or decrease changes the density of crustal rocks, thus changing their buoyancy. Phase changes in minerals that reflect changing pressures or temperatures alter buoyancy contrasts with adjacent rocks. Change to denser minerals decreases rock volume, which also tends to directly lower land-surface altitudes.

Pure strike-slip faulting does not raise or lower a landscape, but major horizontal shifts of mountain ranges and crustal blocks may alter regional distributions of isostatic forces. Many plate-boundary strike-slip faults have cumulative displacements of more than 50 km, so this style of tectonic deformation may change the crustal loads on opposite sides of a fault sufficiently to cause isostatic re-adjustments. This important aspect of strike-slip faulting deserves its separate box within pure uplift tectonics in Figure 1.4. Changes in altitude that occur at bends and sidesteps of strike-slip faults, are included in the local-

tectonic-displacement box because they are classified as scrunch and stretch tectonics. Transpressional or transtensional components of most plate-boundary fault zones also are best considered as part of scrunch and stretch tectonics.

Tectonic processes and isostatic uplift may increase land-surface altitudes, but landscape altitudes also change because of several geomorphic processes. We have already mentioned the surficial processes of fluvial and volcanic deposition. Another is fluvial erosion, which tends to lower hills and streams. Both sets of processes affect crustal weight, and when sufficient may cause isostatic adjustments.

Diagenesis of recently deposited basin fill tends to lower land-surface altitudes. Compaction of saturated clayey, silty beds in a sedimentary basin is analogous to crustal changes that produce denser minerals. It is pure vertical subsidence. Bulk density increases as water is gradually expelled from sediments by the weight of the overlying stratigraphic section, plus several hydrodynamic forces. The resulting decrease in bed thickness lowers the overlying strata and the land surface.

Ground water derived from infiltrating rain and snowmelt dissolves minerals. Solution is a greatly different geomorphic process than landsliding because it is not visually conspicuous. It occurs below the land surface and the resulting ions are invisible in emerging clear springs that nourish streamflow. But substantial mass is removed over Quaternary time spans at depths that range from surficial soil profiles to more than 1 km.

The net surface uplift resulting from all Figure 1.4 processes is an algebraic sum.

Surface uplift = Rock uplift + Geomorphic Processes (1.3)

The sum of geomorphic processes has feedback loops to isostatic uplift and tectonic deformation.

Stretching and scrunching are important tectonic processes that lower or raise landscape altitudes. Most importantly, they (not isostatic uplift) initiate the creation of mountain ranges. Let us think of these as being *tectonic denudation* (Fig. 1.6) and *tectonic burial* (Fig. 1.8). Both are common, and operate at a variety of spatial scales. I'll focus mainly on scrunch processes because local uplift may appear anomalous when it is ten times the expected regional uplift. Also, it seems that tectonic denudation processes are already nicely discussed in the literature of the past two decades.

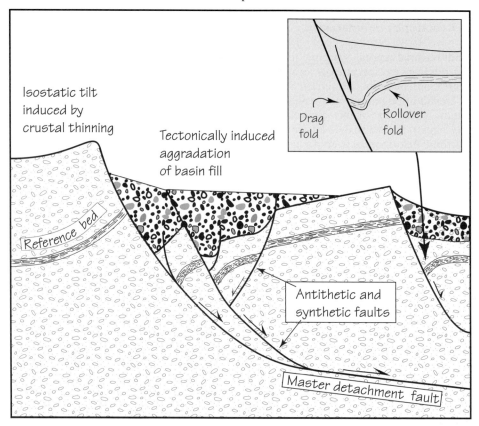

Figure 1.6 Diagrammatic sketch of extension associated with normal faulting that causes tectonic denudation and crustal thinning. Rollover folds form where gravitational collapse progressively increases closer to the normal fault. Frictional resistance during displacement of the hanging-wall block generates the shear couple responsible for drag folds next to the fault.

1.2.2 Stretch and Scrunch Tectonics

Tectonic stretching (Fig. 1.4) is important. The resulting tectonic denudation is widely recognized, and generally is thought of as normal faulting that thins the crust (Armstrong, 1972; Davis and Coney, 1979; Shackelford, 1980; Spencer, 1984; Coney and Harms, 1984; Pain, 1985; Wernicke, 1992; Dickinson and Wernicke, 1997; Burbank and Anderson, 2001, p. 149-151). Normal faulting also occurs locally in compressional settings (Molnar and Lyon-Caen, 1988; Gammond, 1994; Eusden et al., 2005a). England and Molnar (1990) combined tectonic denudation and surficial erosion into a single process called "exhumation". Low angle detachment faulting (Lister et al., 1986; Bradshaw and Zoback,

1988; Lee and Lister, 1992; Dokka and Ross; 1995; Bennett et al., 1999) can efficiently remove large amounts of bedrock, thereby promoting isostatic rebound (Wernicke and Axen, 1988). Normal faulting in the Basin and Range Province of the western United States has resulted in extension of more than 250 km (Wernicke and Snow, 1998), with a crust that has thinned to about 30 km (Jones et al., 1992). The lower crust of the Basin and Range province should behave as a viscous fluid (Bird, 1991; McCarthy and Parsons, 1994), tending to fill voids created by tectonic extension.

Stretch tectonics has distinctive features and resulting landforms (Fig. 1.6). The footwall block typically has minimal secondary faulting, but tensional forces create a myriad of antithetic and synthetic faults in the hanging-wall block. These result mainly from removal of vertical support. Complex structures are induced in hanging-wall blocks where normal fault dip becomes less with depth below the surface to create *listric* faults. This promotes fault-bend folding. Gravitational collapse is greatest near master detachment faults to create rollover folds (Hamblin, 1963, 1965). These tectonic processes lower surface altitudes. Local vertical displacements, such as range-bounding faults, create space that allows deposition of alluvial fans and other basin fill. Such aggradation raises surficial altitudes so deposition of basin fill is a process that partially offsets tectonic lowering, perhaps by a factor of half. Upper crust thinning enhances the potential for upwelling and isostatic uplift. Crustal rigidity extends this isostatic rebound into the footwall block at the left side of Figure 1.6, in an exponentially decreasing manner with increasing distance from the range-bounding fault. Spatially variable isostatic rebound tilts the land surface.

Tectonic denudation caused by a variety of stretch processes thins the upper crust. These reduce crustal loading, and together with an increase in geothermal gradient and lithospheric upwelling promote isostatic uplift that partially offsets the stretch-induced subsidence (Bird, 1991). This self-arresting feedback mechanism is opposite of that caused by tectonic scrunching.

The style of normal faulting affects the behavior of fluvial systems. The example used here examines stretch-tectonics controls on the thickness of piedmont alluvial fans. Continuing lowering of a valley and/or uplift of the adjacent mountains creates the space for new increments of piedmont deposition. The resulting alluvial fans reflect the style and rate of tectonic deformation. Prolonged displacement on a range-bounding normal fault can result in fan deposits more than 1,000 m thick. Alluvial-fan deposits are thickest where basins quickly drop away from the mountains, such as the high-angle normal faults of the Basin and Range Province of the western USA. Fan deposits are much thinner where tectonic displacements occur on low-angle faults. Examples include where thrust-faulted mountain fronts are shoved up and over adjacent basins along low-angle faults that dip back into the mountains (Section 3.2.3). I use many examples from the Death Valley region of southeastern California in subsequent chapters, so introduce an

interesting example of stretch tectonics here. The locale is the western flank of the Panamint Range.

Low-angle normal faults have played an important role in both tectonic extension and landscape evolution of the Death Valley region. Style of alluvial-fan deposition varies with type of fault. Debate continues as to how important such detachment faults are as compared to normal faults that dip steeply at 45° to 65° (Wernicke, 1981, 1995; Walker et al., 2005). Cichanski (2000) made a detailed study of the curviplanar low-angle normal faults on the west flank of the Panamint Range that were first noted by Noble (1926) and Maxon (1950). As a geomorphologist, I have no doubt that normal faults that dip only 15° to 35° had substantial slip during the late Cenozoic. My premise is based on the idea that changes in the kinematics of faulting change the landscape.

The evidence is the contrasting styles of alluvial-fan deposition. One would expect different types of alluvial fans resulting from low-angle and high-angle normal faulting. Adjustments of fluvial systems to movements on 60° and 25° normal faults are much different (Fig. 1.7). Slip on either steep or gentle fault surfaces causes fluvial systems to cut down into the footwall block and to deposit a new increment of detritus on the hanging-wall block. Part of the newly exposed fault plane is subject to the initial stages of dissection by water flowing in rills, and part is quickly buried by the newest increment of alluvial-fan deposition.

Fan slope is also a function of magnitude and type of streamflow events, and the amount and particle-size distribution of the entrained sediment (Bull, 1962; Hooke, 1967). Although many alluvial fans in the Basin and Range Province slope less than 10°, steeper fans are common. Most fans along the Lost River fault zone near Borah Peak in Idaho have fanhead slopes of more than 20° (Section 3.3.4). An assumed fan slope of 20° for the ancestral fans along the western flank of the Panamint Range seems reasonable for this discussion.

Thicknesses of tectonic alluvial fans are a function of fault dip and fan slope. The combination of a 60° normal fault and a 20° fan surface provides ample space for thick deposits to accumulate adjacent to the footwall block. Fan thickness in the Figure 1.7B example is 40 m, and would be the maximum of 50 m if the range-bounding fault were vertical. Steep faults are sites of thick fans of small areal extent. Extension on high-angle normal faults also favors incision of deep valleys in the footwall block.

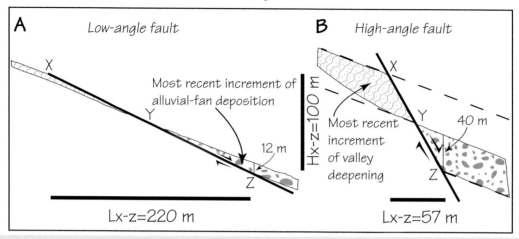

Figure 1.7 Diagrammatic sketches showing how change from low-angle to high-angle normal faulting changes landscape characteristics. H is slope fall, and L is slope length horizontal distance. Vertical tectonic displacements, Hx-z, total 100 m in both cases as the footwall block slips from X to Z. Horizontal tectonic displacements, Lx-z, of 57 and 220 m are a function of normal fault dip. B is the present threshold-intersection point where erosion changes to deposition, assuming that the increase of relief of the footwall block is distributed evenly between alluvial-fan deposition and valley deepening.
A. 25° normal-fault dip and a 20° fan slope.
B. 60° normal-fault dip and a 20° fan slope.

In contrast, only thin veneers of deposits accumulate on a 20° sloping fanhead in response to movements on a 25° low-angle normal fault. Fan thickness in the Figure 1.7A example is only 12 m, but the width of the newest increment of onlapping fan deposits is 110 m – four times that of the high-angle fault example. Such low-angle faults are sites of thin fans of large areal extent. An emphasis on horizontal instead of vertical displacement also inhibits erosion of deep valleys in the footwall block. These shallow valleys are part of a diagnostic landscape assemblage suggestive of low-angle normal faulting, as are the smooth sloping hillsides that resemble the carapace of a turtle, the "turtlebacks" of Wright et al. (1974).

The thinnest deposits near the intersection point (where erosion changes to deposition) are readily removed by fluvial erosion after deposition ceases. Such erosion may have occurred along the west flank of the Panamint Range, and elsewhere in the Death Valley region. Initiation of steep range-bounding faults in the Pleistocene that cut the now inactive low-angle faults (Cichanski, 2000) would stop deposition of the ancestral fans and begin the process of eroding them. The combination of incremental exposure of

the plane of a low-angle fault while it is active and subsequent partial stripping of a thin mantle of fan deposits results in spectacular rilled fault planes (Fig. 2.19A).

Scrunch deformation is everywhere in hanging-wall blocks of thrust faults. In addition to synthetic and antithetic faulting, scrunch processes include folding, flexural-slip faulting along bedding planes, and shoving of wedges of crumpled, brittle rocks up gently inclined fault planes. Scrunch style tectonics may dominate locally to the extent of raising surface altitudes an order of magnitude faster than regional uplift rates.

It makes for pretty messy earth deformation, but adds much variety to rock uplift (Fig. 1.8). The belt of former piedmont terrain between the two thrust-fault zones is called a piedmont foreland, the topic of Section 3.2.3. Bedrock uplift resulting from scrunch tectonic processes increases landscape altitudes and relief of mountains, thus accelerating erosion that partially offsets regional uplift.

Scrunch processes may promote lithospheric downwelling opposite in style to the mantle upwelling described for tectonic stretching. Deposition in tectonic basins raises altitudes. Scrunching and

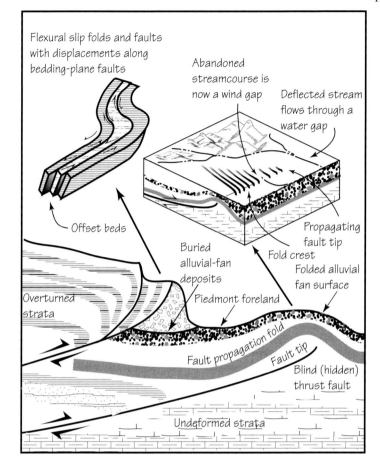

Flexural slip folds and faults with displacements along bedding-plane faults

Abandoned streamcourse is now a wind gap

Deflected stream flows through a water gap

Offset beds

Buried alluvial-fan deposits

Propagating fault tip

Fold crest

Folded alluvial fan surface

Overturned strata

Piedmont foreland

Fault propagation fold

Fault tip

Blind (hidden) thrust fault

Undeformed strata

Figure 1.8 Diagrammatic sketch of types of contractional faulting and folding associated with tectonic shortening that causes burial and crustal thickening. Overturned strata may suggest displacement by a normal fault. Displacements along bedding-plane faults occur where planes between beds are relatively weak; note rock flowage into fold axes. Thrust faulting buries the apex of the piedmont alluvial fan, and a younger fault folds the fan surface.

deposition thicken the crust, thereby promoting iso-static subsidence that partially offsets concurrent rock uplift. Geothermal gradients become cooler where the crust is thickened from the surface down, and the relatively cooler rocks have a lesser potential for iso-static uplift.

Tectonic burial has not received as much attention in the literature as tectonic exhumation, so I use Figures 1.9-1.16 to illustrate the diversity and importance of scrunching. Creation of a fault zone causes more than just uplift, because thrust faults are not vertical. Horizontal rock displacement is a major consequence of scrunching. The hanging wall block is raised as it is shoved up the incline of a gently dipping thrust fault (Fig. 1.9A). The horizontal component of displacement increases local crustal thickness. Amounts of horizontal displacement are a tangent function of fault-plane dip: 100 m of vertical dis-

placement is accompanied by only 27 m of horizontal displacement for a 75° dipping fault. This increases to 100 m for 45° and to 373 m for a fault with a 15° dip. Mass is added to the footwall-block terrain by tectonic conveyance and deposition of sediments eroded from the newly raised block as the fault trace advances in an incremental manner.

In the best of all worlds, tectonic geomorphologists would use planar or conical landforms as time lines passing through tectonically deforming landscapes. Dating of faulted alluvial geomorphic surfaces can provide valuable information about late Quaternary uplift rates. However, estimation of tectonic displacement rates of faulted stream terraces probably is more reliable for stretch than for scrunch tectonics.

The fan surface upslope from the scarp crest in Figure 1.9B is no longer linear. Its undulations sug-

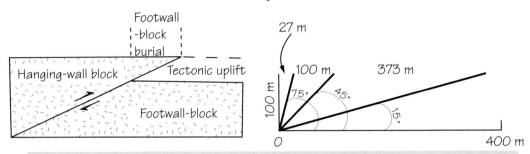

Figure 1.9 Tectonic uplift and burial induced by thrust faulting.
A. Diagram showing components of uplift and burial created by movement along a thrust fault. Both processes thicken the crust and are functions of fault-plane dip.

gest complicated tectonic deformation. A common first impression is that scarp height is indicative of the magnitude of tectonic throw, but scarp height exceeds true displacement where sloping alluvial surfaces are ruptured. A closer approximation can be obtained by noting the vertical separation of projections of the tectonically undeformed fan surfaces upslope and downslope from the fault zone. But this Cucamonga Canyon alluvial fan has a slope that decreases downfan resulting in lack of parallelism of the projected surfaces. A mean apparent throw of 9.3 m based on maximum and minimum displacements is triple the deformation attributable to scrunching. These apparent displacements need to be corrected for the dip of

the faults, which is unknown. A complete discussion is deferred until Section 3.3.4, which describes how to estimate throw for normal-fault scarps on alluvial fans. The interpretation shown in Figure 1.9B is that several synthetic thrust faults ruptured the surface, during several Holocene earthquakes (Morton and Matti, 1987). Another possibility is that the hummocky terrain is nothing more than piles of debris near the fault tip that have been bulldozed by thrust faulting along a single thrust fault. Third, compression may have folded the surficial materials. Most likely, the scrunched material resulted from several processes. Holocene bedrock uplift varies from point to point, but approximates the sum of the vertical

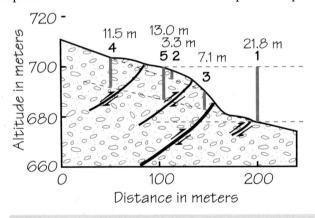

Apparent vertical tectonic displacements (throw)

1 Scarp height
2 Scrunching uplift component
3 Minimum fan surface offset
4 Maximum fan surface offset
5 Mean rock uplift

Figure 1.9 Tectonic uplift and burial induced by thrust faulting.
B. Inferred thrust faults along cross section based on topographic profile. All estimates of displacements are apparent, and except for scarp height are based on projections of adjacent undeformed alluvial-fan surfaces upslope and downslope from the fault zone. Cucamonga alluvial fan, San Gabriel Mountains, southern California.

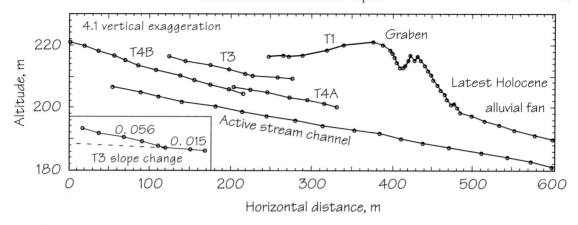

component of thrust-fault displacement, folding, and other scrunching that results from compressional deformation of the wedge of material above the thrust fault. The magnitude of horizontal displacement determines the amount of tectonic burial. The algebraic sum of these processes equals the changes of surficial altitudes because this young alluvial fan is virtually uneroded. Isostatic adjustments are not likely at this small scale.

Tilted stream terraces are sure to catch the attention of the tectonic geomorphologist, especially when alluvium deposited with a 3° downvalley dip now slopes 2° to 5° upvalley (Fig. 1.10). A splay of the Hope fault that bounds the Seaward Kaikoura Range of New Zealand ruptured the Waimangarara River stream terraces. The two oldest, late-Holocene, stream terraces, T1 and T2, have the same backtilt, so the tectonic deformation is younger than the T2 terrace-tread age. Terrace T1 is 5 m above T2. Terrace tread age was estimated with weathering rind analyses, a surface-exposure dating method (Whitehouse and McSaveney, 1983; Whitehouse et al., 1986; Knuepfer, 1988). Analysis of boulders on the T2 tread implies a late Holocene age (Fig. 1.11). This tuffaceous greywacke sandstone does not have nice, sharp weathering rinds, and rind thickness ranges from 1 to 4 mm. I used the McSaveney (1992) procedure. A peak at ~2.5 mm dates as 2,200 ± 300 years before present. Even a 4 mm peak would date to only ~4,700 years B. P.

Terrace T3 is not backtilted but has a four-fold decrease in slope as it approaches the deformed older stream terraces (Fig. 1.10). So it appears that the range-bounding fault ruptured between T2 and T3 time, and again since T3 time.

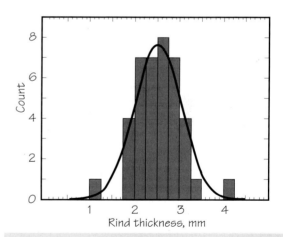

Figure 1.11 Distribution of Waimangarara River T2 stream terrace tread weathering rinds in cobbles of greywacke sandstone deposited before the older of two recent surface-rupture events. Normal distribution curve has been added. 0.25 mm class interval. n = 40.

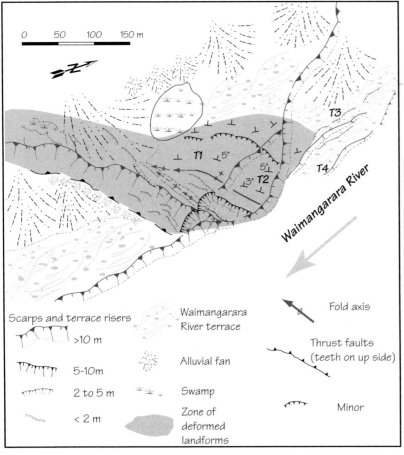

Figure 1.12 Geologic structures, landforms, and tectonically deformed stream terraces at the mouth of the Waimangarara River, Seaward Kaikoura Range, New Zealand as mapped by Jarg Pettinga, University of Canterbury.

Scarps and terrace risers

>10 m

5-10m

2 to 5 m

< 2 m

Waimangarara River terrace

Alluvial fan

Swamp

Zone of deformed landforms

Fold axis

Thrust faults (teeth on up side)

Minor

Scarp height is an impressive 18 m. Vertical offset of 7 m is a minimum value because T1 and T2 have been buried by an alluvial fan downstream from the fault scarp. The prominent graben at the folded scarp crest (Fig. 1.12) can be used to postulate locations of antithetic and synthetic faults above the master thrust fault, which is presumed to dip less than 50° (Van Dissen, 1989).

I suspect that neither the large scarp height nor upvalley stream-terrace tilt is indicative of slip rates on this segment of the Hope fault. The T1 fault scarp on the other side of the river is only about 3 to 4 m high, which is a more reasonable offset for two surface-rupture events. Adjacent segments of this mountain front lack high fault scarps that date to the most recent event. Dip and style of faulting may change within short distances, and subsurface exploration techniques are needed here to fully appraise two possible scenarios. The geologic map (Fig. 1.12) portrays a zone of deformation that tapers towards the southwest, seems to be diffuse on the north side, and is abruptly terminated by the range-bounding thrust fault on the south side. A model of imbricate thrust faulting (Fig. 1.13A) can account for the width of the deformation zone, and synthetic and antithetic faults could produce grabens.

If one uses the critically tapered wedge model of Davis and Namson, (1994), the scrunching shown in Figure 1.13B reflects a fault-kinematic equilibrium. Wedge shape would influence dip of the basal detachment surface, synthetic and antithetic fault movements, and thickness of scrunched rock and alluvium.

So, much of the rock uplift here may be the result of tectonically induced scrunching processes of folding and bulldozing. Brittle fractured greywacke sandstone under low confining pressures may behave like loose boulders. More coherent bedrock slabs may

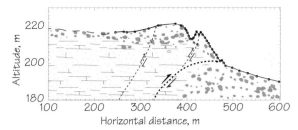

Figure 1.13 Models for deformation of the
Waimangarara River terraces.
A. Fault steepening towards the surface
rotates the stream terraces, creating the
backtilting of T1 and T2. Grabens at scarp
crest record antithetic and synthetic faulting.

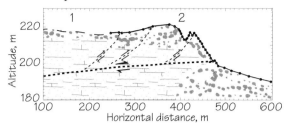

Figure 1.13 Models for deformation of the
Waimangarara River terraces.
B. Critical wedge model in which movement along
gently dipping thrust fault has bulldozed and/or
folded the fractured greywacke sandstone.

fail by rupture along secondary faults. The systematic deformation shown in Figures 1.12 and 1.13B could result from folding instead of haphazard bulldozing processes. Scarp-crest grabens would result from tensional stresses at the crest of an anticline in a folding-dominant model.

The Waimangarara River has frequent large flow events that deposit bouldery alluvium on the adjacent piedmont. Erosional widening of the bedrock valley floor in the mountain-front reach may have thinned the slab above the range-bounding thrust fault prior to the recent surface rupture events. Reduction of rock mass strength below a critical-tapered-wedge threshold would have favored tectonic scrunching processes in the broad valley floor upstream from the fault trace, but not along the adjacent parts of this steep mountain front. Rock uplift, r_u, at location 1 in Figure 1.13B is mainly a function of magnitude of slip along the fault plane, D, and dip of the thrust fault, α.

$$r_u = sin\alpha\,D \qquad (1.4)$$

Rock uplift at location 2 in Figure 1.13B could be largely bulldozed materials above the plane of the thrust fault where scrunch rock uplift, sr_u, has occurred at several fault splays.

$$r_u = sin\alpha\,D + sr_u\Sigma(1, 2, 3, 4) \qquad (1.5)$$

The longitudinal profile of the Waimangarara River reflects several possible tectonic inputs. The stream changes its vertical position in the landscape in response to bedrock uplift. However, fluvial adjustments to rock uplift in the longitudinal profile do not distinguish between regional isostatic uplift, slip on thrust faults, folding, and local bulldozer scrunching of fractured greywacke sandstone. I conclude that the Waimangarara River stream terraces are not ideal time lines passing through a tectonically deforming landscape. The deformed stream terrace treads are good reference surfaces for describing the complicated total bedrock uplift, but should not be used for estimating fault slip rates. Thrust-fault displacement has a vertical component, but secondary folding and crushing is largely a function of horizontal displacement. Both contribute to rock uplift. This local increase in crustal loading due to scrunching is too small to overcome lithospheric rigidity, so let us examine an example that is sufficiently weighty to influence isostatic processes.

Erosion becomes ever more important with increase in spatial extent and steepness of a landscape, longer time spans, and decrease of rock mass strength. Erosion rates increase exponentially with hillslope steepness (Ahnert, 1970), so relief that is increased by scrunching accelerates the denudational processes that tend to lower a mountain range that is being created by tectonic forces. One impressive example of large scale thrust faulting, and rapid erosion during the past 1.5 My, is the Salt Range in Pakistan (Burbank and Anderson, 2001). Potential tectonic burial by a tectonically translocated mountain-range size block that is 3 km high and 18 km long (Fig. 1.14) never transpired because erosion occurred as rapidly as scrunching raised poorly consolidated fluvial sediments of the Siwalik Formation

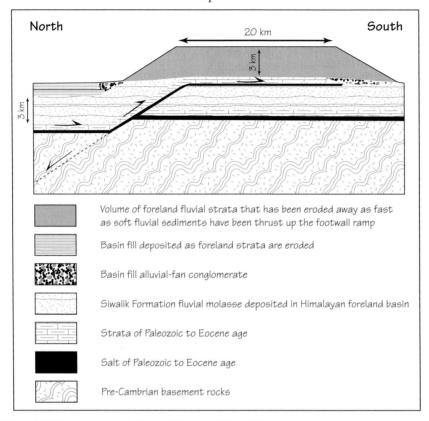

North   South

Volume of foreland fluvial strata that has been eroded away as fast
as soft fluvial sediments have been thrust up the footwall ramp

Basin fill deposited as foreland strata are eroded

Basin fill alluvial-fan conglomerate

Siwalik Formation fluvial molasse deposited in Himalayan foreland basin

Strata of Paleozoic to Eocene age

Salt of Paleozoic to Eocene age

Pre-Cambrian basement rocks

Figure 1.14 Potential large scale tectonic burial that has been offset by erosion of the Salt Range, Pakistan. From Burbank and Anderson, 2001, Figure 7.5, and Burbank and Beck. 1991.

up the ramp of the footwall block. Note the lack of a tapered wedge of mountain range relief away from the top of the ramp. The Salt Range tectonic setting appears to represent a case where regional fluvial erosion balances the tendency for uplift to increase relief. Estimated rates of erosion are 2 m/ky over the large area of 1500 km². Tectonic loading has been largely offset by concurrent erosion, which increases crustal thickness elsewhere in depositional basins.

Active folding provides extreme examples of scrunch-induced bedrock uplift even where rates of regional uplift are modest. Spatial contrasts in uplift rates generally are gradual for active folds and abrupt for active faults. Horizontal strata under a constant rate of tectonic shortening are folded upward, but the crest of the resulting anticline does not rise at a uniform rate. Rockwell et al. (1988) show that uplift for a single, simple fold quickly accelerates to a maximum, and then slows to zero despite unabated com-

pression (Fig. 1.15). About 36% of potential uplift has occurred after only 4% shortening of a horizontal bed, a situation where a slow rate of horizontal displacement causes remarkably rapid uplift of the fold hinge. But there is a limit to how much uplift can be produced by contraction (shortening) of a single fold. Continued scrunching creates faults and new folds.

Anticlines in fold and thrust belts commonly have thrust faults in their cores (Fig. 1.8), which further complicates assessment of bedrock uplift. Folding may be largely replaced by tilting after a thrust fault propagates through to the land surface. We should expect the landforms and geomorphic responses to tectonic deformation to vary along the trace of fold created by a propagating thrust fault (Fig. 1.8).

Spatial variations of local tectonic deformation should reflect the cumulative displacements of individual earthquakes. Level-line surveys of recent historical earthquakes nicely show the contrast in

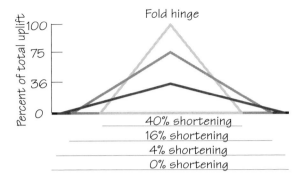

Figure 1.15 Tectonic uplift and burial induced by contractional folding. Deceleration of rates of folding induced uplift, using a model of uniform rates of tectonic shortening (from Rockwell et al., 1988).

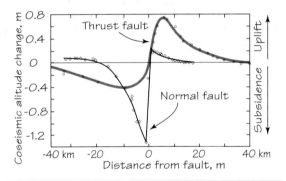

Figure 1.16 Spatial variations in deformation caused by two magnitude M 7.3 earthquakes. The thrust-fault example is the 1952 Kern County, California earthquake. The normal fault example is the 1983 Borah Peak, Idaho earthquake. Note that both subsidence and uplift occur with extensional and contractional earthquakes (from Stein, et al., 1988).

styles of folding and faulting associated with normal and thrust faulting (Stein et al., 1988). Such work reveals that some local uplift occurs during a normal-fault surface rupture and some subsidence occurs with a thrust-fault rupture event (Fig. 1.16). Rebound uplift of the footwall block was about 20% that of the hanging-wall block tectonic subsidence result-ing from the Borah Peak earthquake. Minor faulting and folding is concentrated in the hanging-wall block of normal faults, but the footwall and hanging-wall blocks of an active thrust fault may have similar secondary deformation.

Not all tectonic deformation occurs at the moment of an earthquake due to the response time needed for subsequent mantle upwelling. Fault creep and gradual folding may also deform the land surface. Modeling done by Freed and Lin (2002) links tectonic deformation to post-seismic relaxation of viscous lower crust and/or upper mantle – a process that continues for decades. Although some folding or warping during a particular earthquake event occurs as post-seismic deformation, few studies have the data to assess both pre- and post-seismic folding rates over time spans of centuries.

One such investigation uses stream terraces formed as a result of downcutting induced by folding and faulting of an anticline. Streams flowing across rising mountains incise bedrock and tectonically induced downcutting is proportional to bedrock uplift rates. Nicol and Campbell (2001) estimated uplift rates for an anticline by measuring the heights of terrace treads above the active channel, and by using weathering-rind and radiocarbon methods to date abandoned floodplain remnants.

The scene is a young fold-and-thrust belt in the foothills of the Southern Alps of New Zealand. The Waipara River has a watershed area of 950 km² where it cuts though Doctor's Anticline. The Karetu thrust fault in the core of the anticline has broken through to the surface. Regional tectonically induced downcutting has been subtracted from the total tectonic displacement measured in the anticline reach to produce the graph of Figure 1.17. The highest terrace has been raised 23 m relative to the active channel but is only about 600 years old. Downcutting curves for two reaches of the stream channel show that accelerated downcutting occurred between 0.6 and 0.2 ka. The mean local bedrock uplift rate was an astonishing 52 m/ky! This example of extreme scrunching is 50 times the estimated Holocene uplift rate for this tectonic province.

Nicol and Campbell also use the terrace ages and heights (Fig. 1.17) to assess the temporal distribution of uplift before and after an earthquake on the Karetu thrust fault that occurred 350 ± 50 years ago. Maximum uplift occurred in that century-long time span. The steep sections of the tectonically induced downcutting plot between 0.6 ka and 0.35 ka and

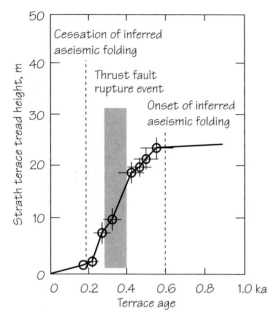

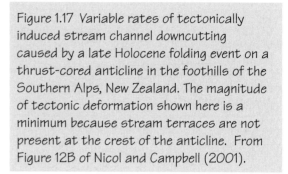

Figure 1.17 Variable rates of tectonically induced stream channel downcutting caused by a late Holocene folding event on a thrust-cored anticline in the foothills of the Southern Alps, New Zealand. The magnitude of tectonic deformation shown here is a minimum because stream terraces are not present at the crest of the anticline. From Figure 12B of Nicol and Campbell (2001).

Marine terraces provide an opportunity to assess horizontal as well as vertical movements of rocks. Pazzaglia and Brandon (2001) provide an elegant example in their discussion of the tectonic landforms of the Cascadia forarc high (Fig. 1.18). They examine coastline landforms where the Juan de Fuca plate is converging with the North American plate at 3.6 m/ky at a bearing of 54 °.

between 0.35 ka and 0.2 ka are inferred to be the result of aseismic folding.

These several examples provide interesting food for thought about scrunching that results in vertical tectonic displacements but tell us very little about magnitudes of horizontal earth deformation.

Shore platform-sea cliff landform couplets generally are created only at times of prolonged sea-level highstands, such as the present. The horizontal and vertical distances between modern and ancient inner edges of marine terraces differ as a function of the horizontal and vertical rates of tectonic displacement. Sea level has also varied in a *eustatic* (world

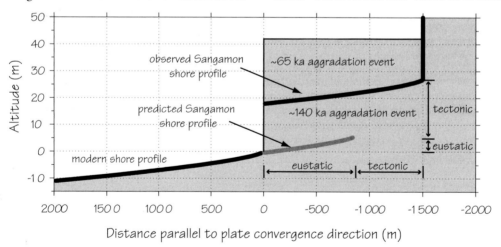

Figure 1.18 Horizontal and vertical displacement of marine-terrace landforms on a subducting plate boundary. Schematic cross section across the modern shore platform, sea cliff, 122 ka Sangamon shoreline, and a partially buried sea cliff. Mouth of the Queets River, Olympic Mountains of northwestern Washington, USA. Figure 17 B of Pazzaglia and Brandon (2001).

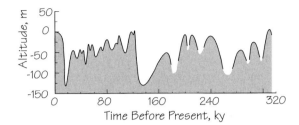

Figure 1.19 Fluctuations of global sea level since 330 ka (from Chappell and Shackleton, 1986).

1.3 Landscape Responses to Regional Uplift

Streams incise ever deeper as bedrock is raised into the powerful buzz saw of stream-channel downcutting. Amounts and rates of tectonically induced downcutting are functions of vertical tectonic displacement rates, excess unit stream power, and resistance of earth materials to degradation. Downcutting by small streams flowing over resistant welded tuff may be unable to match a bedrock-uplift rate of 0.1 m/ky; such reaches erode continuously. Downcutting by perennial rivers flowing over soft rock easily keeps pace with bedrock uplift of 5 m/ky. But stream-channel downcutting occurs only during appropriate climatic and tectonic conditions.

The tendency of streams to cut down to the minimum gradient needed to transport their sediment load has been a long standing fundamental concept in fluvial geomorphology (Powell, 1875; Mackin, 1948; Leopold, Wolman, and Miller, 1964; Leopold and Bull, 1979; Bull, 1991). Headwater reaches of streams in rising mountains tend to stay on the degradational side of the threshold of critical power, but downstream reaches, with their greater unit stream power, are more likely to attain the base level of erosion through the process of *tectonically induced downcutting*.

Gregory and Chase (1994) minimize the influence of base level in a diametrically opposite model. They conclude that Cenozoic canyon cutting in the Front Range of the Rocky Mountains in Colorado resulted entirely from climatic changes that increased stream power. Molnar and England (1990) also favor dominance of climate-change causes of stream-channel downcutting in this same region. The resulting isostatic uplift would promote further canyon downcutting. Zaprowski et al., (2005) prefer a model where climatic changes would increase the concavity of the longitudinal profiles of rivers crossing the western Great Plains. Greater concavity would require more intense rainfalls and larger, more frequent flood events in the Quaternary than during the Pliocene.

However, analysis of gradient changes of rivers flowing east from the Rocky Mountains (Figs. 1.20A, B) reveals that flexural isostatic rebound of the lithosphere due to Cenozoic erosional unloading accounts for only 20% of the concurrent increase of relief (McMillan et al., 2002). Therefore, tectonic uplift is necessary in order to explain the Front Range canyon cutting and the concurrent deepening of val-

wide) sense (Fig. 1.19) due to changing volumes of glacial ice and ocean temperatures (Shackleton, 1987). The Sangamon sea-level highstand at about 124 ka was about 5 m above the present ocean level. This means that even under tectonically inactive conditions the Sangamon shoreline for this gently sloping coast should be 5 m higher and quite far inland.

Marine terraces were created at times of globally synchronous sea-level highstands (Chappell, 1983; Chappell and Shackleton, 1986; Lambeck and Chappell, 2001). The ~124 ka terrace has been uranium-series disequilibrium dated using coral from New Guinea, New Hebrides, Barbados, Haiti, the Mediterranean Sea, Hawaii, Japan, and California. Bloom et al. (1974) estimated the altitudes of many sea-level highstands. These were brief time spans of unchanging terminal base levels for rivers, much like the past 6 ky. Remnants of shore platforms created at a variety of sea levels indeed are useful time lines passing through tectonically active landscapes. Rapid sea-level changes between the highstands raised and lowered the mouths of streams but this is not the same type of base-level fall as faulting of a streambed (Sections 2.2.4, 2.6).

The Figure 1.18 analysis assumes similar geomorphic processes and alluvium mass strength (gravels) at Sangamon time as compared to the present. Because of the higher eustatic sea level the Sangamon highstand sea cliff would have formed at 945 ± 145 m inland of the modern sea cliffs. Instead it is located an additional 505 ± 150 m farther inland. This suggests a horizontal tectonic displacement of about 450 m during the past 122 ka or a mean horizontal tectonic velocity of 3.7 ± 1.1 m/ky. The eustatic component is much smaller than the tectonic component for uplift during the same time span and is simply the change in world-wide sea level.

leys across the adjacent Great Plains. McMillan et al. conclude that post-depositional changes in slope of the stream channels in the western Great Plains of Wyoming and Nebraska since 18 Ma are the result of broad-wavelength tectonic uplift centered under the Rocky Mountains. Uplift began during deposition of braided-stream gravels of the Miocene Ogallala Formation. Tectonically induced downcutting has continued to lower the active stream channels relative to the strath beneath the basal Ogallala fluvial gravels.

It is not easy to discern uplift in landscapes that lack obvious Quaternary faulting and folding. So this book emphasizes tectonic influences on the landscapes of individual watersheds, preferably where tectonic controls are obvious such as active range-bounding faults. McMillan et al. were able to estimate regional tectonic influences on landscape evolution with a combination of paleohydrologic, stratigraphic, and geophysical analyses involving a spatial scale of 250 km and a time span of ~15 My. This challenging project produced some interesting results.

The post Ogallala time span coincides with the gradual northward extension of the Rio Grande rift from New Mexico; and it seems reasonable that the accompanying regional tectonic uplift also decreased towards the north. Tectonic rock uplift was followed by an episode of erosion-induced isostatic uplift that began when the rivers of the region ceased deposition and began 5 My of fluvial degradation.

Leonard (2002) analyzed the larger valleys draining the eastern flank of the Rocky Mountains. Uplift caused tectonically induced downcutting, which promoted isostatic uplift. He assumes that the base of the Ogallala formation was planar and tilted eastward. The Arkansas River valley in southeastern Colorado was eroded to deeper levels than the valley of the North Platte River in southeastern Wyoming. Maximum warping of the Colorado piedmont occurred near the Arkansas River. Leonard's modeling suggests that the isostatic component of rock uplift (Fig. 1.4) accounts for 50% of the total rock uplift with the remainder being tectonic uplift. About 540 m more uplift occurred along the Figure 1.20C transect at the Arkansas River valley than at the valley of the South Platte River. Leonard's results approximate the lesser uplift amounts suggested by McMillan et al., (2002) along the valley of the North Platte River.

These thoughts about the diverse character of uplift will be used when we explore uplift of specific mountainous landscapes. Chapters 3 and 4 go into more detail regarding the complications that arise when one attempts to determine how fast the mountains are rising. Figure 1.4 is a rudimentary summary. It hits the main points, but the influences of many of the variables are not easily constrained to the tidy boxes of this simple model. Deep seated crustal flow (Zandt, 2003) is largely ignored. The isostatic component of rock uplift is the algebraic sum of many processes. It is a function of crustal temperature or mineralogy changes, crustal subduction, spreading and flexing, strike-slip fault loading changes, tectonic denudation or burial, faulting and

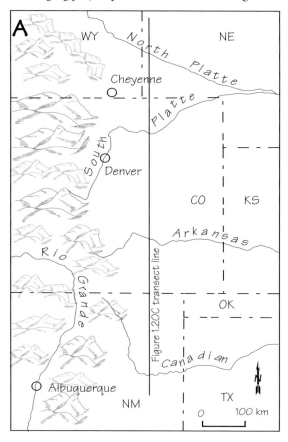

Figure 1.20 Late Cenozoic tectonic and isostatic uplift of the Colorado piedmont east of the Rocky Mountains.
A. Trunk channels of the major rivers flowing eastward across the western Great Plains. Rio Grande is in a rift valley that has propagated northward during the Cenozoic.

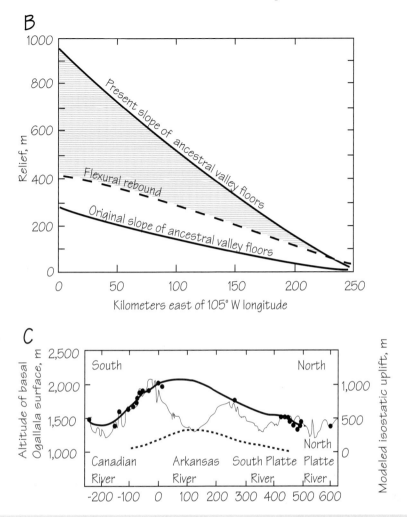

Figure 1.20 Late Cenozoic tectonic and isostatic uplift of the Colorado piedmont east of the Rocky Mountains.

B. Post-depositional changes in slope of fluvial gravels in the western Great Plains of Wyoming and Nebraska, USA. The "present slope" is the strath at the base of the 18 Ma Ogallala Group fluvial gravels. The estimate of the original depositional slope – paleosurface slope – during gravel deposition was calculated using an equation (Paola and Mohrig, 1996) relating size of gravel in a braided stream to depth and slope of flow. Dashed line shows the modeled uplift of the Ogallala strath caused by isostatic rebound due to erosion. Change in relief is relative to fixed hinge point at eastern edge of study area. Flexural rigidity used in model is 1024 N·m. From Figure 4 of McMillan et al. (2002).

C. Polynomial fit of basal Ogallala Formation surface (heavy solid line) based on reference points (solid circles) projected into transect line of Figure 1.20A. Modeled flexure due to erosional unloading (dashed line) used a flexural rigidity of 1024 N·m. Topographic profile shown by thin solid line. Transect line is along 103° 50' W meridian. From Figures 2 and 3 of Leonard (2002).

folding that changes crustal thickness, and the geomorphic processes of erosion, deposition, solution, and compaction.

The bedrock-uplift concept modernizes the ways in which we study tectonics of mountain ranges on active or passive plate margins. Geomorphology now plays a major role in studies of earth history because of the need to understand landscape responses to uplift caused by either tectonic or isostatic uplift. Temporal and spatial scales of study dictate research objectives and procedures. Evaluation of sediment flux from continental landmasses to ocean basins uses different time spans, areas, and geomorphic processes than local erosion and deposition associated with a single-rupture event fault scarp. Conceptual geomorphic models that seem ideal for their formative study area and dataset may become rather tenuous when applied to different spatial, tectonic, and climatic settings.

Fortunately, individual drainage basins are the basic components of mountain-range landscapes and even for rivers that flow across continents. This book appraises tectonics and topography at the scale of moderately small watersheds with an emphasis on processes that shape hills and streams. Response times for drainage nets and their adjacent hillslopes are best studied when tectonic perturbations are nearby. Different rock types and climates add spice to landscape studies. These important variables need to be added to the recipe if you want to more fully understand the tectonic geomorphology of mountains. All these aspects are accommodated by the basic theme of fluvial-system behavior.

Now that we have outlined the essence of bedrock uplift, let us examine constraints on the limits of how fast and far streams can cut down into landscapes at the watershed scale. The next chapter is about specific processes and landforms that help us understand how active faulting and folding shapes the hills and streams of fluvial systems.

Chapter 2

Concepts for Studies of Rising Mountains

Apulse of uplift along a range-bounding fault is transmitted to all parts of a fluvial landscape. How does this occur? Streams are the connecting link between the different parts of watersheds that we treat here as fluvial systems. Is this connecting link equally strong in all humid and arid watersheds? How long does it take for a tectonically steepened reach to migrate upstream to the headwaters? Surely large rivers respond more quickly than small streams. Hillslope-erosion processes don't even start to feel the effects of increased relief until the upstream migrating steeper stream reach arrives at their footslopes. Response times to a surface-rupture event on a range-bounding fault vary greatly with drainage-basin area, climate, and rock type. In contrast, the impacts of a seismic-shaking event are felt quickly throughout modest-size watersheds as changes in sediment yield and mass-movement processes.

Let us discuss standard and new ways to study fluvial-system behavior with an emphasis on responses to tectonic perturbations. Chapter 2 concepts will help you evaluate and explore new and diverse approaches in tectonic geomorphology. They are my foundation for understanding interrelations between tectonics and topography.

Aerial view of Laguna Salada, Mexico. An active normal fault separates landscapes characterized by erosion and deposition. Photograph by Peter L. Kresan ©.

2.1 Themes and Topics

Chapter 2 concepts focus on rates and styles of geomorphic processes in diverse tectonic, climatic, and lithologic settings. Continental landscapes of planet Earth are formed in large part by interactions of tectonic and fluvial processes, which are modulated by pervasive late Quaternary climate changes. *Tectonics* is the study of crustal deformation: the evolution of geologic structures ranging from broad transition zones between crustal plates to small faults and folds. *Geomorphology* is the study of landscapes and the processes that shape them. The influences of vertical and horizontal earth deformation on fluvial, coastal, and glacial processes and the resulting landscapes comprise the domain of *tectonic geomorphology*.

Many processes shape the surface of planet Earth, but the action of running water is responsible for most subaerial landscapes. This book is about *fluvial systems* – hilly to mountainous source areas that supply water and sediment to streams, which convey their load to depositional basins. Sustained uplift along active faults and folds may create mountain-front escarpments. Tectonically active mountain fronts appeal to tectonic geomorphologists, because uplift steepens stream gradients, which accelerates *watershed* (synonymous with the term *drainage basin*) erosion by making hillslopes steeper.

Climatic changes may create landforms that approximate time lines passing through tectonically deforming landscapes. Examples include shoreline marine terraces from polar to tropical realms, and abandoned flood plains rising like flights of stairs above rivers. Both tectonic landforms are vital sources of information about local uplift history and regional isostatic adjustments.

Tectonic geomorphology has two facets – basic research to better understand landscape evolution and practical applications. Basic research is applied to define potential hazards posed by active tectonics processes (Hecker, 1993) and to diminish risk to people and engineering structures. Engineers and planners need knowledge gained from landscape studies (Fig. 2.1) in order to predict *earthquake hazards* (frequency and magnitude of surface ruptures, seismic shaking, and coseismic tsunamis, landsliding ,and flooding) to minimize *earthquake risk* (loss of life and property).

Chapter 1 discussed several processes that raise and lower land surfaces. This chapter examines how far streams can cut down into bedrock – their base level limit as they erode mountain ranges into drainage basins. We also explore fluvial-system behavior to lithologic and climatic controls in different tectonic settings in the context of response times, the threshold of critical power, and tectonically induced downcutting. These concepts will give you the necessary foundation for perceiving the nuances of how hills and streams respond to mountain-building forces.

2.2 The Fundamental Control of Base Level

2.2.1 Base Level

Studies of tectonics and topography use base level as a reference datum for rivers. John Wesley Powell (1875, p. 203–204) introduced the term *base level* as the altitude below which a stream cannot downcut. The ocean is regarded as a general, or *ultimate, base level* even though Quaternary sea levels have fluctuated 130 m (Chappell and Shackleton, 1986: Chappell, 2001). Fluvial processes cease where rivers flow into lakes or the ocean, because the hydraulic gradient is reduced to zero and potential energy is not further transformed into kinetic energy. The concave longitudinal stream profile that typically develops upstream from a base level reflects adjustments between hydraulic factors that G.K. Gilbert referred to as "an equilibrium of action" (1877). Anomalously steep and narrow reaches may reflect lithologic controls on a longitudinal profile (Kirby et al., 2003).

Many *local base levels* occur between the headwaters of a stream and the terminus in an ocean, lake, or basin of internal drainage (Fig. 2.2). Resistant outcrops are considered local base levels because a stream is unable to lower its bed as easily as through relatively softer materials in adjacent upstream and downstream reaches. Downcutting promoted by uplift is reduced or delayed (sometimes greatly; see Sections 2.5.1, 2.5.2) where resistant outcrops create relatively stable reaches. Such local base levels are temporary compared to the relative permanence of the oceans. Mean sea level may change slowly but short-term fluctuations of sea level are as much as 130 m in 15 ky. Downstream parts of fluvial systems at 20 ka now are deep under the sea. Alluvial reaches of streams may be regarded as an even more temporary category of local base levels. Indeed, each point along a stream, be it underlain by rock or alluvium, is part of a continuum of streambed altitudes. Each short reach of a stream exerts a base-level control on adjacent reaches that partly determines the longitudinal profile and stream-channel patterns such as meandering and braided.

Base levels for adjacent reaches of a stream can be raised or lowered. The *base-level processes* of aggradation and degradation may be caused by either tectonic or climatic perturbations, or they can result from internal adjustments initiated by changes in the hills and streams of a fluvial system. The exciting challenge for the tectonic geomorphologist is to recognize and interpret key features of tectonically controlled aggradation and degradation within a fluvial system that provide clues about styles and rates of earth deformation.

2.2.2 Base-Level Change

The spatial consequences of base-level change emanate both upstream and downstream from a base-level perturbation. *Base-level fall*, such as tectonic lowering of a streambed downstream from a fault or fold axis, is readily transmitted upstream by creation of a short reach of increased gradient. This local increase in stream power tends to initiate degradation. The degrading reach propagates upstream as headcuts, waterfalls, and rapids that become smaller as the perturbation migrates away from its tectonic origin. An important consequence of accelerated stream-channel downcutting is the increase of sediment yield caused by undercutting that steepens adjacent hillslopes. The result is an increase in bedload transport rate

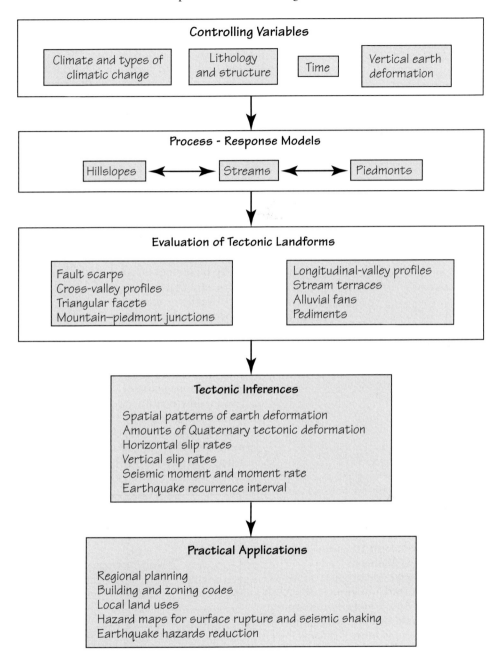

Figure 2.1 Flow-chart checklist for tectonic-geomorphology studies.

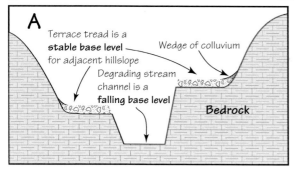

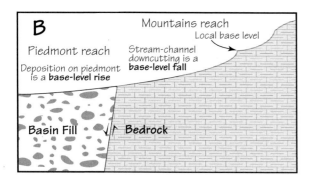

Figure 2.2 Reaches of stable, falling, and rising base level along a hypothetical fluvial system in an arid region. Mountains are being raised relative to the basin on a normal fault so the stream is degrading.
A. Cross-valley topographic profile.
B. Longitudinal stream topographic profile. Waterfall knickpoint is a local base level that separates stream-channel reaches and hillslopes with different characteristics.

that tends to offset the effects of an anomaly created by a tectonically steepened gradient.

Base-level fall affects reaches that are downstream from a tectonic perturbation much differently than upstream reaches. Active range-bounding normal faults typically separate eroding mountains from aggrading basins. An example from Mexico is shown in the photo on the title page for this chapter. A base-level fall that accelerates degradation upstream from an active fault also creates the space for accumulation of thick basin fill downstream from the fault.

The effects of *base-level rise* seem to be minimal in a variety of settings. An aggrading playa lake bed affects only short terminal reaches of streams. The

effects of base-level rise are not transmitted nearly as far upstream as those of a base-level fall. Construction of a dam across a stream is a local base-level rise that is propagated upstream from the reservoir (Leopold and Bull, 1979; Leopold, 1992; Gellis et al., 2005). The longitudinal profile of the newly created valley fill would be the same as before the base-level rise if equal thicknesses of new alluvium extended to the headwaters. This does not occur. Instead, a wedge of alluvium is deposited that extends only as far upstream as needed to maintain the slope at 50 to 70% of its original value. Other variables, such as hydraulic roughness, change concurrently as gradient is decreased. The affected reach reestablishes a new unchanging (equilibrium) configuration through a different set of interactions between variables. The sediment wedge does not migrate farther upstream. So, reaches upstream from the small wedge of new valley fill are not affected by the base-level rise. One important conclusion is that aggradation of river valleys over long reaches is primarily the consequence of climatic-change impacts over the hillslopes of an entire watershed. Climatic perturbations influence bedload supply from hills and transport capacity of streams in ways that maintain the process of deposition for the duration of an aggradation event.

Propagation of the effects of a local base-level rise in downstream reaches is minor. Local aggradation along a mountain stream creates a patch of alluvium consisting of a gentler upstream reach and steeper downstream reach (Fig. 2.3). The locations where the stream changes its mode of operation from degradation to aggradation (or vice-versa) are

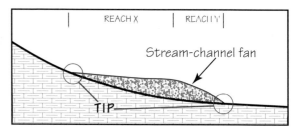

Figure 2.3 Longitudinal profile of a valley floor showing adjacent alluvial reaches that are more gentle (X) and steeper (Y) than the slope of the bedrock channel prior to temporary deposition of an alluvial channel fan. Threshold-intersection points are shown by TIP.

threshold-intersection points. The steepened reach is inherently unstable. Increase of stream power favors entrenchment, leading to the development of an incised channel that concentrates flow, thereby initiating a self-enhancing feedback mechanism that favors removal of the patch of alluvium. Patches of alluvium deposited in active channels tend to be temporary. Localized brief episodes of accelerated deposition also occur in depositional environments such as deltas and alluvial fans where sediment tends to subsequently be redistributed. Aggradation upstream from lakes and the ocean is different because it occurs in the terminal reach of a stream.

2.2.3 The Base Level of Erosion

I now define and use a valuable equilibrium concept that describes adjustments between the hierarchy of adjacent reaches in a drainage net. The *base level of erosion* is the equilibrium longitudinal profile below which a stream is unable to degrade and at which neither net vertical erosion nor deposition occurs (Powell, 1875, p. 203-204; Barrell, 1917). A stream cannot permanently degrade below its base level of erosion and maintain the gradient needed to transport its bedload. A reach of stream at the base level of erosion has achieved a time-independent configuration of its longitudinal profile that is maintained as long as the controlling variables do not change in an average sense over suitably long time spans. This notion integrates the system, equilibrium, and base-level concepts. It considers the longitudinal profile spatially as being an infinite sequence of adjacent base levels (Gilbert, 1877), and temporally as capable of being reestablished at multiple positions within a rising landscape. Changes in any of the variables affecting stream power or resisting power (terms are defined in Figure 2.13) may change the base level of erosion.

The importance of base level in studies of landscape evolution can be illustrated by the simple example of a stream flowing into a lake, or the ocean. The terminal reach of the stream cannot erode below the lake level because it will no longer have sufficient gradient (stream power) to transport the bedload supplied to it by upstream reaches. So the reach can only cut down to the minimum gradient needed to convey its sediment load with the prevailing stream discharge. Of course, both of these dependent variables of fluvial systems changed whenever late Quaternary climate changed.

An upstream migration of the base level of erosion begins at the lake base level. Assume that next reach upstream from the terminal reach is steeper than is needed to convey the bedload and that resistance of rocks to erosion remains constant. The excess stream power will degrade the streambed until it too reaches the base level of erosion. *Unit stream power* (discharge per unit width of streamflow width) becomes progressively less in consecutive upstream reaches, so more time is needed to downcut to the base level of erosion. This attainment of equilibrium conditions progresses spatially at exponentially slower rates because stream discharge decreases upstream in an exponential manner. Attainment of the base level of erosion in a sequence of upstream reaches of a stream initiates spatially consecutive pulses of landscape change that migrate up the adjacent valley sides.

Bedrock reaches in the headwaters of a watershed never achieve the base level of erosion. Matmon et al. (2003) came to the same conclusion for the southern Appalachians where they evaluated bedrock-erosion rates based on [10]Be analyses of river sediment samples. "... it appears that Hack's dynamic equilibrium might never be achieved at the scale of headwater streams" of <50 km[2]. This is because equilibrium longitudinal profiles are never attained in the headwater extremities of the drainage basin. The time needed to achieve the base level of erosion stretches to infinity for the mere trickles of headwater streamflow. Rock mass strength becomes progressively more important upstream.

Eaton and Church (2004) used a stream table to examine the relation between equilibrium stream channel morphology and discharge, bedload supply, and valley slope. Streambed and channel morphology changes were minor, except for channel slope. "... the system tends to move toward the minimum slope capable of transporting the sediment supply".

Attainment of the stream channel base level of erosion affects hillslopes too. A valley floor at the base level of erosion is a stable base level for the adjacent hillslopes in much the same way as a lake is the base level for the terminal reach of the stream.

Hillsides adjacent to stable valley floors typically have three segments. The *footslope* is the concave base of the hillslope. It is a surface of detrital conveyance where geomorphic processes cannot further lower a hillslope graded to broad valley floor. The *crestslope* is the convex erosional topographic profile descending from the ridgecrest. The *midslope* is where hillslope topography changes from convex to concave and may be long and straight. Concave footslopes are most common in downstream reaches of a watershed and midslopes and footslopes may not be present in upstream reaches if rapid stream-channel downcutting generates hillslopes that are convex from ridgecrest to valley floor.

The base level of erosion concept is important because it defines equilibrium conditions that favor beveling of broad bedrock valley floors (Hancock and Anderson, 2002). Previous valley floors may be preserved above an active channel as flights of stream terraces when valley-floor downcutting by the stream is renewed at times of excess stream power. Several types of stream terraces are defined in Section 2.4.1. The point here is that straths are surfaces beveled by streams while at the base level of erosion. Return to similar climatic conditions promotes development of parallel strath terraces. Vertical separations between successive straths are a measure of stream-channel downcutting induced by bedrock uplift. Each longitudinal terrace profile represents a similar base level of erosion that has been raised tectonically relative to the active channel.

A reach of a stream at the base level of erosion is easily identified in the field as a surface of detrital sediment transport. Bedload is moved across the beveled bedrock surface leaving a veneer of gravel on top of a roughly planar bedrock surface.

Repeated aggradation events during the Pleistocene were followed by stream-channel down-cutting. The past 10 to 15 ky of net degradation has allowed even small streams of humid regions to catch up with their base level of erosion. This is why the present is a time of strath formation for many streams, for example the Ventura River (Rockwell et al., 1984) of southern California and the Reno River of Italy (Fig. 2.4). This position remains the same in tectonically stable landscapes (Fig. 2.5), but will be at progressively lower positions in landscapes of rising mountain ranges, including both of the above examples.

The time needed to achieve the base level of erosion is a function of 1) available unit stream power in excess of that needed to transport sediment and overcome hydraulic roughness, and 2) the resistance to erosion of materials beneath the streambed. Duvall et al. (2004) analyze the mutual influence of uplift rate and rock type on longitudinal profiles of streams in the California Coast Ranges. Streams downcut and then bevel straths quickly in reaches underlain by soft materials such as unconsolidated mudstone.

Streams of tectonically inactive regions also achieve the base level of erosion but flights of strath terraces are not the typical landform of these fluvial systems. Valley-floors may be buried instead of being raised above the active channel and preserved as strath terraces. Note the similar depths of incision shown in Figure 2.5. Even the depth of incision of the late Pleistocene valley was at the same base level of erosion in this tectonically inactive setting, despite changing streamflow characteristics as a result

Figure 2.4 Strath forming under the active channel of the Reno River at Marzabotto, northeastern Italy. The 4,597 km² watershed is characterized by frequent large floods. Large unit stream power allowed the river to remain at the base level of erosion during the late Holocene. Lack of gravel in the active channel may in part be the result of prolonged mining of the streambed for concrete aggregate. Photo provided courtesy of Frank Pazzaglia.

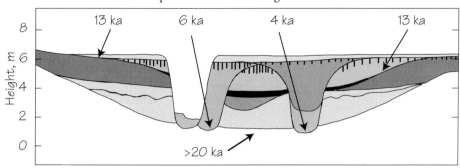

Figure 2.5 Cross section at Curry Draw, southeastern Arizona showing three Holocene episodes of arroyo cutting and backfilling since deposition of the Coro Marl at 13 ka. Vertical hachures indicate the relative ages of soil-profiles and the numbered arrows date times of maximum channel down-cutting for three prehistoric valley-floor degradation events. Note the similar 4 to 4.6 m depths of the modern, 4 ka, and 6 ka arroyos. Similar levels of backfilling were attained by the intervening aggradation events. Stratigraphic cross section is from Haynes (1987). Figure 44 of Bull (1997).

of Pleistocene–Holocene climatic change. Stream behavior changed after the Pleistocene–Holocene transition, and many streams in the American southwest underwent repeated episodes of arroyo cutting (Bull, 1997). This unchanging base level of erosion clearly reveals lack of significant uplift during the late Quaternary. Neither isostatic nor tectonic uplift occurred here so former valley floors were not isolated by the process of tectonically induced downcutting.

The base level of erosion helps us understand why stream channels, although occupying only a miniscule percentage of mountain-range area, are so important in controlling bedrock erosion rates. The base level of erosion for each consecutive reach restrains degradation of the adjacent hillslopes. Consequently the regional drainage net controls the tectonic geomorphology of mountainous landscapes.

2.2.4 The Changing Level of the Sea

We use sea level as the ultimate base level for specific rivers and as the reference framework for altitudinal positions within fluvial landscapes. Rotation of the Earth causes the radius at the poles to be about 21 km shorter than at the equator, and gravity varies on the surface of this oblate spheroid because of spatial variations in the mass of Earth's crust and mantle. Earth's mean sea level, although perfectly level at each point, has minor bumps and hollows as large as 106 m (Heirtzler and Frawley, 1994). So the *geoid* is that

particular equipotential surface of the Earth's gravity field which best fits, in a least squares sense, global mean sea level. In contrast to minimal rates of non-climatic change of this level surface, the altitude of a river mouth varies greatly over short time spans as sea level rises and falls.

A key question concerning the importance of the ocean as a base level is whether or not entrenchment of coastal streams occurred during times of declining sea level as continental glaciers expanded. The answer lies in the slope of the new reach of a river that is created as sea-level decline exposes part of the continental shelf. Assume that no changes occur in discharge of water and sediment from the upstream reach. The continental shelf for the first 100 m below present sea level commonly is gentler than the slope of the adjacent upstream reach (Fig. 2.6A). This gentler new terminal reach would tend to aggrade because of a decrease in stream power (lesser slope). The opposite situation would occur where 100 m of sea-level decline dropped a river into the head of a submarine canyon, or onto the steeper foreset part of a delta. The new terminal reach would be steeper than the adjacent upstream reach and would tend to degrade (Fig. 2.6B). In contrast to the aggradation case, such channel incision is a base-level fall that can be propagated far upstream.

The depth to the edge of the continental shelf and its distance offshore play important roles in the responses of rivers to decline in sea level during the late

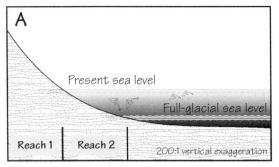

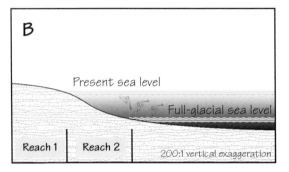

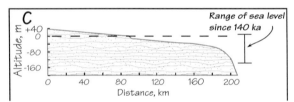

Figure 2.6 Changes in gradients of terminal reaches of coastal rivers caused by sea-level fall or rise.
A. Diagrammatic sketch of 100 m of sea-level fall that decreases fluvial hydraulic gradient. Decrease of gradient in reach 2 favors aggradation.

This model of effective base-level rise also applies to basins of internal drainage where lakes become playas when the climate changes.
B. Diagrammatic sketch of 100 m of sea-level fall that increases fluvial hydraulic gradient. Increase of gradient in reach 2 favors stream-channel entrenchment of both reaches when sea level falls.
C. Topographic and bathymetric profile of the Colorado River of Texas where it enters the Gulf of Mexico. From Figure 1 of Talling (1998). See Figure 1.19 for sea-level curve since 140 ka.

Quaternary. Talling (1998, Fig. 1) shows topographic and bathymetric profiles for river mouths in a variety of plate-tectonic settings. An example from a passive margin setting illustrates some key points (Fig. 2.6C). About 40 m of sea-level decline would extend the mouth of the Colorado River off the present coastline of Texas by almost 100 km but the gradient would remain essentially unchanged. Aggradation or degradation of this reach would therefore be in response to changes in discharge of water and/or sediment load. The major sea-level decline of the latest Quaternary represented a lowering of base level of about 130 m below the present level of the sea. This resulted in sufficient extension of the mouth of the Colorado River so that it dropped over the outer edge of the continental shelf. The result was a major base-level fall that propagated far upstream (Blum and Valastro, 1994). The Mississippi River also drains into the Gulf of Mexico but underwent a different response to the same sea-level fall. Talling (1998) notes the 200 times greater rate of sediment delivery to the mouth of the Mississippi River as compared to the Colorado river. The result is that the Mississippi River delta has been built out to the edge of the continental shelf, so even moderate sea-level fall substantially steepens the terminal reach of the river. But stream-channel entrenchment does not necessarily result from this base-level fall.

Times of maximum sea-level decline also were times of large increases in sediment load in the drainage basins of many mid-latitude rivers as a result of increased glacial erosion and periglacial processes on hillslopes. Thus it is common for fluvial aggradation of some river valleys to coincide with times of low sea levels (Bull, 1991, Section 2.1.3). The great influx of bedload may exceed the concurrent increase in stream power where a falling sea level steepens the terminal reach of a river. Deposition would result. Postglacial alluviation resulting from Pleistocene glacier retreat is one of several *paraglacial* processes (Ballantyne, 2002 a,b).

Large rivers may entrench and backfill their terminal reaches. In studies of the lower Mississippi River, Fisk described 100 m of late Quaternary channel entrenchment (Fisk, 1944, 1947; Saucier, 1994, 1996; Albertson and Patrick, 1996). Degradation was in part due to increased stream power as sea-level decline exposed the foreset beds of the delta and created a steep terminal reach. Entrenchment propagated upstream, but was followed by aggradation when influx of glacio-fluvial detritus overwhelmed

Figure 2.7 Shore platform at Kaikoura Peninsula, New Zealand beveled during the past 6 ky of stable sea level. Rocks are folded and faulted mid-Tertiary mudstone and limestone. Waves keep the platform swept clean of littoral deposits. Two persons near base of sea cliff for scale.

the effects of coastal base-level fall and backfilled the entrenched channel. The terminal reach of the river became progressively gentler during the ensuing sea-level rise, which promoted further deposition that continues presently.

Streams do not permanently degrade below their ocean base level. It is unlikely that a river will scour 80 m below sea level even temporarily during major floods if a terminal reach has an altitude of only 20 m. However changing the total relief from 10 to 120 m, by 100 m of sea-level decline, might allow 80 m of stream-channel entrenchment because of diminished constraint of the ocean as a "permanent" base level. Such entrenchment could occur only when discharges of sediment and water favored a degradational mode of operation – for example, during an increase in water discharge and a decrease in bedload transport rate.

Two conclusions emerge. First, sea-level position constrains the amounts of possible channel downcutting. Second, changes in sea level may not be as important as changes in streamflow characteristics in determining aggradation and degradation modes of stream operation in terminal reaches of rivers.

Marine base-level controls also include lateral erosion into sea cliffs during sea-level highstands to create shore platforms. The sea has remained at about the same level during the past 6 ka (Lajoie, 1986), which has favored retreat of rocky coastlines (Fig. 2.7) that would not undergo much erosion if only briefly exposed to the force of waves.

Consider the interactions of a stable sea-level highstand and rivers along two steep coastlines with hills of soft rock; one coastline is tectonically stable (Fig. 2.8A) and the other is rising (Fig. 2.8B). With no uplift and a stable sea level for 6 ky, shore platforms become wider as sea cliffs retreat. The rate of widening of the shore platform decreases with time, because wider platforms are more effective in dissipating wave energy. Shore-platform widening causes a *lateral erosion induced base-level fall* where streams enter the ocean. This base-level fall can be large for steep streams. The river cuts down to a new equilibrium profile, similar to the former longitudinal profile, as the consecutive base-level falls migrate upstream. The new profile is offset horizontally by a distance equal to the amount of sea-cliff retreat, and is lower than the profile before the sea-level rise by an amount equal to the lateral erosion induced base-level fall (Fig. 2.8A).

Much different interactions prevail for a rising coastline and a stable sea level (Fig. 2.8B). Sea-cliff retreat could have occurred during all of the past 6 ka for the stable coastline (Fig. 2.8A), but only for about 1 to 3 ky for a coastline rising at more than about 1 m/ ky. Furthermore, much of this brief period of coastal erosion would have occurred prior to attainment of the 6 ka sea-level highstand. Optimal conditions for shore-platform development occur when the rates of sea-level rise and uplift are similar. Assume that the shore platform began to form at 8 ka in the Figure 2.8B example. Coastline retreat, by 5 ka, would be much less than for the tectonically inactive coast, but other processes create wide shore platforms. Rock that is raised into the surf zone is beveled too. The inner edge of the shore platform coincided with the river mouth at 8 ka in the hypothetical situation portrayed in Figure 2.8B. Then 8 ky of uplift raises the longitudinal profile of the river to a higher position. The river continues to cut down in response to a base-level fall that is the sum of the uplift and lateral erosion induced base-level fall.

Erosiveness by a variety of shore platform processes (Kirk, 1977: Stephenson and Kirk, 1998, 2000a, 2000b, 2001) and erodibility of the rock type largely determine whether or not shore-platform degradation equals bedrock-uplift rate. The platform enlarges mainly in a seaward direction as rock is raised into the surf zone after initial sea-cliff retreat at the beginning of a sea-level highstand. Wave abra-sion may be more important on the outer edge than on the inner edge where salt weathering and biota are key factors of bedrock erosion. Shore-platform degradation has kept pace with bedrock uplift along parts of the New Zealand coast (Fig. 2.9).

Landscape evolution at the mouths of rivers also is a function of available stream power. Small streams may be unable to achieve the base level of erosion. Downcutting in response to uplift of their coastal reaches steepens their longitudinal profiles. Such streams undergo a net increase in altitude while continuing to maintain a concave longitudinal pro-file. Powerful rivers generally have rates of down-cutting that equal bedrock-uplift rates. They tend to remain at the same position, relative to sea level, and have similar longitudinal profiles throughout a sea-level highstand. The position of the longitu-dinal profile in the landscape is progressively lower

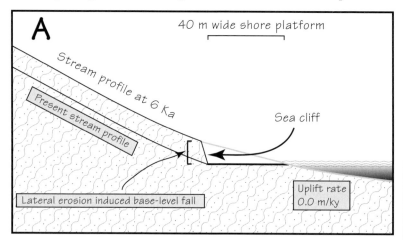

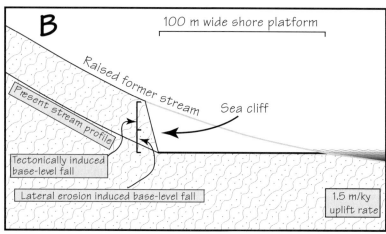

Figure 2.8 Influences of a 6 ky of stable sea level on the longitudinal profiles of powerful rivers.

A. Lateral erosion induced base-level fall along a tectonically stable coast.

B. Elevation of a 6 ka stream profile along a rapidly rising coast.

Figure 2.9 Shore platform at Kaikoura Peninsula, New Zealand that has not been widened by trimming of the sea cliff since a mid-Holocene coseismic uplift event. Late-Holocene increase of shore-platform width is the result of ~0.5–1.0 m/ky of rock uplift that has raised soft rocks into the surf zone.

than before the sea-level rise by an amount equal to the sum of the lateral erosion induced base-level fall, plus the amount of bedrock uplift. Coastal erosion base-level fall creates mountain fronts with characteristics that are similar to faulted mountain fronts (Fig. 2.10). Lateral erosion induced base-level fall affects hillslopes as much as river valleys. Larger streams of the Seaward Kaikoura Range in New Zealand deliver so much sediment to the coast that deltas are deposited. Such depositional base-level rise keeps the waves away from the mountain front.

Apparent lateral erosion can also be the result of tectonic horizontal displacements. Pazzaglia and

Figure 2.10 Truncated coastline where the Seaward Kaikoura Range, New Zealand has an abrupt, straight escarpment, deeply incised V-shaped canyons, and triangular facets.

Brandon (2001) describe how a shoreline that was tectonically translated towards the ocean results in apparent uplift of marine terraces.

2.2.5 Spatial Decay of the Effects of Local Base-Level Changes

Exponential equations may be used to describe either the curving longitudinal profiles of streams, or the decreasing influence of a pulse of uplift generated by a range-bounding fault with increasing distance upstream. Hack (1957) defined the exponential nature of longitudinal profiles of streams. Morisawa (1968) described the relation between decrease in altitude, H, and distance in the downstream direction, L, as

$$H = be^{-mL} \qquad (2.1)$$

where b and m are constants.

The effects of vertical tectonic displacement of a streambed decrease exponentially upstream and downstream from the surface rupture. If P_o is the magnitude of the base-level fall (Fig. 2.11), and P_L is the magnitude of change in streambed altitude at distance L from the fault, then:

$$P_L = P_o e^{-K_{bl}L} \qquad (2.2)$$

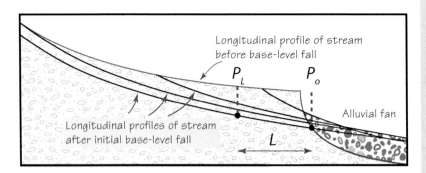

Figure 2.11 Decreasing effects of a local base-level fall upstream and downstream from a hypothetical lateral erosion induced base-level fall. P_o is the magnitude of the perturbation, and P_L is the effect of the base-level perturbation at an upstream distance L.

Equation 2.2 describes spatial adjustments by the processes of tectonically induced downcutting or aggradation. The base-level-reaction constant K_{bl} is

$$K_{bl} = \frac{ln\, P_o - ln\, P_L}{L} \qquad (2.3)$$

K_{bl} is also a function of discharge, structure and lithology, and climate and vegetation–all of which influence the rate of tectonically induced downcutting. Values of K_{bl} are small because the numerator of equation 2.3 is divided by the horizontal distance from the tectonic perturbation.

Strath terrace longitudinal profiles commonly are parallel. Values of K_{bl} approach 0.0 for pairs of terraces where minimal net change in the longitudinal profile has occurred after re-establishment of equilibrium conditions.

Maximum values of K_{bl} occur where vertical tectonic displacement is large relative to distance from the active fault zone, and where stream power or time has been insufficient for the stream to return to a longitudinal profile similar to that present before vertical tectonic deformation. When used in this context, K_{bl} describes the degree of departure, from pre-base-level fall conditions.

Both erosion and deposition in fluvial systems respond to elevation by mountain-building forces. In 1887 A.D., a small ephemeral stream crossing the Pitaycachi fault in northeastern Sonora, Mexico was displaced 3 m vertically by a Magnitude 7 earthquake (Bull and Pearthree, 1988). A major event, this is the longest known surface rupture by a normal fault. This large tectonic perturbation greatly affected fluvial processes in the reaches upstream and downstream from the surface rupture. The soil profile on a pre-

1887 alluvial fan downstream from the fault zone has characteristics that correlate with the soil profile on an alluvial fill terrace upstream from the fault zone. Both were part of a continuous alluvial surface (Fig. 2.12A) deposited during a late Pleistocene aggradation event. Minor entrenchment of the stream channel terminated the aggradation event and allowed the distinctive late Pleistocene soil to form on the stable terrace tread.

A 3 m offset in 1887 of the active channel and the Pleistocene fill terrace caused accelerated stream-channel entrenchment upstream and alluvial-fan deposition downstream from the fault. Deposition of a new alluvial fan began, not adjacent to the fault, but 30 m downstream. This base-level rise caused partial backfilling of the stream channel entrenched into the fill terrace. In 1987, the first 16 m downstream from the fault trace was still entrenched, but the mode of operation had switched to aggradation in the next 14 m of the partially backfilled fanhead trench.

Equation 2.2 was used to describe the spatial changes in channel depth below the late Pleistocene fill terrace tread in a 12.4 m long reach upstream from the fault. The rupture plane is exposed in a 0.3 m high dry waterfall in weathered quartz monzonite. The amounts of tectonically induced downcutting decrease in an exponential manner with increasing distance upstream from the fault. The value of K_{bl} for this reach is 0.106. A regression analysis of the effects of the 1887 surface rupture (Fig. 2.12B) predicts a stream-channel depth at the fault of 2.82 m; the actual value is 2.74 m.

Tectonically induced aggradation of the downstream reach of the fanhead trench also can be described by equation 2.2 (Fig. 2.12C, D). The

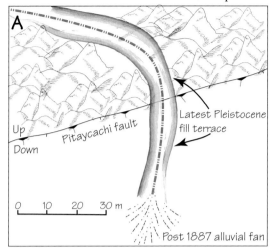

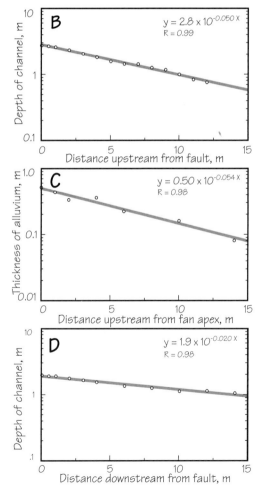

Figure 2.12 Spatial decay of an uplift perturbation, for a 100-year time span, along
a stream that was displaced 3 m vertically in 1887 by rupture along the Pitaycachi normal fault, Sonora, Mexico.
A. Diagrammatic sketch of a small Pitaycachi fault study area.
B. Decreasing amounts of tectonically
induced downcutting in the reach upstream
from the fault.
C. Decreasing amounts of tectonically
induced aggradation in the reach upstream
from the apex of an alluvial fan that is 30 m
downstream from the fault because of pre-
1887 stream-channel downcutting.
D. Decreasing amounts of tectonically
induced downcutting in reach downstream
from the fault.

1887 uplift event increased the relief and sediment yield of the small watershed. Stream power increased because of the steeper gradient, but resisting power increased even more (terms are defined next in Figure 2.13). Aggradation in the fanhead trench reach is the combined result of increased watershed sediment yield and base-level rise caused by alluvial-fan deposition. The resulting post-1887 tectonically induced aggradation was described by measuring thicknesses of modern gray alluvium above reddish brown weathered Pleistocene deposits along the 14 m of the partially backfilled entrenched stream channel. Spatial changes in thickness of aggradation are systematic (Fig. 2.12C). The K_{bl} value is 0.124. Channel depth decreases systematically in the entrenched stream channel (Fig. 2.12D); K_{bl} is 0.224.

We continue by introducing the topics of stream power, resisting power, and thresholds in geomorphic processes. These concepts provide a base to examine topographic profiles of stream channels that have attained equilibrium configurations.

2.3 Threshold of Critical Power in Streams

Altitude and topographic relief increase where bedrock uplift exceeds surficial erosion (equation 1.3). Bedrock uplift increases valley-floor slope, and thereby the stream power available to transport bedload and to erode valley floors. Bedrock uplift is an *independent variable* of fluvial systems where it is not affected by other geomorphic variables. Slopes of hillsides and streambeds are *dependent variables* because they are affected not only by tectonic processes, but also by independent variables such as climate and rock type, and by dependent variables such as soils, plants, and erosional processes. Perhaps we should use the less restrictive term "controlling" instead of "independent" for time spans of more than 1 My. For example, uplift of the Himalayas during the past 35 My has controlled climate on a global scale (Raymo and Ruddiman, 1992). Lithologic control of erosional landforms is associated with spatial variations in structure, fabric, and mineralogy of rocks (Weissel, and Seidl, 1997).

The time span being considered influences how we classify variables (Schumm and Lichty, 1965). Climate should be considered as an independent variable for short time spans, but over long time spans orographic influences of bedrock uplift make watershed climate dependent on bedrock uplift.

Climatic control of fluvial systems is so pervasive that a companion book is devoted to the fascinating subject of Geomorphic Responses to Climatic Change (Bull, 1991). Changes to plant communities are central to climate-change impacts (Vandenberghe, 2003; Flenley and Bush, 2006). Late Quaternary climatic changes commonly overwhelmed the effects of concurrent bedrock uplift by abruptly changing the amounts of water and sediment supplied to streams. As a tectonic geomorphologist, you need to separate tectonic from climatic influences on landscape evolution. Hemisphere-scale climate variations affect both styles and rates of erosion of mountain ranges, such as the Andes (Montgomery, 2002).

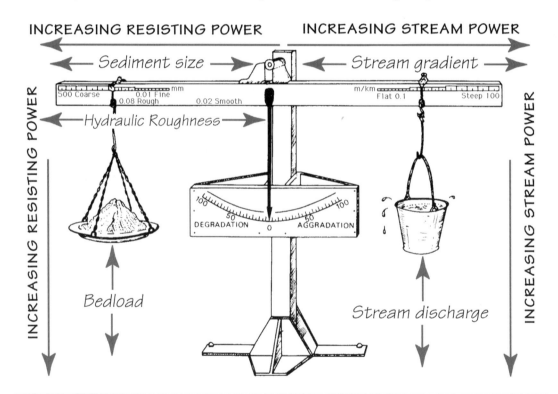

Figure 2.13 Schematic balance between modes of aggradation and degradation in streams; zero is the threshold of critical power. Increases or decreases of one or more variables may cause the mode of stream operation to depart markedly from a threshold condition. (Originally from notes of E.W. Lane; modified from Chorley et al., 1984 and Bull, 1991.)

Streamflow behavior may be regarded as a delicate balance between several controlling factors – stream slope and discharge, sediment size and amount, and hydraulic roughness (Fig. 2.13). The threshold of critical power separates two disequilibrium modes of operation in streams, degradation and aggradation. This threshold is defined as a ratio (Equation 2.4) where the numerator consists of those variables that if increased tilt the balance in favor of degradation, the mode of operation that lowers the altitude of a reach of a stream by fluvial erosion. The denominator consists of those variables that if increased tilt the balance in favor of aggradation. Sustained aggradation raises the altitude of a reach of an active stream channel by selective deposition of bedload.

$$\frac{Stream\ Power\ (driving\ factors)}{Resisting\ Power\ (prohibiting\ factors)} = 1.0 \quad (2.4)$$

The available **stream power** to transport bedload (Ferguson, 2005) as defined by Bagnold (1977) is

$$\Omega = \gamma QS \quad (2.5)$$

where Ω is the total kinetic power, or in terms of power per unit area of streambed,

$$\omega = \frac{\gamma QS}{W} = \gamma QSv = \tau\omega \quad (2.6)$$

where γ is the specific weight of sediment-water fluid, Q is stream discharge, S is the energy slope, w is streambed width, d is streamflow depth, v is mean flow velocity, and $\tau\omega$ is the shear stress exerted on the streambed (Baker and Costa, 1987). Unit stream power is a measure of the power available to do work in a reach of stream.

The denominator of equation 2.4, **resisting power**, becomes greater with increases in hydraulic roughness, and the amount and size of bedload. When resisting power exceeds stream power, aggradation of bedload occurs as the suspended load continues to be washed downstream. The deposits of most stream terraces and many small alluvial fans consist largely of gravelly bedload materials, as compared to the silty suspended-load overbank deposits of large rivers.

Both bedrock uplift and climatic change profoundly affect the components of the threshold of critical power (Starkel, 2003). Bedrock uplift increases watershed sediment yields by increasing landscape

relief and steepening valley floors. Climatic changes during the Pleistocene and Holocene changed discharges of both water and sediment. Climate change may increase bedload production so much that aggradation prevails along nearly all streams of a region, including reaches with tectonically active faults and folds. Watersheds in each climatic setting have different styles and magnitudes of geomorphic responses to changes from glacial to interglacial climates (Bull, 1991). Aggradation events in New Zealand occurred during full-glacial times. In the Mojave Desert of California they occurred during the transition to interglacial climates. Use of the threshold of critical power model emphasizes rates of change of the variables affecting a geomorphic process, thus encouraging you to consider the relative importance of many interrelated factors.

2.3.1 Relative Strengths of Stream Power and Resisting Power

Relative strengths of stream power and resisting power vary spatially in drainage basins (Graf, 1982, 1983). Consider the hypothetical longitudinal profile for the tectonically active arid watershed depicted in Figure 2.14. Stream power continually exceeds resisting power in the headwater reaches. Slope remains excessively steep, especially where resistant rock types and small streamflows near drainage divides favor slow degradation of valley floors. Bedrock uplift of the mountain block occurs as displacements along the range-bounding normal fault zone. This tectonic perturbation to the entire fluvial system occurs in a narrow linear zone crossing the canyon mouth. Episodic bedrock uplift increases the slope component of stream power. Increases of orographically-induced precipitation also occur in rising mountains over time spans of 100 ky to 10 my, so local climate and daily weather patterns commonly are linked to long-term tectonic controls. In marked contrast, late Quaternary climatic changes over time spans of 0.1 to 3 ky were relatively abrupt. They greatly influenced both stream power and resisting power and simultaneously affected all the watersheds of a mountain range. Numerical modeling suggests vastly longer watershed response times to surface ruptures (Allen and Densmore, 2000).

The situation depicted in the aggradational reach of Figure 2.14 is opposite to that of the headwaters. Deposition has been continuous because resist-

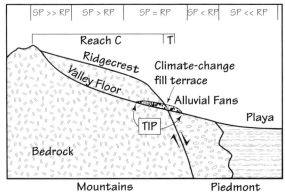

Figure 2.14 Relative strengths of stream power (SP) and resisting power (RP) along a hypothetical fluvial system in an arid closed basin. Tectonic perturbations are initiated by ruptures on a normal fault in reach T and climatic perturbations are initiated in reach C (all the drainage-basin hillslopes). TIP is threshold-intersection point.

ing power has persistently exceeded stream power. Such reaches are particularly obvious in basins of internal drainage where accumulation of playa and associated alluvial-fan deposits constitute a base-level rise that gradually tends to decrease overall piedmont slope and thereby stream power. Tectonic subsidence of a basin – stretch tectonics – tends to offset the base-level rise caused by aggradation. Examples of depositional basins include Great Salt Lake in Utah, the San Joaquin and Death Valleys in California, and lake basins in North Africa and the Middle East.

The mountain-front reach of Figure 2.14 will be the most likely to change its mode of operation as a consequence of climatically induced changes in hillslope water and sediment yield. This is the most sensitive part of a fluvial system to climatic and tectonic perturbations because it is where stream power may exceed, equal, or be less than resisting power for a given streamflow event. Relative discharges of water and bedload are especially important because moderate changes in resisting power may initiate episodes of aggradation or degradation.

2.3.2 Threshold-Intersection Points

Threshold-intersection points are spatial crossings of the threshold of critical power along longitudi-

nal profiles of valleys (Fig 2.3). Aggradation in a mountain-front reach that formerly was downcutting into bedrock moves the threshold-intersection point upstream. Renewed stream-channel downcutting through the new valley fill and fan deposits then shifts the threshold-intersection point downstream leaving the recently elevated valley floor as the tread of a fill terrace (Figure 2.14).

Threshold-intersection points may shift in a great variety of time scales. More than 2 ky may be needed for a shift such as that depicted in Figure 2.14. In contrast, sediment discharge commonly peaks before water discharge during a flood event. This causes the threshold-intersection point to migrate upstream and then downstream as the ratio of stream power to resisting power changes during the flow event. Threshold-intersection points shift more rapidly along the valleys of streams in humid regions than in arid regions, because wet climates provide greater annual stream power to do the work of transporting the sediment load imposed by weathering and erosion of hillslopes.

2.4 Equilibrium in Streams

Those who seek tectonic information from landscapes need to know how tectonic perturbations change the behavior of fluvial systems. How do landscapes evolve when perturbed by vertical and horizontal earth deformation? Do they tend towards unchanging configurations? Do some landforms achieve such a steady state (equilibrium) sooner than others? Vital skills include being able to assess how far removed a landform is from a steady-state condition, and how quickly part of a fluvial system can achieve a new equilibrium. Depositional landforms, such as growing deltas and alluvial fans, do not even tend toward steady-state conditions (Bull, 1977a, 1991, Section 1.6.2).

2.4.1 Classification of Stream Terraces

Some former valley floors are equilibrium reference surfaces that record tectonic deformation. Let us classify types of stream terraces, noting their suitability for tectonic studies. Two types of equilibrium, and the crossing of the threshold of critical power, can be recognized by the presence of distinctive types of stream terraces. We are particularly interested in stream terraces resulting from tectonic and climatic perturbations, but acknowledge that common minor terraces (called autocyclic) form because of responses to fluctuations of geomorphic processes within a fluvial system.

Alternating aggradation and degradation events are common in fluvial systems that are sensitive to climatic changes. Termination of aggradation and degradation events creates key landforms. A degradation event is an interval of net lowering of a valley floor by fluvial erosion. Minor stream terraces form during pauses in the downcutting process. A degradation event commonly is terminated by a prolonged episode of lateral beveling when the stream stays at the same low position in the landscape.

Different climatic controls promote an aggradation event, which tends to backfill valleys (Barnard et al., 2006) and commonly is terminated by deposition of a broad alluvial surface. Aggradation tends to bury fluvial landforms created by the preceding degradation event. Such burial is continuous as shown by lack of buried soil profiles in the stratigraphic section of an aggradation event. Fortunately, net stream-channel downcutting occurs in tectonically active landscapes. Watersheds with infrequent aggradation events are less likely to conceal landforms of the pre-

ceding degradation event when tectonically induced downcutting lowers the active channel below the former valley floors represented by stream terraces. First let us get acquainted with standard stream-terrace terminology (Fig. 2.15A).

Terraces may be paired or unpaired. *Unpaired terraces* (Davis, 1902) typically occur on the insides of meander bends along a stream that is steadily downcutting, but sometimes occur where a stream temporarily erodes laterally into a bedrock hill. A *paired terrace* consists of remnants of a former floodplain that when connected describe a former longitudinal profile of the stream. They occur on both sides of a valley where not removed by erosion. An important characteristic of a paired terrace is that it has continuity along a valley.

Flights of stream terraces can be likened to flights of stairs; both consist of a sequence of alternat-

Figure 2.15 Tectonic, climatic, and internal-adjustment terraces of the Charwell River, New Zealand.

A. Sketch of basic definitions. A strath beveled at the lowest possible position in a landscape is the tectonic reference landform. A climate-change induced aggradation event buries the tectonic strath. The tread of the resulting fill terrace records a time of change from aggradation to renewed degradation. Pauses in the subsequent stream-channel downcutting that generate stream terraces may be due mainly to adjustments caused by variable hillslope water and sediment yield.

B. Multiple strath and aggradation surfaces. Geomorphic responses to late Quaternary climatic changes and tectonic perturbations, and internal adjustments of the fluvial system, are recorded as a flight of stream terraces in this rising landscape. The Flax Hills and Stone Jug aggradation surfaces are fill-terrace treads that record terminations of climate-change induced aggradation events. Internal adjustments are preserved as minor strath, fill-cut, and fill terraces that record pauses in degradation from the aggradation surface to the younger broad tectonic strath terrace at the base level of erosion.

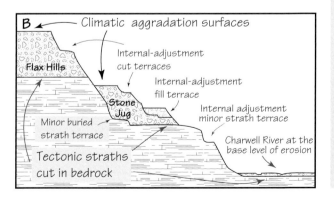

ing treads and risers. A *terrace tread* is remnant of a former valley floor that has been abandoned by the stream as a result of stream-channel incision. The corresponding *terrace riser* is a scarp created by fluvial lateral erosion above the tread.

Former levels of streams may be classed as fill, fill-cut and strath terraces (Leopold & Miller, 1954; Howard, 1959; Bull, 1990). A *fill terrace* is formed by aggradation of a valley floor and subsequent channel incision into the alluvium that leaves remnants of the former active channel as the tread of a paired fill terrace. The tread represents a time of crossing of the critical power threshold when the mode of operation switches from aggradation to degradation. Longitudinal profiles of fill terraces generally do not represent attainment of equilibrium, because this threshold crossing tends to be abrupt and may occur at different times along a valley.

A *strath terrace* is genetically the same as a *fill-cut terrace*: straths are surfaces beveled in bedrock and fill-cut surfaces are beveled in alluvium. This generally brief equilibrium (pauses in a degradation event) is followed by renewed stream-channel downcutting that preserves remnants of the beveled surface, and gravel being transported at the time of formation, as the terrace tread. In this case, the riser (or streambank) was formed shortly before initiation of renewed stream-channel downcutting. Strath and fill-cut terraces differ from fill terraces in that only thin layers of stream gravel cap their erosion surfaces. These thin deposits may be regarded as lag deposits or gravelly cutting tools that most likely would have been entrained and redeposited by the next large flood.

Erosional lowering of a valley floor does not proceed at a constant rate. Stream-channel downcutting rates vary with unit stream power and bedload transport rate. Pauses in valley-floor degradation allow fill-cut and strath terraces to form. Brief reversals in degradation of valleys caused by deposition of bedload result in minor fill terraces.

Typically, both of these brief, minor, types of events create *internal-adjustment terraces.* In contrast to tectonic and climatic stream terraces, they result from temporal variations of dependent variables in a fluvial system. Such stream terraces have been studied in experimental (sandbox) models. They are called "complex responses" (Schumm, 1973; Parker, 1977; Schumm et al., 1987) and are referred to as "auto-cyclic" landforms by Hasbargen and Paola (2000). These authors studied internal-adjustment terraces

in sandbox models by initiating upstream migration of knickpoints that locally increased sediment yield from hillslopes. Although internal-adjustment terraces are important for understanding adjustments in fluvial systems, they do not record the start or end of climate-change induced aggradation, nor can they be used to assess rates of valley floor downcutting that are related to bedrock uplift.

Many other processes cause perturbations of water and sediment discharge that favor formation of internal-adjustment terraces; fires and landslides are examples. A kinematic wave of bedload increase sweeps down trunk stream channels only to be followed by stream-channel incision to produce an internal-adjustment fill terrace. Aggradation may last only a few years or decades. Dating flights of internal-adjustment terraces in adjacent drainage basins usually reveals a lack of synchroneity (Bull, 1991). Each watershed behaves differently.

In contrast, tectonic and climatic stream terraces tend to be synchronous throughout much of a mountain range. Examples of synchronous terraces range from a brief episode of valley backfilling caused by an earthquake-induced sediment-yield increase to regional aggradation of longer duration caused by stripping of the hillslope sediment reservoir during the Pleistocene–Holocene climatic change.

Important terraces form at the end of major aggradation and degradation events. Thick alluvial fill deposited in response to regional climatic change raises valley floors and alluvial-fan surfaces. The amount and rate that valley floors are raised by such deposition are mainly a function of the time span of the climatic perturbation and magnitude of departure from the threshold of critical power. A fill-terrace tread represents the maximum altitude attained by backfilling a valley floor, and records the end of persistent aggradation that commonly lasts 1 to 15 ky. This type of aggradation surface is the fundamental climatic stream-terrace landform. The tread is the only landform created that has temporal significance if the riser above the tread is merely a valley side. Aggradation event surfaces in adjacent watersheds are synchronous where response times to climatic perturbations are similar.

There are limits to the depth of erosion of any valley floor, and the termination of a degradation event sets the stage for creation of an important type of strath terrace.

Lateral erosion that bevels bedrock and widens an active channel at the end of a degradation event may persist for 1 to 10 ky. It creates a strath that is the fundamental tectonic stream-terrace landform (Fig. 2.4). This alluvium–bedrock interface records the lowest possible longitudinal profile for a particular tectonic setting. Strath age is defined as the end of an interval of strath formation. This occurs when renewed, persistent deposition raises the streambed and buries the beveled bedrock surface. Termination usually requires a climatic change. Alternatively, valley floor bedrock beveling can be terminated by base-level fall induced by a large surface rupture in a downstream reach. This increases stream power sufficiently that the threshold of critical power is crossed to create a tectonic strath terrace landform.

Multiple late Quaternary tectonic strath terraces record important times of attainment of equilibrium along the streams of rising mountains. In contrast to the generally minor internal-adjustment strath terraces, tectonic straths tend to be broader, and are more likely to be correlated between adjacent reaches. Reaches in adjacent drainage basins with similar bedrock resistance, discharge, and bedload transport rates generally complete a degradation event at about the same time. If so, time spans of tectonic strath formation may be synchronous for a mountain range. Adjustments of reaches of streams that achieve such equilibrium conditions are discussed in the next two sections.

2.4.2 Feedback Mechanisms

The threshold approach to geomorphic investigations emphasizes how far removed a system is from stable conditions. Thresholds are essential for studying interactions between geomorphic variables that do not tend towards an unchanging "steady-state" condition. So, we need to consider such relations in a bit more detail.

Interactions between geomorphic variables may tend to create a landform, or landscape, that does not change with the passage of time. Alternatively, such interactions may result in an opposite tendency away from equilibrium conditions.

Key to understanding both modes of landscape evolution are feedback mechanisms between dependent variables that drive the system toward or away from steady-state conditions. Studies of both *self-enhancing* and *self-arresting feedback mechanisms* are encouraged by the threshold approach to geomorphology, whereas studies only of self-arresting feedback mechanisms are encouraged by the equi-

librium approach to geomorphology. The behavior of streams is one of the better examples of how self-arresting feedback mechanisms promote equilibrium. Landslides represent the opposite situation where self-enhancing feedback mechanisms drive the process towards its culmination rather than toward a balance between variables.

Progressive accumulation or erosion of hillslope colluvium are examples of self-enhancing mechanisms. Colluvium deposited on bare rock increases infiltration capacity, thereby providing both water and soil to support vegetation. The vegetation traps additional colluvial materials from upslope sources, thereby furthering the tendency for accumulation of more colluvium. For a reversal in the mode of operation to occur, a threshold must be passed that separates tendencies for progressive accumulation from those of progressive erosion of colluvium. Climate-change perturbations cause this system to alternate between two modes of operation and both landforms and landscape evolution cannot tend towards unchanging conditions. Such nonequilibrium modes of operation may involve time spans of decades or millions years.

Progressive increase in areas of massive granitic outcrops in the Sierra Nevada of California over several million years may be considered an irreversible change. The self-enhancing feedback mechanism of rapid runoff from bare rock continues to erode the soil produced by weathering of granitic rocks (*gruss*) at the margins of hillslope outcrops. Clyde Wahrhaftig (1965) made a classic study of such a landscape evolution. He sought to learn more about the formation of stepped topography developed in the massive granitic rocks of the western slopes of the Sierra Nevada. Bedrock exposed by streams expanded in area to become the dominant hillslope landform. Such *topographic inversion* – valley floors eventually become ridgecrests – requires that the drainage net, as well as the hillslopes, undergo progressive change. Thus Wahrhaftig's study provides a nice example of the importance of self-enhancing feedback mechanisms. The steady-state model is inappropriate because a key independent variable – erodibility of surficial materials – changes in both time and space.

A different example of self-enhancing feedback mechanisms is the progressive development and ultimate collapse of a slump rotation block in massive sandstone cliffs (Schumm and Chorley, 1964). A seemingly insignificant tension crack at the top of the cliff signals initiation of a process that may require thousands of years. Insufficient lateral support for the cliff face allows the minute crack to gradually widen at an exponentially faster rate, thereby enhancing several processes that culminate in failure of the sandstone monolith when it collapses into a pile of rock-fall blocks. Water and ice accumulate in the ever widening crack, rotation shifts the center of gravity of the monolith, and block rotation fractures the lower portion of the sandstone pillar thereby causing a progressive decrease in rock mass strength.

Now, let us examine the opposite type of feedback mechanism. The adjustment of a stream to an increase of gravel eroded from the hillslopes of its fluvial system is an example of self-regulation by self-arresting feedback mechanisms. Large increases of bedload change all hydraulic variables in meandering stream channels, which in turn results in a decrease in channel sinuosity, so that more bedload can be conveyed. In braided channels, increases in bedload may cause aggradation of the valley floor with maximum alluviation in the upstream reaches. This increase in streamflow gradient is the principal way stream power is increased. In both meandering and braided streams, adjustments of the hydraulic variables will continue as long as the stream is supplied an excess of bedload (Paola and Mohrig, 1996). Changes in flow characteristics or stream-channel characteristics that are greater than needed to transport the bedload will result in counterbalancing adjustments that will decrease the transporting capacity and competence of the stream.

A key tectonic landform is the longitudinal profile of a stream channel (Merritts et al., 1994). We consider present and past longitudinal profiles as potential reference datums passing through tectonically deforming landscapes. So, it is worth our while to consider several types of equilibrium in such streams, and the landscape characteristics associated with each type.

2.4.3 Dynamic and Static Equilibrium

The base level of erosion concept describes reaches of streams that have achieved one of two types of equilibrium–static equilibrium and dynamic equilibrium (which can be further classified as type 1 or type 2). We define and discuss these terms here in order to clarify how climatic controls and internal adjustments in fluvial systems lead to static equilibrium, and how tectonic controls create situations of dynamic equilibrium in landscape evolution. Unless classed as static or dynamic, the general term "equilibrium" refers to reaches that have attained the base level of erosion.

Static equilibrium is attained briefly when bedload derived from the watershed hillslopes is transported through a reach with neither aggradation nor degradation of the streambed (Leopold and Bull, 1979). One example is a gradual, instead of abrupt, switch from aggradation to degradation at the conclusion of an aggradation event. Bull (1991, Section 5.4.4.1) describes field identification of an aggradation surface that records a period of static equilibrium. A stream that pauses briefly during a degradation event creates minor fill-cut or strath terraces that also record temporary static equilibrium.

John Hack used the ideas of G.K. Gilbert when he applied the concept of *dynamic equilibrium* (steady state) to open fluvial systems of mountains. Hack believed that a steady-state landscape configuration would develop between processes that tend to raise and denude mountains. All elements of a landscape in dynamic equilibrium are mutually adjusted to each other. The resulting landscape assemblage downwastes at a uniform rate, and its configuration does not change (Gilbert, 1877; Hack, 1960, 1965; Carson, 1971). Thus, dynamic equilibrium is independent of time and acknowledges that climatic and tectonic energy is continually entering and leaving the system. Adjustments to perturbations restore a dynamic steady state without significant delays. The steady-state model of landscape evolution continues to be highly regarded by many geomorphologists.

The Gilbert-Hack model is heuristic, because we have yet to obtain field evidence for attainment of true steady state for an entire landscape (Bull, 1977a). Important tectonic and climatic perturbations occur frequently during time spans of 0.1 to 100 ky and affect denudation rates that vary with both time and space within drainage basins. In my opinion, response times of geomorphic processes are sufficiently long for hillslopes to prevent attainment of steady state for either watersheds or mountain ranges. Models using non-steady state assumptions are preferable for studies of landscape evolution.

Sólyom and Tucker (2004) nicely describe obvious examples of why steady state is inappropriate for modeling short-term behavior of entire watersheds by emphasizing interactions between the variables of storm duration, basin shape, and peak discharge. Central to their argument is that precipitation-runoff inputs are not uniformly distributed in a spatial sense, creating disparities that increase with drainage-basin area and with increasing watershed aridity. Nonuniform inputs of precipitation typically create situations of partial-area contributions of runoff from a watershed (Dunne and Black, 1970; Yair et al., 1978), which lead to lesser concavity (rate at which a longitudinal profile becomes flatter downstream).

Lack of a vigorous proof for dynamic equilibrium for entire watersheds does not detract from the value of the base level of erosion concept, and the tendency of specific landscape assemblages toward time-independent shapes. One nice application of the dynamic equilibrium model is the way in which powerful streams respond to bedrock uplift.

Two categories of dynamic equilibrium in streams may be defined in terms of length of attainment of the base level of erosion. Hack (1973, 1982) defined steady-state adjustment between variables as those longitudinal profiles that plot as straight lines on semi-logarithmic graphs because stream discharge increases logarithmically downstream (Wolman and Gerson, 1978). Two types of dynamic equilibrium can be recognized in the field.

Diagnostic landforms for attainment of *type 1 dynamic equilibrium* include a strath beneath the active channel and a valley floor that is sufficiently wide for preservation of strath terraces. Type 1 dynamic equilibrium is present when stream downcutting and bedrock uplift rates are the same. An analogy would be the constant shape and position of a rotating circular saw (the stream) as a log (the mountains) is raised into it. The resulting sequence of longitudinal profiles represents an infinite number of tectonic base levels of erosion as the stream re-establishes equilibrium conditions after each tectonic perturbation in the rising landscape.

Climatic perturbations in real world fluvial systems do not allow perpetual uniform downcutting in response to bedrock uplift. Instead, climatic factors modulate the discharge of water and bedload so as to vary the rate of valley-floor downcutting. The result is a series of closely spaced minor strath terraces, each created briefly at an appropriate time as dictated by changing climatic constraints.

Type 2 dynamic equilibrium is present in streams with a strong tendency toward, but lack of sustained attainment of, the base level of erosion. Diagnostic landforms include narrow valley floors whose longitudinal profiles (like those of type 1 stream reaches) plot as concave lines on arithmetic graphs and as straight lines on semi-logarithmic graphs. Straths and remnants of strath terraces generally are not present.

Recent studies of bedrock stream-channel longitudinal profiles and stream-channel characteristics are a worthwhile departure from the previous emphasis on less confined, easily eroded alluvial streams (Baker and Kochel, 1988; Seidl and Dietrich, 1992: Wohl, 1994; Montgomery et al., 1996; Baker and Kale, 1998; Hancock et al., 1998; Pazzaglia et al., 1998; Sklar and Dietrich 1998; Tinkler and Wohl, 1998a, b; Massong and Montgomery, 2000; Wohl and Merritt, 2001, Whipple, 2004).

Irregular longitudinal profiles are characteristic of *disequilibrium streams* upstream from most type 2 equilibrium reaches where interactions between variables have yet to approximate the base level of erosion. Landforms in disequilibrium reaches include convex valley sideslopes plunging into V-shaped canyons with waterfalls and rapids, and convex longitudinal profiles, even when plotted on logarithmic graphs.

Stream-channel widths for degrading streams tend to be narrower than for streams in equilibrium. Consider the case where renewed degradation converts an equilibrium reach characterized by strath cutting (active-channel width is not confined) to confined streamflow that is typical of downcutting reaches. Streams degrading into bedrock may establish a self-enhancing feedback mechanism that accelerates downcutting until equilibrium conditions are approached. Narrowing of streamflow tends to increase unit stream power (equation 2.6).

Millions of years may be needed for ephemeral streams flowing over resistant rocks to approach their base levels of erosion. Some response time is needed for large rivers to adjust to a pulse of uplift even in humid mountain ranges. The time spans needed for landscape adjustments typically are longer than intervals between uplift. Upstream reaches and adjacent hillslopes will have longer response times. So a model where interacting landscape elements continue to adjust makes more sense than proclaiming still another example of steady-state conditions. One alternative is the allometric change conceptual model (Bull, 1975a; 1991, Section 1.9) that describes orderly interactions between dependent variables in a changing landscape before or after the base level of erosion is attained.

Allometry in biology describes relative systematic changes in different parts of growing organisms (Vacher, 1999). Allometric change in geomorphology describes orderly behavior in nonsteady state fluvial systems. Hydraulic geometry of stream channels (Leopold and Maddock, 1953) is a good example. Increase of flow width with increasing discharge at a gauging station is static (at a single location in a watershed) allometric change. Increase of flow width with downstream increase of discharge (a sequence of locations) is dynamic allometric change. The allometric change model emphasizes the degree of interconnectivity of different geomorphic processes and landscape characteristics. Allometric change emphasizes system behavior instead of attainment of steady state and is mentioned here to alert readers that most regressions between geomorphic variables in this book imply nonsteady-state assumptions.

Perennial streams of humid regions flowing over soft or highly fractured rock are better suited for studies of responses of streams to bedrock uplift during the past 40 ky. One such stream is the Charwell River, whose main branch drains 30 km² of the Seaward Kaikoura Range of New Zealand. Times of aggradation events were estimated by radiocarbon dating of wood in the fill-terrace deposits of the region. The Charwell River is the type area for the Stone Jug, Flax Hills, and Dillondale climate-change induced aggradation events, which have synchronous counterparts elsewhere in New Zealand (Barrell et al., 2005; Litchfield and Berryman, 2005). Terrace tread ages were estimated with weathering rind analyses (Knuepfer, 1988). Strath ages were estimated by radiocarbon dating of tree trunks, in growth position, just above the straths. These age estimates indicate termination of intervals of major strath beveling at about 40 and 29 ka.

Tectonic strath terraces are those that record times of attainment of the base level of erosion (type 1 dynamic equilibrium) in rivers that are being tectonically elevated. The Charwell River has both internal-adjustment and tectonic strath terraces. The piedmont reach extends 12 km downstream from the highly active range-bounding Hope fault, has local faulting and folding, and has experienced episodes of climate-change induced aggradation. Bedrock uplift ranges from 0.2 to 1.3 m/ky. Successive tectonic strath terraces are at lower positions in the landscape because the Charwell River lowers its channel in response to continued bedrock uplift (Fig. 2.15B). Each broad tectonic strath was created at a time of interstadial or interglacial climatic conditions, such as the present Holocene climate. Each episode of tectonic strath formation ended at the time of initiation of an aggradation event.

The long-term tendency of rivers to cut down into a rising landmass is temporarily reversed during aggradation events, which in humid, mesic to frigid climates coincide with times of full-glacial climates (Bull, 1991, Chapter 5; Wegmann and Pazzaglia, 2002). Rivers then switch to degradation and may catch up to, and re-establish, type 1 dynamic equilibrium during times of interglacial climates. This is when streams have excess stream power, relative to resisting power. Rapid stream-channel downcutting (Bull, 1991, Fig. 5.18) began at 14 to 16 ka after termination of the latest Pleistocene aggradation event, and ended at about 4 ka. By then, the Charwell River had attained its tectonic base level of erosion downstream from the range-bounding Hope fault. Bedrock uplift has continued to raise the stream during the past 4 ky, and the active channel has cut down into soft piedmont-reach lithologies at a rate that matches the local bedrock-uplift rate. Fluvial erosion switched from largely vertical to mainly horizontal and the stream beveled a strath as wide as 200 m (discussed and shown later in Figure 2.26).

Degradation stream terraces also record times of attainment of the base level of erosion. The flight of Charwell River stream terraces provides examples of how self-arresting feedback mechanisms result in temporary static equilibrium. Latest Pleistocene aggradation buried the 29 ka tectonic strath and raised the streambed altitude 40 m. Then rapid degradation of the bouldery valley fill occurred. The stream paused a dozen times as it cut down before it caught up to its long-term (tectonic) base level of erosion (Fig. 2.16). Tectonic and climatic controls on other Marlborough streams follow a similar pattern but magnitudes of change in positions of valley floors are functions of the amounts of latest Quaternary aggradation and uplift. Mean rates of post-glacial incision for the Saxton River on the Awatere fault are 1.4 m/ky compared to 5.1 m/ky for the Charwell River (Fig. 2.16). The shapes of the two incision curves are the same, and both streams have returned recently to the base level of erosion where the rate of stream-channel downcutting matches the uplift rate.

Each pause was a time of attainment of static equilibrium that was recorded by a minor terrace. Stream-channel entrenchment into poorly sorted sandy gravel promoted selective entrainment and transport of the finer portion of the valley fill. Accumulation of residual boulders on the streambed

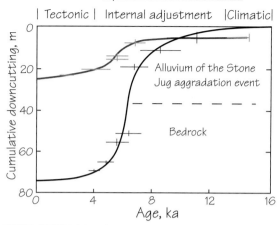

Types of stream terraces

Figure 2.16 Tectonic, climatic, and internal-adjustment terraces of the Charwell River: variations of stream-channel downcutting rate since 16 ka. Uncertainty cross estimates are for terrace ages and distances below the aggradation surface reference level. From Figure 6 of Bull and Knuepfer (1987). Gray plot is for Saxton River on the Awatere fault. From Figure 13 of Mason et al. (2006).

created a lag deposit that armored and protected the streambed from further degradation by 1) increasing the shear stresses needed to initiate movement of streambed gravel, and 2) by increasing the hydraulic roughness. Riparian vegetation grows beside streams with armored beds, further increasing hydraulic roughness. Neither net aggradation nor net degradation occurs, even though the active channel is still far above its tectonic base level of erosion.

These brief episodes of static equilibrium end abruptly when a 1-ky flood event entrains and smashes streambed-armor boulders, and destroys riparian vegetation. Both changes reduce hydraulic roughness and the shear stresses needed to entrain bedload in the normal range of stream discharges. Renewed streambed degradation and winnowing then create a new lag gravel at a lower altitude. Remnants of the former streambed, with the characteristic streambed armor preserved as a capping layer of boulders, remain as treads of fill-cut or minor strath terraces.

Thus without invoking either climate-change or tectonic perturbations, a self-arresting feedback

mechanism can occur repeatedly during the formation of a flight of degradation terraces. These are regarded as internal-adjustment terraces because they are not the result of changes in the independent variables of this fluvial system. Each records a pause in valley-floor degradation caused by temporary increases of resisting power.

Different geomorphic responses to bedrock uplift in the reach upstream from the Hope fault have resulted in narrow V-shaped canyons (Bull, 1991, Fig. 5.7), convex footslopes, and waterfalls. The fractured greywacke sandstone is more resistant to erosion than the soft Cenozoic sediments down-stream from the fault. Bedrock-uplift rate is about 3.8 ± 0.2 m/ky based on altitudinal spacing analysis of uplifted marine-terrace remnants. Degradation by small headwater streams is unable to keep pace with bedrock uplift, but some mid-basin reaches of the Charwell River have sufficient unit stream power to approximate type 2 dynamic equilibrium. The reach just upstream from the Hope fault has remnants of two fill terraces, perhaps reflecting repeated episodes of base-level rise caused by rapid deposition of 30–50 m of latest Pleistocene valley fill in the adjacent reach downstream from the Hope fault.

Many possible interactions between variables can produce equilibrium in streams. Increased slope and reduced stream-channel width may be important in achieving equilibrium after vertical offset of a streambed by an earthquake surface rupture. Changes in hydraulic roughness and stream-channel pattern may occur after a landslide increases bedload. Powerful streams may maintain equilibrium by adjusting interactions between variables, within a limited range, as short-term climatic change influences unit stream power and bedload transport rate. Perturbations that force a reach of a stream beyond its capacity to maintain equilibrium conditions initiate aggradation or degradation events.

Sections 2.6 and 2.7 provide several examples of how strath terraces can be used by tectonic geomorphologists. But first, let us outline the characteristics of response times of fluvial systems to climatic and tectonic perturbations.

2.5 Time Lags of Response

Several concepts, aggradation and degradation events, attainment of the base level of erosion, and time lags of response can be summarized with a simple diagram (Fig. 2.17) *Reaction time* is the delay before

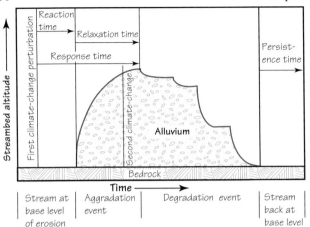

Figure 2.17 Anatomy of an aggradation–degradation event illustrating the
concepts of reaction time, relaxation
time, and response time after a climate-
change perturbation causes an episode
of alluviation of a valley floor.

streambed aggradation begins: for example, the interval when hillslope plant cover decreases thus increasing sediment yield sufficiently that the stream can no
longer maintain equilibrium conditions. *Relaxation
time* is the time span needed to complete the aggradation event. *Response time* is shown as the total
elapsed time from the time of a climatic perturbation to end of the aggradation event. It is the sum of
reaction and relaxation times (Allen, 1974; Thornes
and Brunsden, 1977; Brunsden and Thornes, 1979;
Brunsden, 1980). The ensuing degradation event
returns the stream to the same base level of erosion in
this case. This is indicative of a lack of uplift during
the time span represented by the aggradation–degradation event. Fluvial systems also respond to being
elevated. *Persistence time* is the time span during

which fluvial system behavior is constant, which here
is a condition of type 1 dynamic equilibrium.

2.5.1 Responses to Pulses of Uplift

The concept of response time in a fluvial system to
tectonic inputs can be illustrated with a threshold-
equilibrium plot (Fig. 2.18). Reactions of a stream
to two hypothetical perturbations are illustrated for
a reach that is 2 km upstream from the mouth of
a watershed and initially in type 1 dynamic equilibrium. The first perturbation is a 2 m surface rupture on the range-bounding normal fault. Tectonic
lowering of the reach immediately downstream from
the fault, relative to the mountain block, creates a
knickpoint. This short tectonically steepened reach

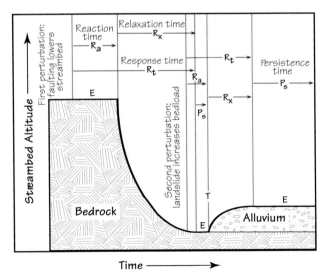

Figure 2.18 Hypothetical threshold-equi-
librium plot showing the components of
response time, R_t, for a reach 2 km upstream from a normal fault. Delayed
responses are depicted for two perturbations; a 2 m surface rupture on the
range-bounding fault, and a landslide upstream from the study reach. Response
time is the sum of the reaction time, R_a,
and relaxation time, R_x. P_s is the time of
persistence of new equilibrium condition,
and T and E are the times of threshold
and equilibrium conditions respectively.

migrates upstream. A knickpoint may be a waterfall initially but commonly becomes rapids – a *knick-zone* – where rock mass strength is low. Processes of upstream migration are different for these two types of stream-channel perturbation. Knickpoints retreat mainly by undermining and collapse and knickzones by streambed abrasion and plucking by fast moving saltating bedload. Both may be significant local departures from the typically concave longitudinal profile indicative of a reach in equilibrium. As such they can inhibit continuity of fluvial systems (Section 2.5.2) to the extent that waterfalls separate relict landscapes in a dramatic fashion (Crosby and Whipple, 2006).

Rates of incision into bedrock are considered to be proportional to shear stress exerted by streamflow (Howard and Kerby, 1983; Tucker and Slingerland, 1994; Snyder et al., 2000, 2003). Some geomorphologists prefer to emphasize total stream power (Seidl and Dietrich 1992), bedload transport rates relative to the threshold of critical power (Kooi and Beaumont, 1994), bedload abrasion potential (Sklar and Dietrich, 1998), or unit stream power (Whipple and Tucker, 1999). The interval is quite short for the first persistence time of Figure 2.17.

Migration of the hypothetical knickzone (Figure 2.17) still further upstream undercuts a hillside and triggers a landslide – the second perturbation. The result is an increase in bedload input to the stream that causes minor aggradation in the study reach, followed by a new equilibrium condition. Note that equilibrium is regarded as an interval with no change of streambed altitude, whereas the threshold is depicted as a point in time when the system reversed modes of operation (for example, from degradation to aggradation). The first two equilibrium intervals are examples of type 1 and type 2 dynamic equilibrium. The third is an example of static equilibrium, because the stream is above its long-term (tectonic) base level of erosion.

Reaction time is a measure of the sensitivity of a fluvial system to a perturbation, and relaxation time is a measure of how efficiently a geomorphic system adjusts to a perturbation. The reach immediately upstream from a surface rupture reacts quickly as waterfalls and rapids are created. Hillslopes in the headwaters of the same stream react slowly, and flat summits and plateaus have such exceedingly long reaction times that they are essentially isolated from downstream perturbations. Resistance of rocks may be the same throughout a drainage basin, but stream power decreases exponentially in the upstream direc-

tion. Spatial variations in response time inhibit attainment of steady-state conditions for entire drainage basins.

Outputs of fluvial systems, such as sediment yield, integrate response times from all parts of a drainage basin. Renewed degradation of a valley floor in response to a range-front uplift event provides a new source of sediment as streamflow downcuts into reaches previously at equilibrium. Stream-channel entrenchment steepens adjacent hillsides, which promotes mass movements; both processes increase sediment yield. Tectonically induced increases of watershed sediment yield increase resisting power and tend to accelerate deposition in reaches downstream from the active fault. Chapter 4 has many examples of the abrupt initiation of deposition of alluvial fans that coincide nicely with this tectonic base-level fall. An aggradation-rate increase on the alluvial fan reach of a fluvial system alluvial fan occurs as quickly as upstream channel downcutting is initiated, but maximum rate increase may be delayed substantially until tributary streams and their large areas of hillslopes are affected by an episode of range-front faulting. So reaction time is short for the depositional reach, but relaxation time may be extended until the effects of a tectonic perturbation have spread throughout much of the drainage basin.

The response-time model described above applies only to those fluvial systems with continuity. Both erosional and depositional reaches of some streams can have characteristics that inhibit continuity of fluvial-system behavior.

2.5.2 *Perturbations that Limit Continuity of Fluvial Systems*

Waterfalls are dramatic landforms that typically isolate upstream reaches from the effects of surface rupture on downstream fault zones. These streambed cliffs effectively *decouple* upstream from downstream reaches. Decoupling isolates the reaches in the headwaters of a watershed, which may show minimal consequences of incremental increases in watershed relief emanating from an active range-bounding fault. Decoupling inhibits stream-channel downcutting in a trunk valley from migrating up tributary valleys. Instead, such upstream reaches are graded to the top of a waterfall, or to a rapid with a drop sufficient to dissipate energy in a hydraulic jump. Rapids and falls are local base levels (Fig. 2.2B).

Geomorphic responses to climatic changes may control the magnitude, time of formation, and position in a watershed of a waterfall (Sections 2.6.2, 2.7). Episodes of uplift along a range-bounding fault increase the height of a bedrock fault scarp even when deep beneath the gravels of a climate-change induced aggradation event. Age controls for the Charwell River landscape evolution allow estimation of process rates (Bull and Knuepfer, 1987). The period between about 26 and 14 ka ago was a time of valley floor backfilling. Episodic vertical movements along the Hope Fault during this time span continued to displace both bedrock and valley fill. Streamflows would quickly eradicate scarps formed in unconsolidated fill, but a sub-alluvial bedrock fault scarp would have become progressively larger until Holocene degradation was sufficient to expose bedrock once again at about 9 ka. The Hope Fault separates the mountain and piedmont reaches; both are rising and relative uplift is about 2.5 m/ky. Thus, late Quaternary climatic perturbation modulated tectonic processes to create and then exhume a prominent bedrock waterfall about 40 m high. After exposure it would retreat upstream as a knickpoint or steeper reach. The present anomalously steep reach 1300 m upstream from the Hope Fault departs from a smooth profile by about 40 m.

Cliffs prevent continuity of geomorphic processes on hillslopes too. One example is the free face on a young fault scarp. This is why we can't do diffusion-equation modeling for a fault scarp that still has a free face. Another common example of lack of fluvial system continuity is the precipitous drop below a tabletop upland or mesa. Flux of sediment and water through cliffy landscapes is interrupted in a system that otherwise would behave systematically enough to be described by a single set of equations. Summit uplands are subject to such different processes that they may be eroding at a miniscule rate relative to the valley floors and hillslopes downvalley from the cliffy terrain that decouples the headwaters from the rest of the fluvial system. Remnants of shore platform–sea cliff landscapes of raised marine terraces (Fig. 2.19A) are another example of decoupled terrain.

Reaches of active deposition may decouple fluvial systems as efficiently as waterfalls. Consider the case of a downcutting stream that emerges from a rising mountain range, crosses an aggrading alluvial fan, and joins a trunk river. A base-level fall on the trunk river will migrate upstream as a headcut but cannot be transmitted through the alluvial-fan reach as long as it remains on the aggradational side of the threshold of critical power. Fans decouple fluvial systems by not allowing base-level falls further downstream to influence stream channels and hillslopes in their source areas.

The Carrizo Plains adjacent to the Temblor Range of California have ephemeral streams that respond to tectonic and climatic perturbations (Ouchi, 2005). Sieh and Wallace (1987) note that the fanhead trench of Wallace Creek has been incised for about 3.7 ky and records 128 m of right-lateral displacement by the San Andreas fault zone during this time span (Fig. 2.19B). Alluvial-fan deposition has subsequently occurred on the fanhead of the unnamed creek, whose young surface has undergone minimal stream-channel offset. Both streams have undergone arroyo cutting (Bull, 1964a, b) during the past 200 years. Erosional knickpoints during this climate-change induced stream-channel entrenchment can easily move up the continuous channel of

Figure 2.19 Decoupled hillslopes, piedmonts, and stream channels.
A. Slowly eroding upland in the Kelly Range, Southern Alps, New Zealand changes abruptly to cliffy headwater reaches of rivers that are downcutting rapidly in response to bedrock uplift of 5 m/ky (Bull and Cooper, 1986). Uplands are shore platform(s) of uplifted marine terraces (SP). Degraded former sea cliff at SC. The foreground cliffs decouple the upland terrain from the adjacent downstream reaches of these fluvial systems. Hut at H is 110 m below shore platform at SP.

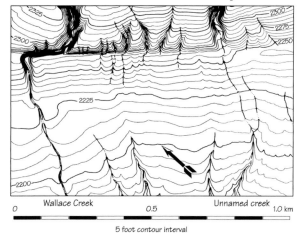

5 foot contour interval

Figure 2.19 Decoupled hillslopes, piedmonts, and stream channels.
B. Topographic map of Wallace Creek crossing the San Andreas fault that shows piedmont stream channels that are decoupled from their source areas by intervening reaches of alluvial-fan deposition. Wallace Creek is the only stream channel with continuity that permits upstream migration of knickpoints. Trace of the San Andreas fault is at base of the escarpment. From Figure 3 of Sieh and Wallace (1987).

Wallace Creek. The 400-m long aggradational reach stops knickpoint migrations in the unnamed creek. The fluvial system of the unnamed creek will remain decoupled until all threshold intersection points are eliminated so all reaches are on the erosional side of the threshold of critical power.

Even big rivers can be decoupled. The Grand Canyon reach of the powerful Colorado River illustrates a third style of fluvial decoupling. This reach lacks the smooth concave longitudinal profile indicative of attainment of equilibrium conditions that one would expect for a big river. Instead, the profile is convex with numerous abrupt steps (Fig. 2.19C). Steps coincide with tributaries from cliffy

watersheds whose frequent debris flows deliver 1 to >3 m boulders to small fans (Fig. 2.19D). The river quickly removes the sand fraction from such debris fans, and can shift some of the smaller boulders a short distance downstream during flood discharges (Webb et al., 1999, Figure 41). Repeat photography (Webb et al., 1999, Figure 14) shows the same large boulders in the Lava Falls rapids after a moderate flow of 6,200 m³/s (220,000 ft³/s).

Encroachment by debris fans controls the longitudinal profile of the river and the hydraulic behavior of the rapids (Dolan et al., 1978; Howard and Dolan, 1981; Kieffer, 1985; Webb et al., 1989; Melis and Webb, 1993; Webb, 1996; Griffiths et al., 1996, 2004). The Grand Canyon reach does not have the typical pool-riffle sequence characteristics of bedload streams (Webb et al., 1996, p. 152) because of the size and flux rate of boulders from tributaries.

The result is constricted, steep reaches with near-critical streamflow characteristics and upstream migrating hydraulic jumps. Half the drop in altitude in the Grand Canyon reach occurs in short rapids (Leopold, 1969). Each bouldery obstruc-

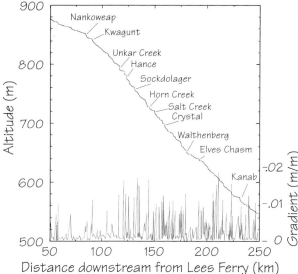

Figure 2.19 Decoupled hillslopes, piedmonts, and stream channels.
C. Longitudinal profile and gradients of the Grand Canyon reach of the Colorado River, Arizona between 50 and 250 km downstream from Lees Ferry. Steep reaches coincide with rapids (only a few are named here) resulting from bouldery debris flows derived from cliffy tributary streams. 1927 data set supplied courtesy of Robert H. Webb, U. S. Geological Survey.

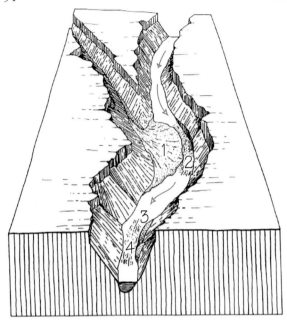

Figure 2.19 Decoupled hillslopes, piedmonts, and stream channels.

D. Diagrammatic Grand Canyon debris fan and rapids. 1 is debris-flow fan from tributary watershed. 2 is constricted river flow plunging down a rapid with large immobile boulders. 3 is debris bar of cobbles and small boulders derived from debris fan. 4 is secondary rapid caused by debris bar. From Griffiths et al., 1996 as modified from Hamblin and Rigby, 1968.

As elsewhere in the American Southwest (Bull 1991, Chapter 2) the change to Holocene climates resulted in much more frequent debris flows. Summer monsoon-type rainstorms and winter cyclonic storms, such as cutoff lows off the coast of Southern California and Mexico, that are sufficiently warm to produce thunderstorms were scarce or absent during times of cooler full-glacial climates. So, the influence of debris flows on the Colorado River longitudinal profile may not have been important before the change to a Holocene monsoonal climate.

Pazzaglia (2004, p. 268) concludes "There remains no good single explanation for why dramatically steep slopes on the Great Escarpments of passive margins erode at slow rates approaching 5 m/My whereas equally steep slopes in tectonically active settings may erode at rates three orders of magnitude faster approaching 5000 m/My." The concept of impediments to continuity in fluvial systems may explain this paradox. Disconnected landscape elements cannot transmit base-level falls in a manner resembling watersheds where the trunk stream channel is the connecting link between different parts of an integrated fluvial system. Numerical analyses that fail to recognize the presence of two separate, adjacent, landscape systems might use a process-response model that does not represent the real world.

Conceptually, the ancient Great Escarpments of South Africa resemble the sandstone spires of Monument Valley, Arizona. Both are spectacular weathering-limited cliffs whose rates of denudation have little relation to base-level changes in the streambeds of nearby fluvial systems.

2.5.3 Lithologic and Climatic Controls of

Relaxation Times

The independent variables of lithology and structure, and of climate, largely determine the types of processes that create hillslope landforms, and the time needed for streams to achieve equilibrium conditions. Joints and fractures and petrologic fabric beneath a watershed change little with time, whereas climate change is ubiquitous. Climate also varies between north- and south-facing slopes, and with altitude. Lithology also varies with space, and even monolithologic watersheds – those drainage basins underlain by a single rock type such as quartz monzonite or greywacke sandstone – typically have a highly variable density of joints, fractures, and shears.

tion acts as a local base level for the upstream pool reach in much the same manner as first described by John Wesley Powell (1875, p. 203–204). Continued influx of debris-flow boulders is sufficient to maintain rapids as significant local base levels. Lava that flowed into the canyon was a local bedrock base level that was removed by the river. Bouldery debris fans are a different type of perturbation. They are renewable base-level controls that define the character of the Grand Canyon reach. Human impacts are increasing the importance of this perturbation. Heights of some rapids are increasing since dam construction eliminated large annual floods (Graf, 1980; Melis and Webb, 1993; Webb et al, 1996).

Precipitation		Temperature	
Mean Annual (mm)		**Mean Annual (°C)**	
Extremely arid	< 50	Pergelic	> 0
Arid	50 - 250	Frigid	0 - 8
Semiarid	250 - 500	Mesic	8 -15
Semihumid	500 - 1,000	Thermic	15 - 22
Humid	1,000 - 2,000	Hyperthermic	> 22
Extremely humid	> 2,000		
Seasonality Index (Sp)*		**Seasonality Index (St)****	
Nonseasonal	1 - 1.6	Nonseasonal	< 2
Weakly seasonal	1.6 - 2.5	Weakly seasonal	2 - 5
Moderately seasonal	2.5 - 10	Moderately seasonal	5 - 15
Strongly seasonal	> 10	Strongly seasonal	> 15

* Precipitation seasonality index (*Sp*) is the ratio of the average total precipitation for the three wettest consecutive months (*Pw*) divided by the average total precipitation for the three consecutive driest months (*Pd*).

$$Sp = Pw/Pd$$

* Temperature seasonality index (*St*) is the mean temperature of the hottest month (*Th*) minus the mean temperature of the coldest month (*Tc*) in °C.

$$St = Th - Tc$$

Table 2.1 Classification of climates.

Relaxation times after fault ruptures of streambeds are short for large rivers flowing on soft rocks, and long for ephemeral streams flowing on hard rocks. Locally massive, hard rocks greatly increase the lifespan of a waterfall that decouples upstream from downstream reaches. Multiple surface-ruptures produce knickpoints that may migrate upstream only to increase height and permanence of waterfalls.

Table 2.1 defines the climatic terms used in this book. Each category, including extremely humid and extremely arid, is characterized by major differences in geomorphic and pedogenic processes. The temperature terms are from Soil Taxonomy (Soil Survey Staff, 1975). Mean annual air temperature at a site approximates soil temperature at a soil depth of 50 cm.

Climatic constraints affect the time needed for fluvial processes to shape a given landform by at least an ***order of magnitude*** (a ten-fold variation). Consider the triangular facets shown in Figure 2.20A. Weak to moderately resistant rocks and an arid, thermic, strongly seasonal climate are responsible for Saline Valley triangular facets with minimal dissection.

The lower portion of the rilled facet approximates the plane of an exhumed 35° to 40° range-bounding normal fault such as those described by Cichanski (2000). One might conclude that bedrock uplift must be rapid to form such dramatic triangular facets. The bedrock-uplift rate probably is typical of other rapidly rising mountain fronts in the Basin and Range Province, most likely being 0.3 to 1.0 m/ky. The minimal degradation seen here is in large part the result of the arid, thermic to hyperthermic climatic setting.

A much different climate influences the triangular facets of Figure 2.20B. The northwest front of the Southern Alps of New Zealand is being raised along the oblique-reverse range-bounding Alpine fault that dips under the range. Quartz–biotite schist offers little resistance to erosion after being weathered in the extremely humid, mesic, weakly seasonal climate. Deep valleys dissect the triangular facets. One might erroneously conclude that this landscape is indicative of a slow bedrock-uplift rate. Instead, this is one of the fastest rising major mountain fronts in the world – rock uplift and ridgecrest uplift is about 5 to 8 m/ky

Figure 2.20 Lithologic and climatic control of tectonic landforms illustrated by a comparison of triangular facets.

A. Mountain front along the southwest side of arid Saline Valley in southeastern California. The mountain–piedmont junction coincides with a normal fault. The slightly rilled lower surface, just above the mountain–piedmont junction, has a homogeneous appearance because it is fault gouge. Contrasting lithologies are obvious higher on the slope where the thin layer of gouge has been removed by erosion. Local patches of colluvium and alluvium cling to the fault plane such as at the top of the waterfall at the left side of the view.

B. Mountain front along the northwest side of extremely humid southern Alps of New Zealand. The dense rain forest provides little protection against rapid erosion of schist. The mountain–piedmont junction coincides with the oblique-reverse Alpine fault.

(Bull and Cooper, 1986; Yetton and Nobes, 1998). Valley-floor surface uplift is < 1.0 m/ky because these big rivers with large annual stream power have impressive stream-channel downcutting rates. Increases of bedload size and amount may have contributed to modest long-term increases in stream-channel gradient in reaches upstream from the range-bounding Alpine fault. The main divide of the Southern Alps is high partly because sustained rapid erosion of deep valleys promotes isostatic compensation that further increases the altitudes of peaks and ridgecrests.

The factors that influence surface uplift (Fig. 1.4) can be used to elaborate on the usefulness of tectonic landforms such as triangular facets. Surface uplift more closely approximates bedrock uplift in Saline Valley because erosion is mini-

mal. This favors preservation of triangular facets as a tectonic landform. However, rapid erosion of weathered rocks creates tectonic landforms suggestive of relatively less bedrock uplift in the Southern Alps. The magnitude of climate-controlled erosion is large enough to affect styles of crustal faulting (Koons, 1989; Norris and Cooper, 1995). Average surface uplift for Southern Alps watersheds surely is reduced by rapid erosion of the landscape, but still exceeds that of less tectonically active Saline Valley.

The importance of climatic control on landscape evolution demonstrated by this comparison underscores the difficulty of using landforms for quantitative estimates of bedrock or surface uplift rates. Alternatively, one can rank qualitative classes of surface uplift based on assemblages of tectonic landforms within a given climatic province (Chapter 4).

The concept of relaxation time also applies to the consequences of Pleistocene–Holocene climatic change. Reaction times typically are brief when protective plant cover is changed and hillslope soils undergo net erosion instead of net accumulation. The pulse of valley-floor alluviation caused by stripping of the hillslope sediment reservoir has a relaxation time of only a brief 1 to 3 ky in hot deserts (Bull, 1991), but is much longer for vegetated hillslopes of humid regions. Density of hillslope plant cover is not changed as much, and the volume of soil and colluvium is an order of magnitude greater. The relaxation time of Japanese watersheds to the Pleistocene–Holocene climatic change exceeds 10 ky and may be a factor in the present high watershed sediment yields (Oguchi, 1996).

2.5.4 Time Spans Needed to Erode Landforms

Tectonic geomorphology studies focus mainly on the past 10 to 100 ky in areas of accelerated landscape evolution (rapid bedrock uplift, soft rocks, and extremely humid climate) and on more than 10 My in slowly changing pedimented landscapes of some arid regions. Hills and streams continue to change after tectonic uplift of mountains has virtually ceased. The time needed for erosion to create landforms indicative of tectonic stages of landscape evolution ranges from less than 1 ky to more than 1,000 ky.

The time span needed for each landform noted on the left side of Figure 2.21 is a function of uplift,

rock resistance, and volume of material to be eroded after cessation of uplift. Only a short time is needed for the concentrated power of a stream to remove a small volume of unconsolidated alluvium to create a fanhead trench. Immense time spans (>10 My) are needed to consume the last vestiges of an uplifted planar surface. Such escarpment retreat is accomplished by gradual weathering of bedrock and slow erosion of hillslopes, and the volume of rock to be removed is huge.

Isotopic ages allow rough estimates of the times needed to erode landforms in the Mojave Desert and the Coast Ranges of California. Potassium-argon ages for volcanic materials in mountains and basin fill range from 0.5 to more than 5 Ma. Granitic and metamorphic rocks predominate in the arid Mojave Desert, and soft mudstone and sandstone predominate in the semiarid to subhumid Coast Ranges.

Erosion rates vary with climatic setting by at least two orders of magnitude (Fig. 2.20). The Mojave Desert and Coast Range plots on Figure 2.21 are separated by approximately an order of magnitude. The sheared granitic and metamorphic rocks of the semiarid to subhumid Transverse Ranges occupy an intermediate position. Estimates of denudation rates based on amounts of sediment trapped in 450

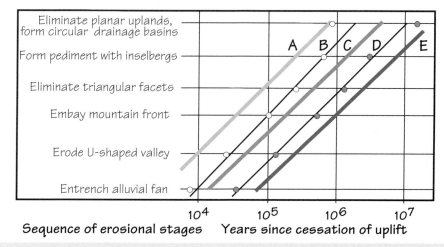

Figure 2.21 Diagram comparing estimated times needed for changes in landforms after cessation of active uplift for different climates and rock types in 10 km² fluvial systems. The Coast Range and Mojave Desert stages are spaced on the ordinate so that most points approximate straight lines. Plots without control points have less dating control. A. Sheared and fractured greywacke sandstone in humid New Zealand. B. Soft mudstone and shale in the semiarid central Coast ranges of central California. C. Sheared and altered granitic and metamorphic rocks of the subhumid San Gabriel Mountains of southern California. Quartz monzonite in the arid Mojave Desert of California. E. Gneissic and granitic rocks in the extremely arid Sinai Peninsula of Egypt. After Bull (1988).

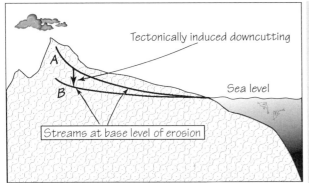

Figure 2.22 Sketch of two longitudinal stream profiles graded to similar sea-level highstands. Rivers erode down into rising mountains and then widen their valley floors by beveling strath surfaces when they are not able to downcut further. In this case tectonically induced downcutting between times A and B has left the downstream reaches of the longitudinal profile as a strath terrace passing through the rising landscape.

debris basins (Scott and Williams, 1978; Brown and Taylor, 1982) suggest rapid denudation of the San Gabriel Mountains at about 1.5 m/ky. Extremes of rates of landscape evolution are represented by the easily eroded fractured greywacke sandstone and schist of the extremely humid Southern Alps of New Zealand and by the extremely arid Sinai Peninsula.

A wide range of spatial and temporal scales of investigation is needed for the overdue incorporation of landscape analyses as an integral component of the plate tectonic paradigm. Process-oriented studies emphasize small spaces and time spans as short as the elapsed time since a recent earthquake (Arrowsmith and Rhodes, 1994). At the other extreme, spaces can be as large as mountain ranges, or entire tectonic provinces, and time spans may exceed 10 My (Davies and Williams, 1978; Ollier, 1982).

2.6 Tectonically Induced Downcutting

Streams incise ever deeper as bedrock is raised into the powerful buzz saw of stream-channel downcutting. Amounts and rates of tectonically induced downcutting are functions of vertical tectonic displacement rates, excess unit stream power (equation 2.6), and resistance of earth materials to degradation. Downcutting by small ephemeral streams flowing over resistant welded tuff may be unable to match a bedrock-uplift rate of 0.1 m/ky; such reaches remain on the erosional side of the threshold of critical power. Downcutting by perennial rivers flowing over soft rock easily keeps pace with bedrock uplift of 5 m/ky. But stream-channel downcutting occurs only during appropriate climatic and tectonic conditions.

The tendency of streams to cut down to the minimum gradient needed to transport their sediment load has been a long standing fundamental concept in fluvial geomorphology (Powell, 1875; Mackin, 1948; Leopold, Wolman, and Miller, 1964; Leopold and Bull, 1979; Bull, 1991). Headwater reaches of streams in rising mountains tend to stay on the degradational side of the threshold of critical power, but downstream reaches, with their greater unit stream power, are more likely to attain the base level of erosion through the process of *tectonically induced downcutting* (Fig. 2.22).

2.6.1 Straths, Stream-Gradient Indices, and Strath Terraces

Many streams return to the base level of erosion after tectonically induced downcutting is interrupted by aggradation events that temporarily raise the streambed. The Charwell River, New Zealand fluvial system (Figs. 2.23, 2.24) is sensitive to both tectonic and climatic perturbations; it has frequent climate-change induced aggradation events, numerous internal-adjustment terraces, and occasional times when the stream bevels its valley floor to create a tectonic landform – a major strath. Prior piedmont valleys, with their flights of Pleistocene stream terraces, have been preserved. Their rich history of landscape evolution has been set to one side as a result of rapid right-lateral displacement of the watershed by the Hope fault, which is at the mountain front–piedmont boundary. The following discussion focuses on the present-day valley and its flight of terraces, whose creation was modulated by several late Quaternary global climatic changes.

The mere presence of either marine or strath terraces has tectonic significance. Only one sea-level highstand was higher than the present high stand during the past 350 ka. It occurred at about 125 ± 5ka

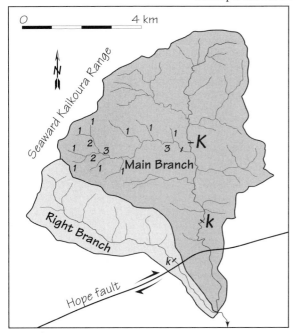

Figure 2.23 Drainage networks of the Main and Right Branches of the Charwell River, New Zealand. k is knickpoint migration from Hope fault since ~9 ka. K is where several knickpoints have accumulated to create a large step in the streambed. Numbers are for Strahler (1952, 1964) stream orders for a third order tributary. From Infomap 260 031, New Zealand Department of Survey and Land Information.

when the oceans were about 5 to 6 m above present sea level (Chappell, 1983, 2001; Chappell et al., 1996; Israelson and Wohlfarth, 1999). So just the presence of coastal shore platforms higher than 6 m shows that the land is rising relative to the sea-level datum. Similarly, the presence of flights of paired strath terraces shows that the terrain is rising, relative to the long-term base level of erosion of the stream. Dating the times of formation, and measuring heights, of either strath or marine terraces provides estimates of bedrock-uplift rates.

Reaches of the Charwell River upstream and downstream from the range-bounding Hope fault have different styles of response to uplift. This part of the Seaward Kaikoura Range is rising three

times faster than the adjacent piedmont reach. Rock mass strength also is much greater in the mountains, where unit-stream power becomes progressively less farther upstream. The longitudinal profile in the mountains has the characteristics of a disequilibrium stream, whereas the river flowing down the piedmont easily achieves type 1 dynamic equilibrium.

Tectonic strath terraces of the Charwell-River reach downstream from the front of the Seaward Kaikoura Range illustrate the importance of this landform to tectonic geomorphologists. The longitudinal profile is much more concave upstream from the range-bounding Hope fault and the average gradient is fivefold less downstream from the fault (Figs. 2.23, 2.24). The marked change in concavity of the two reaches mainly reflects rapid uplift of more resistant mountain bedrock, and pronounced overall widening of active-channel streamflow in the piedmont reach. Downstream increases of discharge and decrease in size of bedload are only moderate in this short distance, so may be less important than lithologic and tectonic controls. Slower uplift and softer rocks in the piedmont reach favor prolonged attainment of the base level of erosion at the conclusion of degradation events that follow pulses of aggradation.

Valley-floor portions of fault zones were buried beneath thick alluvium during aggradation events. The highly irregular longitudinal profile upstream from the Hope fault in part reflects cumulative surface ruptures as much as 40 m that were not able to migrate upstream until bedrock beneath episodic valley fill was exposed to erosion. We need ways to quantify both the irregularities in the longitudinal profile upstream from the Hope fault and degree of smoothness downstream from the fault zone. Stream-gradient indices are introduced as a valuable concept here.

John Hack used characteristics of large rivers in the humid Appalachian Mountains of the eastern United States to define a *stream-gradient index* that describes influences of many variables that influence the longitudinal profiles of stream channels (Hack, 1957, 1973, 1982).

Equilibrium adjustments, termed *hydraulic geometry*, assume orderly interactions between streamflow variables. Hydraulic geometry of stream channels is based on stream-gauging data, and typically has an order of magnitude scatter of data when used in logarithmic regressions of discharge and streamflow characteristics (Leopold and Maddock, 1953; Leopold, Wolman, and Miller, 1964). It defines statistical relationships between streamflow

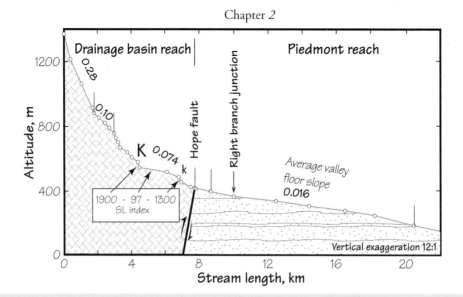

Figure 2.24 Longitudinal profile of the Charwell River from the headwaters to the junction with the Conway River, South Island, New Zealand. SL is stream-gradient index. k is a knickpoint that has migrated upstream from the Hope fault where it originated as a fault scarp beneath alluvium between 26 and 9 ka. K is larger multiple-event knickpoint. From Figure 4 of Bull and Knuepfer (1987).

parameters and channel morphologies. Using the approximate relations provided by hydraulic geometry dispenses with having to measure streamflows in virtually inaccessible localities.

Discharge (Q) from a watershed increases as a power function with drainage-basin area (A_d):

$$Q = cA_d^{\,n} \tag{2.7}$$

Many studies have compared length down a stream channel, L, from the main divide with drainage basin area, A_d, and have found that L increases at least as rapidly as Ad, (exponent is >0.5),

$$L = bA_d^{\,0.6} \tag{2.8}$$

where the units for L are miles and for A_d are square miles. This exponential function is now revered as "Hack's law" and has been the subject of many re-evaluations (Smart and Surkan 1967; Mueller, 1972; Seidl and Dietrich, 1992; Montgomery and Dietrich, 1992; Rigon et al., 1996). Experimental watershed studies by Lague et al. (2005) found that power functions of mainstream length increase almost linearly with drainage area, and that Hack's law is not

significantly dependent on uplift rate. The systematic decrease in slope as described by concave longitudinal profiles of stream channels is nicely described by a power function between the slope of a reach of a stream, S_{sc}, and A_d.

$$S_{sc} = kA_d^{\,\theta} \tag{2.9}$$

Hack used length, L, as a reasonable proxy for discharge, Q. He tested whether streams had achieved equilibrium by analyzing spatial variations in the product of slope of a reach, $\Delta H/\Delta Lr$ (change of altitude/length of reach) and the horizontal length to the midpoint of the reach from the watershed divide, Lsc. This is the "stream gradient index", or SL index, where SL is defined as:

$$SL = \frac{\Delta H}{\Delta L_r} L_{sc} \tag{2.10}$$

Verification of Hack's SL model was achieved when he showed that reaches of Appalachian rivers had fairly constant values of SL for consecutive reaches. The Appomattox River has remarkably constant SL values over a distance of 150 km (Hack, 1982). See the Figure 2.30 example discussed in

Section 2.7. This implies that 1) larger "bankfull" flow events were cumulative responses of the entire watershed, 2) streambed hydraulic roughness is constant downstream, and 3) bedload size and transport rate remained about the same downstream.

Analyses of *stream-gradient index* should be classed as *narrow* or *inclusive*. Narrow pertains to an anomalous stream-gradient index. It typically is only a short reach that describes only one or several contour intervals. It is useful for describing magnitudes of longitudinal profile abnormalities caused by locally high rock mass strength and/or knickzones that have migrated upstream from a source of base-level fall perturbations.

Inclusive pertains to long reaches of a longitudinal profile that have a constant rate of longitudinal-profile decrease in gradient associated with progressively larger streamflows from a headwater divide source area. Adjacent reaches with dissimilar inclusive gradient-indices describe variations in longitudinal profile caused by factors such as adjustments to spatially variable uplift rates (Keller and Rockwell, 1984), changes in rate of downstream increase of stream discharge, change in the direction of a valley, and change in median particle size of gravelly streambeds. Narrow gradient-indices describe disequilibrium reaches of streams. Inclusive gradient-indices can be used to describe situations of type 1 or type 2 dynamic equilibrium.

Narrow *SL* values of 1300, 97, and 1900 for adjacent reaches upstream from the Hope fault (Fig. 2.24) record the inability of the Charwell River to smooth out some irregularities in the longitudinal profile. These anomalies result from frequent large tectonic displacements of the streambed and variable rock mass strength of greywacke sandstone. The heights and present positions of these knickpoints are also a function of late Quaternary climate change.

Stream-gradient index analyses may not apply equally well to all streams. I suspect that this index describing the behavior of an erosional fluvial system should not be used where streams are aggrading. Of course, a brief pulse of deposition that has uniform thickness would not change analysis results because erosional processes prior to deposition created the form of the longitudinal profile.

Application to ephemeral streams should proceed with caution, especially where most convective-storm rainfalls generate flash floods over only part of a watershed. Infiltration of streamflow into a dry streambed results in progressive decrease in discharge, which is opposite to the trend of the large perennial rivers where Hack defined the stream-gradient index. Hack's model assumes that longitudinal-profile concavity results from ever-increasing stream power in the downstream direction. Concentration of sediment load concurrent with decreasing discharge of water can move a degrading ephemeral streamflow closer to, or across to the depositional side of, the threshold of critical power. For such reasons, ephemeral streams typically have longitudinal profiles that are much less concave than for humid region watersheds that generate bankfull discharges of similar size (Wolman and Gerson, 1978). We can expect stream-gradient indices to increase downstream, even in equilibrium reaches, where ephemeral streamflow behavior has constant or even decreasing stream power in consecutive downstream reaches.

Hack also noted that long equilibrium reaches of perennial rivers plotted as a straight line on semilogarithmic regressions of altitude H and ln of L_{sc}. Each long reach can be numerically described by an inclusive gradient-index. Such linear relations are described by

$$H = C - k(lnL) \qquad (2.11)$$

C is a constant and k is the inclusive gradient-index (slope of the regression). The derivative of equation 2.10 with respect to L is streambed slope, S:

$$SL = \frac{dH}{dL} = \frac{d(klnL)}{dL} = \frac{k}{L} \qquad (2.12)$$

The inclusive gradient-index can be estimated by regression analysis or by using data points from the longitudinal profile:

$$k = \frac{(H_i - H_j)}{(lnL_i - lnL_j)} \qquad (2.13)$$

where H_i and H_i are the altitude and distance from the watershed divide for an upstream point on the stream channel, and H_j and L_j are for a downstream point on the longitudinal profile.

Examples from the Charwell River are introduced here. The Right Branch of the Charwell River is presently beveling a strath as it flows from the Hope fault to its junction with the Main Branch.

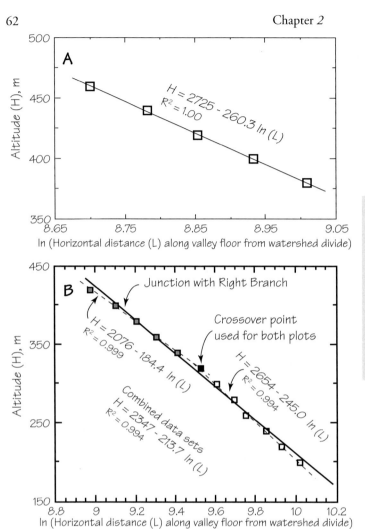

Figure 2.25 Inclusive stream-gradient indices for piedmont reaches of the Charwell River.

A. Right Branch from Hope fault to junction with the Main Branch.

B. Main Branch from Hope fault to the junction with the Conway River. The crossover point is used in both regressions.

A semi-log regression of altitude and distance from the headwater divide (Fig. 2.25A) indeed plots as a straight line. The perfect correlation coefficient is in part due to soft mudstone bedrock beneath the stream channel, and the lack of topographic obstacles. Of course correlation coefficients tend to be high where cumulative altitude is regressed against cumulative distance. The inclusive gradient-index for this fairly small stream is 260.

A similar analysis for the Main Branch also demonstrates attainment of equilibrium conditions. The complete dataset has a correlation coefficient of 0.994 and an inclusive gradient-index of 214, just what one would expect for a stream whose watershed is three times larger than that of the Right Branch.

Alternatively, the Main Branch can be modeled as two reaches with different characteristics. This improves the correlation coefficients slightly. The trend of consecutive points for the reach upstream from the crossover point does not reveal where the Right Branch enters the Main Branch or where the Main Branch narrows where it flows through a gorge cut in massive sandstone.

The upstream reach flows straight down the piedmont at a bearing of 170°. The river impinges on Flax Hills at the location of the crossover point. This topographic obstacle deflects the course of the river by 50°, changing the direction of the valley to a bearing of 230°. The steeper regression trend for the downstream reach has an inclusive gradient-index of

245. The contrast between gradient indices of 184 and 245 supports treating these as two datasets, thus counteracting the initial impression. Evaluation of the equilibrium stream channels of the Right Branch and Main Branch for reaches just downstream from the Hope fault is more appropriate for comparison of gradient-indices for these humid region streams, 184 for the Main Branch and 260 for the Right Branch.

Straths are beveled along reaches of streams at equilibrium – where the inclusive gradient-index remains constant from reach to reach. It is useful to view strath formation in a context of a strath-formation threshold defined as the effective stream power needed to mobilize streambed materials, and above which the stream can do the work of beveling a strath. This threshold is reached more often in downstream reaches of a stream as is suggested by the general lack of straths in the upstream half of most watersheds. Climatic and lithologic variables play critical roles in determining the wide range of conditions affecting the strath-formation threshold for a specific drainage basin. Rainfall-runoff magnitudes and rates greatly affect peaks and durations of streamflows and amount and size range of suspended and saltating sediment. Such interactions between the fluctuating variables over the long term affect the numerical values for the inclusive gradient-index.

A broad modern strath in the reach immediately downstream from the Hope fault has been beveled across soft sedimentary rocks (Fig. 2.26) and is indicative of prolonged attainment of type 1 dynamic equilibrium. This late Holocene valley floor is a nice example of a steady-state landform. Peter Knuepfer (1988) did the weathering-rind dating of exposed

boulders on the treads of the flight of degradation terraces (Fig. 2.16). The lower strath terraces of the flight are shown here. The scarp of the oblique right-lateral Hope fault is the bushy riser at the far end of the pasture. This is a nice example of attainment of equilibrium for a specific reach of the fluvial system (Fig. 2.25B). The Charwell River quickly re-established the base level of erosion here many times after brief departures during the past 4 ka. These minor variations in streambed altitude are merely the product of the normal fluctuations in the spectrum of discharge of water and sediment.

Strath terrace heights in the Flax Hills reach, 7 to 8 km downstream from the Hope fault, were surveyed using the modern strath as a reference level (Bull, 1991, Fig. 5.19). It is 2.5 ± 0.5 m below the active channel of the Charwell River. Seven radiocarbon ages on fossil wood collected from basal aggradation gravels just above several older straths reveal that tectonically induced downcutting in this reach is 0.37 ± 0.03 m/ky. Assuming that this bedrock-uplift rate was uniform during the past 200 ka, one can estimate strath ages by dividing strath height by tectonically induced downcutting rate. For example, the strath presently at 30.3 ± 0.5 m above the modern strath is estimated to have formed at about 82 ka.

$$\frac{30.3\,m}{30.3\,m/ky} = 82 \pm 8\,ka \qquad (2.14)$$

Times of tectonic strath formation occurred at approximately 0, 40, 54, 62, 72, 82, and 114 ka in the Flax Hills reach. Times of strath terrace forma-

Figure 2.26 Reach of the Charwell River downstream from the Hope fault has been at the base level of erosion. The resulting steady-state landform is a 400 m wide strath cut in soft Cenozoic basin fill that is capped with a veneer of stream gravel. This surface of detrital transport since 4 ka continues to be lowered at a rate equal to the rock-uplift rate.

Strath height, m	Marine terrace age, ka	Inferred strath age, ka
0	0	0
*	29	#
14.9	40	40
	44	#
20.1	53	54
23.1	59	62
26.6	72	72
30.3	81	82
	96	#
	100	#
42.1	118	114
	124	
53.7	No match for 54 m strath	145
62.0	176	168

Table 2.2 Relations between tectonically induced valley-floor downcutting and inferred ages of major straths of terraces along the Flax Hills reach of the Charwell River, South Island, New Zealand. Marine-terrace ages are from Chappell and Shackleton (1986); and Shackleton (1987).
* River cut a strath near the mountains, but only incised part way through the Flax Hills aggradation-event alluvium in the study reach.
# Tectonic strath of this age is present in another reach of the River

tion at about 29, 44, 96, and 100 ka were observed in other reaches, but not in the Flax Hills reach because 1) erosion has removed the strath, 2) the strath was not exposed at the time of my survey, 3) or insufficient vertical separation to distinguish between adjacent straths because of the locally low bedrock-uplift rate. Strath terrace age estimates for the Charwell River (Table 2.2) coincide with the ages of dated global marine highstands of sea level (Chappell and Shackleton, 1986; Gallup et al., 1994). The coincidence between the inferred ages of Charwell River straths and the isotopic ages of global marine terraces is the result of similar timing of climate-change modulation of marine coastal and fluvial geomorphic processes during the late Quaternary.

Both the Charwell fluvial system and the coastal marine system are controlled by global climatic changes that fluctuate between full-glacial and interglacial extremes. Times of rapid aggradation of New Zealand valleys occurred at times of maximum accumulation of ice on the continents and lowstands of glacio-eustatic sea levels. Downstream reaches of these streams easily attain type 1 dynamic equilibrium during interglacial climates at times that coincide with the times of maximum melting of continental ice masses and attainment of sea-level highstand. Hillslope plant cover and geomorphic processes in the Charwell River watershed were greatly different for these two regimes (Bull, 1991, Chapter 5).

The 29, 40, and 54 ka tectonic straths of the Charwell and nearby rivers can be identified by radio-carbon and luminescence dating of the adjacent overlying deposits. Together with the modern (0 ka) strath, they provide readily accessible time lines to assess local rates of tectonically induced downcutting of the valley floor. For example, the 0 and 29 ka straths can be used to estimate the bedrock-uplift rate for a reach of the Charwell River that is 1 km downstream from the Hope fault. Analyses of weathering-rind thicknesses on surficial greywacke cobbles by Knuepfer (1984, 1988) were the basis of the 10.8 ± 1.9 ka age estimate for the tread of the fill-cut degradation terrace shown in Figure 2.27. The lack of paleosols or beds of loess in the basal 23 m of uniformly massive sandy gravels is suggestive of a single pulse of aggradation of Stone Jug gravels. The process of returning to the Charwell River base level of erosion involves stream-channel downcutting through the aggradation event gravels and then through a thickness of bedrock equal to the total rock uplift since the time of the pre-aggradation event strath. Holocene degradation of an additional 39 m below the 29 ka strath occurred in this reach. The rate of tectonically induced downcutting is

$$\frac{39\,m}{29ky} = 1.3 \pm 0.1 m/ky \qquad (2.15)$$

Two additional points are illustrated by Figures 2.26 and 2.27. Even the small 30 km² main branch of the Charwell River may easily attain the base level of erosion in downstream reaches for sufficiently long time spans to bevel tectonic straths (type

Figure 2.27 View of 11 ka Charwell River fill-cut terrace. About 23 m of gravel lie on a tectonic strath that formed at about 29 ka. The 39 m between the buried strath and the present tectonic strath reflects the amount of tectonically induced river downcutting since 29 ka: the basis for estimating an uplift rate of 1.3 ±0.1 m/ky.

1 dynamic equilibrium). Second, it is not necessary to preserve a complete section of aggradation gravels in order to identify the aggradation event that buries a tectonic strath.

The excellent agreement between ages of global marine terraces and local tectonic stream terraces ties the Charwell River terrace chronology to a global climatic chronology. Global climatic changes result from variations in the Earth's orbital parameters – the astronomical clock – (Berger, 1988). The similarity of the Table 2.2 pairs of ages may permit assignment of ages for straths with less radiocarbon dating control.

Potential dating uncertainties for straths older than 40 to 50 ka include violation of the assumption that both systems have similar response times to global climatic perturbations. This can happen for aggradation events because watershed characteristics are sufficiently variable to result in crudely synchronous, or even diachronous, aggradation surfaces for a suite of adjacent watersheds. Watershed responses to

climate change in much of the South Island of New Zealand are nicely synchronous because of fairly similar topography, lithology, and humid climate. Brief regional aggradation events can also have nontectonic origins such as regional coseismic landslides (Hancox et al., 2005).

Termination of periods of strath formation is much more likely to be synchronous. Studies of the Greenland ice cores give us a better appreciation of the rapid onset of major climate changes (Alley, 2000; Peteet, 2000). The age uncertainties for isotopically dated times of global marine terraces are less than the ±5 to ±10 ky uncertainties for strath ages that are a function of surveying and uplift-rate-calibration errors (see equation 2.14 for an example).

2.6.2 Modulation of Stream-Terrace Formation by Pleistocene–Holocene Climatic Changes

Times of formation of tectonic landforms commonly reflect other important variables of fluvial systems such as *annual unit stream power* (a measure of a stream's capacity to do work). Times of tectonic strath formation along the Charwell River were largely controlled by the rather overwhelming influence of late Quaternary climatic changes. Climatic-change impacts of watershed geomorphic processes raise and lower the streambed at rates faster than the concurrent bedrock uplift caused by the sum of tectonic forces and isostatic adjustments (Fig. 2.28).

The piedmont reach of the Charwell River was either aggrading or was catching up to new base levels of erosion. This reach was raised by a combination of uniform rapid bedrock uplift and intermittent valley-floor backfilling of 30–60 m.

Aggradation events were the dominant process during the Pleistocene, whereas the Holocene has been characterized by degradation. In order to occasionally catch up to a new tectonic base level of erosion the stream had to degrade through the most recently deposited valley fill, and then through a thickness of bedrock equal to the amount of bedrock uplift since the last time the stream attained type 1 dynamic equilibrium. The Charwell River barely had enough time to bevel a new tectonic strath after attaining the base level of erosion, before the onset of the next aggradation event. These brief episodes of attainment of equilibrium allow comparison between the times of strath cutting with the times of solar insolation maxima and sea-level rise. The agreement

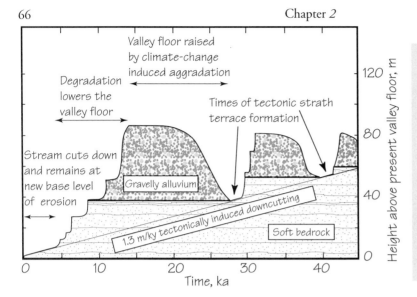

Figure 2.28 Changes in the streambed altitude of the Charwell River, New Zealand reflect the combined influence of tectonic and climatic controls during the past 45 ka. Tectonic strath terraces are created only during brief time spans that follow climate-change modulated episodes of tectonically induced downcutting. Simplified from Figure 5.24 of Bull (1991).

would not be nearly as nice (Table 2.2 and Figure 2.28) for larger streams that may remain at the base level of erosion for 60-90% of the time.

One reason for distinguishing between type 1 and type 2 dynamic equilibrium is that tectonically induced downcutting can be used to estimate uplift rates only when comparing situations of type 1 dynamic equilibrium. These streams have parallel longitudinal profiles of stream terraces that indicate return to similar combinations of variables for reaches where channel width is less than valley-floor width. This is not the case for type 2 dynamic equilibrium streams incising into bedrock. Longitudinal profiles may be concave and exponential, but unfortunately we can only examine the present assemblages of landforms because type 2 streams do not create suitable landforms that are preserved.

Consider the Grand Canyon reach of the Colorado River in northern Arizona. Active normal faulting at the western end of the Canyon during the Quaternary caused tectonically induced downcutting of roughly 0.4 m/ky but this decreased to 0.2 m/ky at 100 km upstream in the eastern reach of the Canyon (where strath terraces are more likely). It appears that this 100 km long reach has been steepened 400 m in the past 2 My. Steeper gradient and narrower channel width are the obvious consequences, but changes in hydraulic roughness may be just as profound. These several alterations do not let us use changes in the altitude of the longitudinal profile to estimate uplift rate. Furthermore, influx of large boulders from cliffy tributary streams does not allow the river to behave

as a system of interrelated reaches (Figs. 2.19C, D). Waterfalls in the Charwell River upstream from the Hope fault also limit use of fluvial landforms to estimate uplift rate.

2.7 Nontectonic Base-Level Fall and Strath Terrace Formation

Not all strath terraces represent time lines in tectonically deforming landscapes. So let us clarify other aspects of this valuable landform with examples of the few exceptions to what might have seemed a general rule in the preceding discussions.

The most obvious nontectonic strath is an unpaired terrace resulting from local lateral migration of a stream into a bedrock hillslope. Such a nontectonic strath could even form while a valley floor is being slowly raised during the terminal stages of an aggradation event.

Pauses in a degradation event may temporarily allow a stream to bevel either fill-cut surfaces in alluvium or strath surfaces in bedrock before the stream has downcut sufficiently to return to a new base level of erosion. These common erosion surfaces are nontectonic internal-adjustment terraces (Charwell River at ~ 14 to 4 ka (Fig. 2.16) for example).

Base-level falls can be induced by climatic perturbations to fluvial systems as well as by uplift. A good example is from the piedmont along the mid-Atlantic coast of the eastern United States. Isostatic uplift continues at a very slow rate in response to gradual erosion of the Appalachians and tectonic uplift generally is so minor as to be trivial. So, this would seem to be an improbable region to observe large amounts of stream-channel downcutting below prominent strath levels. But, beautiful, prominent strath surfaces occur along the lower reaches of large rivers that have cut spectacular bedrock gorges just above their terminal tidewater reaches.

The Great Falls are in the terminal reach of the Potomac River west of Washington, DC (Fig. 2.29). The prominent strath surface was beveled across the highly resistant late Proterozoic sandstone and schist of the Mather Gorge Formation. Cosmogenic [10]Be and [26]Al dating of 18 samples collected from Mather Gorge downstream from the Great Falls of the Potomac River (Bierman et al., 2004; Reusser et al., 2004) indicate rapid incision of the strath began at about 30–32 ka. This strath was beveled during the preceding 50 ky and perhaps during an even longer time span. The 22 m height of the strath terrace shown in Figure 2.29 surely cannot represent the consequences of either tectonic or isostatic uplift in this region in such a short time span.

A detailed study by Bierman et al. (2004) concludes that gorge incision coincides with a 50 m decline in sea level at the beginning of the most recent ice age. Sea level continued to fall to more than 100 m below the present level (Fig. 1.19) – a major climatic perturbation. The regional nature of the perturbation is suggested by synchronous similar straths in other coastal plain rivers such as the lower Susquehanna, Rappahanock, and James Rivers.

Processes of incision by large and small rivers into bedrock intrigue geomorphologists (Tinkler and Wohl, 1998a, b) because the valley floor is the base level for all adjacent hillslopes. River incision rates control the rate of landscape response to changes in rock uplift rate and Quaternary climate change (Howard, 1994). Seidl and Dietrich (1992) assume that incision rate is proportional to stream power, but this model may not explain some variations in incision rates (Stock and Montgomery, 1999) and longitudinal profile shapes (Sklar and Dietrich, 1998). The mix of variables surely has to include rock mass strength (Selby, 1982b, Moon, 1984) and bedload size and amount (Sklar and Dietrich, 2001).

I would add that information about all of the above variables is of little use unless one knows how far removed a reach of a river is from the base level of erosion. Gilbert (1877) knew that streams close to the base level of erosion or on the aggradational side

of the threshold of critical power would be limited in their capacity to incise into bedrock – regardless of unit stream power, rock mass strength, or bedload transport rate – because the valley floor would be mantled with protective alluvium much of the time. However, a streambed far to the degradational side of the threshold will be exposed to valley-floor degradation. The magnitude of departure from the threshold of critical power is the crucial factor and is controlled by both climatic and tectonic factors. Application of such stream power/resisting power ratios for adjacent aggrading and degrading reaches is illustrated in Figure 6 of Bull (1997). Field-work based modeling needs to focus more on the locations and time spans where excess stream power is available to incise valley floors.

The Potomac and other rivers extended downstream into newly exposed reaches of the continental shelf that were steeper than the prior tidal reaches. Terminal reaches either dropped over the edge of the continental shelf, or were steepened by dropping into the head of a submarine canyon (Talling, 1998, Fig. 3). Assuming no change in discharge of either water or sediment, such rivers would then have terminal reaches strongly on the erosional side of the threshold of critical power. The potential for erosion would have been still greater if sea-level fall coincided with either a decrease of bedload transport rate or an increase in stream discharge.

Figure 2.29 Broad strath terrace approximately 22 m above the present floor of the Potomac River at Great Falls National Park, Maryland. Strath incision of this magnitude is anomalous in this tectonically inactive setting.

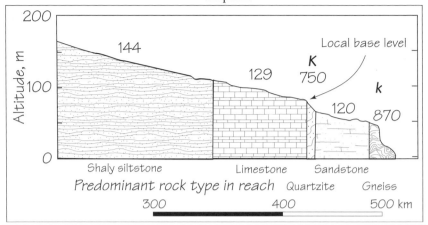

Figure 2.30 Stream-gradient indices for the Potomac River reach upstream from tidal Mather Gorge. Inclusive gradient-indices of the longitudinal profile (low SL values) approximate straight lines on this semi-logarithmic graph have achieved type 1 or type 2 dynamic equilibrium. Stream-gradient index analysis uses a logarithmic scale of distance, in this case from the headwaters of the South Branch of the Potomac River. k is the 30 ka knickpoint of the Great Falls of the Potomac. K is reach with exceptionally resistant lithology that tends to arrest and accumulate upstream migrating knickpoints. Both are narrow gradient-indices. Graph and rock types from Hack (1973).

A second, much different type of climatic perturbation might have played an important role in strath-terrace formation. Reusser et al. (2004) point out that massive loading by nearby continental glaciers would have the effect of depressing the crust beneath the ice, which would tend to create a belt of rock uplift in front of the continental glacier. This creates an ephemeral crustal upwarping – a forebulge. Such raising of the land surface in the terminal reaches of large coastal rivers in front of the ice would initiate stream-channel downcutting that has continued to the present.

Timing of strath incision could also be modulated by local presence of resistant valley floor rocks that could delay the onset of accelerated downcutting. Potomac River incision appears to have migrated rapidly upstream; Bierman et al. (2004) conclude that the initial knickpoint migration required only a few thousand years. The river continues to deepen Mather Gorge at a uniform rate of about 0.8 m/ky since 35 ka.

Similar climatic perturbations may have played a role in forming the classic Schooley and Somerville erosion surfaces of Davis (1890) in the coastal Appalachian region. Stanford et al. (2001, 2002) use the post late-Miocene decline in sea level to account for two episodes of valley-floor incision

and strath-terrace formation. Stepwise glacioeustatic events dominated landscape evolution on this low-relief passive margin. Incision occurred when global ice mass increased, thus causing sea-level decline. Times of formation of broad straths coincided with long periods of stable or rising sea level. The resulting flexural isostatic uplift of the area was the product of both mountain-range denudation and shifting of the resulting sediment to coastal basins of deposition. Estimated regional long-term uplift rates are miniscule, being only about 0.02 m/ky, and in contrast to rapid gorge incision of Mather Gorge between 37 to 13 ka of about 0.8 m/ky (Reusser et al., 2004).

The downstream reaches of the Potomac River have SL indices that clearly show either attainment of equilibrium conditions, or have pronounced knickpoints indicative of obvious disequilibrium (Fig. 2.30). The Great Falls is a late Quaternary feature but the resistant quartzites of Blue Ridge may act as a long-term local base level that accumulates upstream-migrating knickpoints in much the same way as noted for the Charwell River of New Zealand (Fig. 2.24). This lithologic control is a long-term impediment to the continuity of the Potomac fluvial system.

Combination of Cenozoic flexural isostatic uplift and upstream knickpoint migration can temporarily accelerate stream-channel downcutting, thus facilitating preservation of remnants of old erosion surfaces (Pazzaglia and Gardner, 1993, 1994, Pazzaglia et al., 1998; Zaprowski et al., 2001). Strath terraces have formed in tectonically inactive regions such as Australia (Goldrick and Bishop, 1995) and in southern Arizona; both areas have minimal connection to sea-level fluctuations. Such landforms have great antiquity because they are the result of valley floor downcutting in response to isostatic uplift resulting from gradual erosion over millions of years. Bedrock uplift in this case is purely isostatic. Times of strath incision, as for tectonically active settings, occur at times when climatic controls favor strong departure from the threshold of critical power by altering discharge of water and sediment. They also influence the concavity of river longitudinal profiles (Zaprowski et al., 2005). Tectonically inactive Appalachian streams have more concave longitudinal profiles where peak annual discharge is greatest, suggesting that this factor may be more important than watershed size in determining rates of valley floor incision. Location within a fluvial system also is important, so we next turn our attention to a dimensionless way of defining locations within watersheds.

2.8 Hydraulic Coordinates

This concluding section of Chapter 2 introduces a way of locating hydraulic positions within smaller elements of fluvial systems – hillslopes and drainage basins. Use of hydraulic coordinates is a dimensionless format that facilitates comparisons between large and smaller drainage basins.

A good way to relate tectonics to topography is with models that quantify spatial and altitudinal positions of water and sediment flux in fluvial systems. I do this with dimensionless hydraulic coordinates for points on local hillslopes and for watershed locations relative to the trunk valley floor of a drainage net. Active range-bounding faults mark abrupt transitions between the erosional and depositional domains of many tectonically active fluvial systems. Tectonic geomorphologists use hydraulic coordinates to relate the position of an active fault or fold to nearby or distant parts of a fluvial system, and to compare the morphologies of hills and streams in different tectonic and climatic settings.

Two planimetric coordinates are described in terms of ratios of horizontal distances of flow direction down hills and streams. The *hillslope-position coordinate*, H_{pc}, is the planimetric length from a ridgecrest divide to a point on the hillslope, L_h, divided by the total length from the divide to the edge of the valley floor, L_t.

$$H_{pc} = \frac{L_h}{L_t} \qquad (2.16)$$

Hillslope-position coordinates range from 0.00 at the ridgecrest to 1.00 at the base of the footslope.

The *basin-position coordinate*, B_{pc}, is the planimetric length, L_v, from the headwaters divide of a drainage basin along the trunk valley to a streambed point divided by the total length of the valley, L_{vt}, from the headwaters divide to the mouth of the drainage basin. The point in the valley floor should be directly down the flow line from a hillslope point of interest, which commonly is directly opposite the stream-channel point.

$$B_{pc} = \frac{L_v}{L_{vt}} \qquad (2.17)$$

Total length is not measured along a sinuous stream channel, because this is a landform that changes too quickly for our longer-term perspective, nor along a chord between two endpoints. Instead, the basin-position coordinate describes distance along the trend of a valley. Interpretation of the numerical values is fairly straightforward. For example, 0.50R/0.67 describes a point half way down the right side hillslope for a point whose flux of sediment and water is two-thirds of the distance through a drainage basin. Right (R) and left (L) sides are when looking downstream.

The aerial photos used in Figure 2.31 illustrate descriptions of hillslope- and basin-position coordinates. Locations of points along trunk streams are easy to define in low-order basins (Fig. 2.31A); flow proceeds from points 1 to 2 to 3 whose hydraulic coordinates are listed in Table 2.3. Point 2 marks the location where fluxes of water and sediment from a small tributary valley enter the trunk stream. Point 4 is a third of the way down the right-hand side hillslope opposite a basin-position coordinate of 0.48. Note that I have made a subjective decision to regard rills as part of this hillslope instead of as lower order tributary valleys. This type of subjective decision is always present and is largely a matter of scale and mode of depicting drainage nets, and the purposes

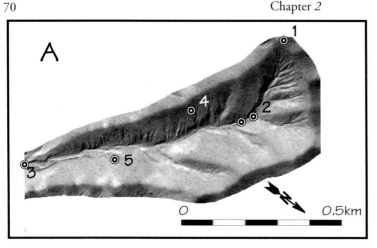

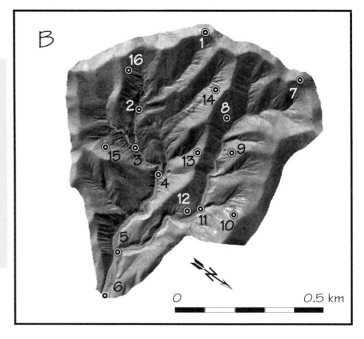

Figure 2.31 Aerial photographs of drainage basins eroded into soft rocks of the eastern Rodman Mountains, central Mojave Desert, southern California. Hydraulic coordinates for each numbered point are listed in Tables 2.3 and 2.4.
A. Low-order watershed with points to illustrate basic concepts of hydraulic coordinates.
B. Complex watershed with points to illustrate numbering for nested basin-position coordinates.

of your investigation. The rills, which would not be apparent on conventional topographic maps, are useful. In this case, they alert us to measure distance along flow lines that are not at right angles to the trunk valley floor in much of this watershed. The left-side position of point 5 is denoted with an L, and this slope also drains directly into the trunk channel.

Basin-position coordinates can also be used to describe points on hillslopes within multiple orders of nested tributary valleys. The first step in defining hydraulic coordinates in more complex watersheds is

to identify the trunk channel of the longest sub-basin, from the headwater divide to the mouth of the drainage basin. We then identify the junctions at which water and sediment from tributaries join this trunk stream. By working our way upstream we go from the mouth of the master stream to the sources of sediment and water. Two large sub-basins are shown in Figure 2.31B. The longest flow path is through the southern sub-basin, which in this case has a larger watershed area than the northern sub-basin. Flow begins at point 1 and leaves the watershed at point

Hydraulic coordinate number	Basin-position coordinate	Hillslope-position coordinate
1	0.00	0.00
2	0.27	1.00
3	1.00	1.00
4	0.48	0.30R
5	0.70	0.72L

Table 2.3 Basin-position and hillslope-position coordinates for points in the low-order drainage basin shown in Figure 2.31A.

6. Points 2, 3, 4, and 5 are the locations of several stream junctions whose tributary valleys contain the hillslope points of Table 2.4. Each stream junction marks an abrupt increase of discharge from tributary watersheds that here are progressively larger in the downstream direction.

Hillslope-position coordinates are the same as for the simple watershed of Figure 2.31A, but basin-position coordinates should include the location of the tributary in the nested hierarchy of stream orders. For example, points 7 through 12 are all located in the tributary that joins the trunk channel at a basin-position coordinate of 0.85, so (0.85) is placed before the basin-position coordinate in the tributary. Point 10 is on a left-side hillslope of a small basin that has a third-order relation to the trunk valley floor. Its location would be described by [0.85, 0.68) 0.70; 0.27L]. The number 0.85 tells us where flow from the large second-order tributary joins the third-order trunk stream. The number 0.68 refers to the fact that this first-order stream joins a second-order stream at a position that is 68% of the distance from the head-water divide of the second-order basin. Flow passing through hillslope-position coordinate 0.27L passes directly to first-order channel at the local basin-position coordinate of 0.70. Similarly, point 13 is 40% down a left side hillslope and 52% down a first-order basin that drains to a second-order basin at 90% of its length, which in turn drains to the third-order stream at 57% of its length: [0.57, 0.90) 0.52; 0.40L]. Assignment of relative planimetric positions is flexible, depending on map and image scales and the needs of the geomorphologist.

Dimensionless numbers can also be used to describe relative relief positions. The *hillslope-fall ratio*, H_f, is the decrease in altitude from a ridgecrest divide to a point on the hillslope, H, divided by the total decrease in altitude from the divide to the base of the footslope, H_{th}.

$$H_f = \frac{H}{H_{th}}$$ (2.18)

The *basin-fall ratio*, B_f, is the decrease in altitude from the headwaters divide along the trunk valley to a point in the valley floor, H_v, divided by the total watershed relief (decrease in altitude from the head-water divide to the mouth of the drainage basin) R.

$$B_f = \frac{H_v}{R}$$ (2.19)

Hillslope and basin-fall coordinates for a sequence of nested drainage basins can be described in much the same way as for hillslope- and basin-position coordinates by using percentages of total relief in each sub-basin draining to the trunk stream.

Dimensionless hydraulic coordinates and ratios minimize the factor of size in comparisons of fluvial landscapes. The basin-position coordinate can be used to describe knickpoint migration in adjacent watersheds of different sizes and stream powers. In Chapters 2 and 3 it will be recommended that a standard basin-position coordinate of 0.9 be used to evaluate valley floor width–valley height ratios in relation to range-bounding faults. Hillslope-position coordinates can be used to standardize survey procedures. Uses include descriptions of features of the relative locations of change from convex to concave slopes, landslide head scarps, and where rills start to incise with respect to their source ridgecrests. I use hydraulic coordinates to examine tectonic signatures in hillslope morphology. I also prefer to use the basin-fall ratio for dimensionless analyses of the stream-gradient index (Hack, 1973) and for semi-logarithmic longitudinal profiles of streams.

The longitudinal-profile of the trunk stream of the Right Branch of the Charwell River (Fig. 2.32)

Hydraulic coordinate number	Basin-position coordinate	Hillslope-position coordinate
1	0.00	0.00
2	0.33	1.00
3	0.46	1.00
4	0.57	1.00
5	0.85	1.00
6	1.00	1.00
7	(0.85) 0.00	0.00
8	(0.85) 0.31	0.67R
9	(0.85) 0.44	0.51L
10	(0.85, 0.68) 0.70	0.27L
11	(0.85) 0.68	1.00
12	(0.85) 0.70	0.81R
13	(0.57, 0.90) 0.53	0.40L
14	(0.57) 0.28	0.52L
15	(0.46) 0.87	0.39R
16	(0.33) 0.45	0.36R

Table 2.3 Basin-position and hillslope-position coordinates for points in the more complex drainage basin shown in Figure 2.31B.

summarizes many Chapter 2 concepts. This is an example of how climate change modulates the timing of an episode of retreat of a knickpoint created by displacements on the range-bounding Hope fault.

Frequent vertical displacements on the range-bounding Hope fault at a basin-position coordinate of 0.71 (1.00 for watershed analyses upstream from the mountain front) are sufficiently strong perturbations to reduce continuity (Section 2.5.2) of mountain and piedmont reaches. The fractured greywacke sandstone of this 10 km² watershed is so sheared and fractured that resistant ledges for prominent waterfalls generally are absent along a very narrow valley floor. Rates of knickpoint retreat and stream channel incision are strongly tied to unit stream power – the mean bed shear stress of Whipple (2004). Nearby smaller and larger watersheds display much different responses to similar Hope fault tectonic perturbations, even though the rock mass strength is similar. Stream power thresholds have to be crossed before significant knickpoint retreat can occur, and total annual stream power becomes progressively less with smaller basin-position coordinates of a knickpoint that is migrating upstream.

Only a few floods in the broad spectrum of stream-discharge events exceed the threshold unit stream power required to incise the bedrock trunk stream channel of the small Right Branch drainage basin. The result is a long response time to the 9 ka Hope fault tectonic perturbation. Sources and rates of introduction of bedload from the hillslope are not variable in these brecciated, fractured greywackes. Such stream-channel abrasive tools are poised to be dumped into the stream channel in all parts of the watershed. The response of the Right Branch to incision is partly a function of being an elongate drainage basin with no major tributaries (Fig. 2.23).

In contrast, the Main Branch of the Charwell River is circular with many abrupt increases in discharge with the addition of major tributary inputs. Knickpoint retreat may slow dramatically after reaching a major tributary junction. Bull and Knuepfer (1987) note "The waterfall . . . occurs just upstream of a junction with a major tributary where the stream also crosses a ridge of more massive sandstone." Thus, lower unit stream power and increased rock mass strength appear to have created knickpoints that have become cumulative in a short reach of a fluvial system.

Linear trends on semi-logarithmic plots for reach B (Fig. 2.32) and for reach D (also see Fig. 2.25B) attest to attainment of stream channel equilibrium. Reach D flows on soft Cenozoic rocks for 2 km before joining the main branch of the Charwell

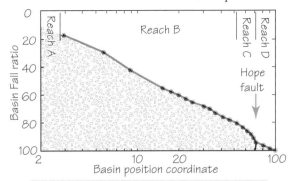

Figure 2.32 Dimensionless semi-log-arithmic profile of the Right Branch of the Charwell River, New Zealand. From Figure 4 of Bull and Knuepfer, 1987.

River (basin-position coordinates of 0.71 to 1.00). It is presently at its base level of erosion and beveling a tectonic strath since 4 ka. Reach D has achieved type 1 dynamic equilibrium as described by the Figure 2.25 equation. Reach B is still actively downcutting, but plots as a straight line on this semi-log plot, even though the stream flows through steep, rugged landscape. The central section of reach B, between altitudes of 800 and 600 m, has an inclusive gradient-index of 250 (same as reach D). So, lacking straths, reach B is a nice example of type 2 dynamic equilibrium between basin-position coordinates of about 0.27 and 0.52. Convex reach C (0.52 to 0.71) is in disequilibrium because of 40 m of uplift along the Hope fault. Post-26 ka reaction time (~17 ky) to surface ruptures is abnormally long. The stream could not degrade into bedrock as long as latest Pleistocene climate-change induced stripping of the hillslope sediment reservoir kept the system strongly to the aggradational side of the threshold of critical power. Exhumation of the sub-alluvial fault scarp that started at about 9 ka initiated knickzone retreat, which has progressed only a few hundred meters upstream. Long-term rock uplift favors maintenance of type 2 dynamic equilibrium conditions in reach B. Response times may be so long that the next climate-change aggradation event may occur before the 40 m of cumulative Hope fault displacements, that occurred between 26 to 9 ka, extends upstream to a basin-position coordinate of 30. Lack of fill terrace remnants in reach B suggests that it was sufficiently steep to remain on the degradational side of the threshold of critical power. Continued stream-chan-nel downcutting might reduce or eliminate convex reach C. Climate-change induced aggradation seems to occur only in reach D, a base-level rise that spreads upstream far enough to bury the Hope fault for the duration of an aggradation event. Headwater reach A (basin-position coordinates of 0.00 to ~0.27) is persistently degrading and unable to achieve equilibrium because of low annual unit stream power relative to the rock mass strength of materials beneath the trunk stream channel.

Surprisingly the Right Branch fluvial system is neither sensitive (long reaction time) nor efficient (long relaxation time) in its response to a large tectonic perturbation emanating from the mountain front. System adjustments here are strongly modulated by climatic and lithologic controls.

Chapter 2 tectonic concepts should be applied in the context of watershed climatic and lithologic controls on geomorphic processes. This helps us better understand the significance of external factors such as late Quaternary global climate change, sea-level fluctuations, and vertical tectonic deformation. The base level of erosion is the reference datum for studies of tectonics and fluvial topography. The threshold of critical power separates degradation and aggradation modes of operation of fluvial systems. It is purposely defined as a multi-variable ratio to remind us not to overemphasize a single variable, such as streamflow gradient, when trying to comprehend fluvial-system behavior. Time lags of response help us focus on the frequencies and magnitudes of tectonic and climatic perturbations, their locations within a fluvial system, and the magnitudes and time spans of departures from equilibrium conditions that such perturbations usually cause.

Bedrock uplift has a major influence on geomorphic processes and landscape evolution. Relief orographically controls precipitation and temperature, and defines potential energy of flowing water even where tectonic elevation of mountains ceased long ago. Increases of fluvial-landscape relief emanate from active geologic structures through the process of tectonically induced downcutting. Streams act as connecting links that transmit tectonic perturbations to upstream reaches. Active faults and folds separate fluvial reaches with vastly different processes and landforms. Degradation changes to piedmont aggradation where a stream crosses an active range-bounding fault. Let us explore how bedrock uplift affects mountain fronts in the next two chapters.

Chapter 3

Mountain Fronts

3.1 Introduction

Mountain-front escarpments have caught the attention of humankind for centuries. Settlers of the American west viewed distant mountains as a change of scene and as an impending challenge. Geologists wonder if active faults and folds separate mountains from lowlands. Active geologic structures in topographic escarpments are zones of concentrated tectonic base-level fall for fluvial systems. One challenge is to discern which mountain fronts have fault zones that are sufficiently active to generate damaging earthquakes.

Progressive urban encroachment onto mountainous escarpments occurs after gently sloping land is occupied or when homeowners seek impressive views from their residences. Residential construction on tectonically active escarpments, such as in Los Angeles, California, Salt Lake City, Utah, and Wellington, New Zealand, increases earthquake and landslide risks. Steep, high mountain fronts can be menacing but the next surface rupture may occur along the more subtle low fronts and scarps. In order to assess potential earthquake-related hazards the paleoseismologist needs to identify and date key tectonic landforms and apply her or his knowledge of landscape evolution. Tectonic geomorphologists are faced with

diverse questions when viewing a sea of suburban development that laps onto foothills of lofty mountains (Fig. 3.1). How old is the escarpment, and what are the past and present rates of uplift? How seismogenic are the pulses of mountain-building uplift? Has a steady-state balance been achieved between uplift and denudation of the mountain slopes? (Lavé and Burbank, 2004), or is a model of continuously changing landscape more appropriate?

Earth scientists, engineers, and planners benefit from geomorphic tectonic assessment of whether or not range-bounding fault zones are active or inactive. How long has it been since the most recent surface rupture, and when will the next one occur? What advice should be given to those seeking to bulldoze low piedmont fault scarps in order to build new housing subdivisions, or to those already admiring

Image showing the proximity of the Los Angeles metropolitan area (lower-right coastal plain) of southern California to the imposing mountain front of the rugged San Gabriel Mountains which are rising >2 m/ky. Pacific Ocean in foreground. Mt. Baden Powell at the right side rises to 2,866 m. Shuttle Radar Topographic Mission perspective view with Landsat overlay; image PIA02779 courtesy of the Jet Propulsion Laboratory and NASA.

Figure 3.1 Urban development encroaching onto a thrust-faulted mountainous escarpment east of Cucamonga Canyon, San Gabriel Mountains, southern California. Both the high mountains and the lower structural bench are being raised along thrust faults. Less obvious active faults rupture the urbanized piedmont alluvial fans.

their views from homes built on the crests of high fault scarps? Clearly there is a need to do more than merely describe the locations and types of faults present. With each passing year Quaternary earth scientists are better able to define the locations and magnitudes of future surface ruptures, and to estimate the rates of uplift along faults associated with low and high escarpments.

The San Gabriel Mountains and the thrust faults along its south flank are associated with a bend in the strike-slip San Andreas fault. Right-lateral movements along this restraining bend cause local crustal shortening, so the thrust and strike-slip styles of faulting are intimately related. It is logical for paleoseismologists to ask "do synchronous surface ruptures of the San Andreas and thrust faults occur as a single mega-earthquake event?" Alternatively, thrust-fault earthquakes occur independently. The hazard implications for the Los Angeles metropolitan region are profound. The San Fernando (U.S. Geological Survey, 1971), Whittier Narrows (Hauksson et al., 1988), and Northridge (Hudnut et al., 1996) earthquakes demonstrate the seismically active nature of the mountain front and adjacent basin (Dolan et al., 1995). We can expect more damaging earthquakes especially if the major range-bounding Sierra Madre-Cucamonga fault (Fig. 3.3) ruptures.

Will 40–90 km of this fault rupture synchronously with the next surface rupture of the San Andreas fault? The result would be a Mw magnitude >8.0 earthquake. Mw is earthquake moment magnitude (Hanks and Kanamori, 1979). Thrust faults in the Elkhorn Hills (Sieh, 1978a), and perhaps elsewhere, ruptured during the 300 km surface rupture of the San Andreas fault in 1857.

An appraisal of the Mw magnitude 8.3 Gobi-Altay, Mongolia 1957 earthquake (Bayarsayhan et al., 1996; Kurushin et al. 1997) serves as a useful prototype. It's surface-rupture length was 250 km. The spatial arrangement of thrust and strike-slip faults is remarkably similar to the Cucamonga and San Andreas faults. They conclude that the probability of such an event is speculative, but "the similarities are too great for the possibility of such an event to be ignored". Although the rapid contractional strain rate between the coast and the San Gabriel Mountains dictates big thrust-fault earthquakes (Dolan et al., 1995), it is rather unlikely that they will occur concurrently with the next San Andreas strike-slip earthquake (Hough, 1996).

Figure 3.2 Image showing the tectonic setting of the San Gabriel Mountains, which appear in the lower right part of the view as a lens of raised terrain caught between the San Andreas fault (prominent diagonal slash at right side) and the range-bounding thrust faults on the Los Angeles side of the mountain range. The Garlock fault at the top right of the view bounds the north side of a wedge of tectonically quiet terrain in the western Mojave Desert. Shuttle Radar Topographic Mission perspective view with Landsat overlay; image PIA03376 courtesy of the Jet Propulsion Laboratory and NASA.

The purpose of Chapter 3 is to review several ways to assess the hazard potential of tectonically active mountain-front landscapes. Mountain fronts are created by diverse styles of faulting and folding, so the overall theme is landscapes that respond to tectonic base-level fall. I apply the conceptual models of Chapter 2 and use fluvial landforms to better resolve

several specific problems. Stream channels, terraces, and faulted alluvial fans are used to determine which thrust faults are capable of producing the next earthquake, and to measure the true throw of normal-fault surface ruptures.

Passive margin escarpments fall outside of the paleoseismology emphasis of this book. Erosion-in-

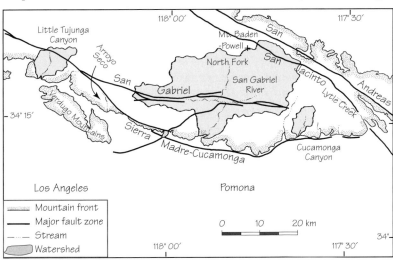

Figure 3.3 Location map for place names and illustrations pertaining to the San Gabriel Mountains.

duced isostatic uplift of tectonically inactive escarpments is fascinating. Examples include Drakensburg in South Africa (Gilchrist et al., 1994; Brown et al., 2002; van der Beek, 2002), Blue ridge in the eastern United States (Spotila et al., 2004), and Western Gnats bordering the western side of the Deccan Plateau in India (Ramasamy, 1989).

The term "mountain front" pertains to more than the topographic junction between the mountains and the adjacent piedmont. A *mountain front* is a topographic transition zone between mountains and plains. This landscape assemblage includes the escarpment, the streams that dissect it, and the adjacent piedmont landforms.

Our discussion starts with the diagnostic landforms of triangular facets, mountain–piedmont junctions, and piedmont forelands of an active mountain range bordering the Los Angeles metropolitan area in southern California (Figs. 3.2, 3.3). Triangular facets evolve during a million years of erosion and episodic uplift of resistant rocks of arid regions (Fig. 2.20A). Formation of mountain–piedmont junctions may be likened to a contest between the relative strengths of uplift along a range-bounding fault zone and fluvial dissection. Locations of active thrust faulting typically shift from mountains into adjacent basins. *Piedmont forelands* are newly raised and deformed blocks between the new and old thrust faults. These low piedmont scarps are easy to overlook, but may eventually rise to become impressive escarpments with the passage of geologic time.

The Gurvan Bogd mountains of the Gobi Altay, Mongolia were formed by a system of strike-slip faults with a reverse component. The magnitude $Mw{\sim}8.0$ earthquake of 1957 has attracted paleoseismologists from around the world to study the marvelous scrunch tectonics of this remote arid region. Low ridges rise through broad piedmonts of >3,000 m high mountains and roughly parallel the older mountain fronts. Florensov and Solonenko (1963) used the term "*foreberg*" for these hills created by the complexities of scrunch thrust faulting (Kurushin et al., 1997). Such folds, antithetic and synthetic faults, and elongated backtilted ridges result from the shortening component on a broad active intracontinental fault zone. They have a common function, which is to broaden the deforming zone by creating new structures that accommodate both strike-slip and scrunch shortening components of tectonic deformation. The landscape further suggests that these new structures evolve by lateral propagation, increase in

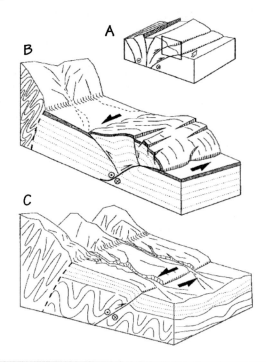

Figure 3.4 Scrunch tectonics of piedmont forelands and forebergs. Illustration and caption are from Figure 5 of Bayasgalan et al. (1999).

A Cartoon of a transpressional "flower structure" adapted from Sylvester (1988).

B. Cartoon of the internal deformation within a foreberg, based on observations at Gurvan Bulag, Mongolia. Note the flattening of the underlying thrust at very shallow depths, which is probably responsible for the collapse of the thrust "nose" by normal faulting (Kurushin et al., 1997) and the left-stepping backthrusts and right-stepping normal faults, which suggest a component of left-lateral slip.

C. Cartoon showing migration of active faulting away from the main range front, leaving uplifted and dissected fans in the hanging wall of the new fault and older, abandoned faults and shear fabrics within the uplifting mountain range.

amplitude, and may eventually merge and form new fault zones of considerable length.

The evolution described here is thus peculiar to strike-slip faults with a reverse component, and can form many of the features of the "flower structures" that are often described in such oblique-shortening zones (Fig. 3.4A). The interplay between the rates of sedimentation and erosion allows some elevated fans between the foreberg ridges and the mountains to be much less dissected than would otherwise be expected, because the rising foreberg is a base-level rise (Fig. 3.4B). Few arid-region streams have sufficient annual unit stream power to accomplish the tectonically induced downcutting needed to cross the rising landform (Owen et al., 1997). The streams of the highly seasonal semiarid San Gabriel Mountains dissect such piedmont forelands easily. Range-bounding faults of the lofty mountain range become much less active as tectonic deformation is transferred to the newest outermost fault zone (Fig. 3.3C).

Piedmont foreland and foreberg shapes result from changes in thrust-fault dips of the underlying thrust faults in the uppermost 200 m and also at depths or more than 1 km. Add features like fault-bend and fault-propagation folds and it is easy to see why each structural geologist devises a different tectonic scenario for a study region. I use the model of Ikeda and Yonekura (1979) and Ikeda (1983) for the San Gabriel Mountains, where characteristic suites of fluvial landforms document shifts in the locations of scrunch tectonics.

Chapter 3 also explores normal-fault landscapes. I evaluate a conceptual model for segmented surface-rupture behavior of active faults, and then apply the fault segmentation model to a normal fault in Idaho. The best way to test the characteristic earthquake model is to make measurements of historical and prehistorical surface ruptures with sufficient precision to define surface rupture behavior in the boundaries between fault segments. Such responses of hills and streams to episodic surface ruptures are then used in Chapter 4 to identify active range-bounding faults and to discern regional patterns of uplift rates of mountain fronts.

3.2 Tectonically Active Escarpments

Hills record long-term interactions between uplift and landscapes. Compared to streams, hills respond slowly to the cumulative effects of many small increments of uplift along active geologic structures. Long response times to uplift are due primarily to huge volumes of rock that have to be weathered into erodible-size fragments before tectonic landforms such as triangular facets can be created (Menges, 1987, 1990a, b). Mountainous topography is the consequence of fluvial erosion initiated by the first pulse of uplift. Mountains continue to evolve for millions of years after tectonic uplift has ceased. However, isostatic uplift (Fig. 1.4) continues in response to denudational unloading. The resulting landscape assemblages record the rates and magnitudes of rock uplift and concurrent fluvial erosion; both processes increase relief.

The topic of mountainous escarpments is part of the much broader subject of hillslope development whose erosion is initiated by base-level fall. Many of the early papers are worth reading. Hillslope processes and forms are reviewed by Young (1972), Carson and Kirkby (1972), and Cooke and Warren (1973); important papers about hillslopes include those by Gilbert (1877), Horton (1945), Strahler (1950, 1957), Schumm (1956), Leopold and Langbein (1962), Hack (1965), Abrahams (1994), and Anderson and Brooks (1996). Significant papers about the effects of uplift on mountainous escarpments include those by Davis (1903), Louderback (1904), Blackwelder (1934), Gilbert (1928), King (1942, 1968), and Wallace (1977, 1978).

3.2.1 Faceted Spur Ridges

The splendid triangular facets of the Wasatch Range escarpment in north-central Utah have been a classic example of a tectonic landform since the time of William Morris Davis (1903). Blackwelder (1934), Hamblin (1976), and Wallace (1978) describe triangular facets as being fault planes that have been modified by erosion, an explanation that seems appropriate for mountains bounded by normal faults.

Triangular facets result from base-level fall, and occur in a variety of tectonic settings. Erosion of facets at the truncated ends of spur ridges may be associated with normal faults (Fig. 3.5A), anticlines (Fig. 3.5B), thrust faults (Fig. 3.5C), and even along escarpments formed by erosional base-level fall (Fig. 3.5D). The overall similarity of the facets shown in Figure 3.5 is suggestive of a more general relation than erosional notching of normal-fault planes.

The faceted ends of the spur ridges are steep hillslopes that reflect recent cumulative range-front uplift. Sharp-crested spur ridges divide an escarp-

Figure 3.5 Triangular facets of different tectonic environments.
A. Spur ridges truncated by a normal fault on the east side of the Toiyabe Range, central Nevada.

Figure 3.5 Triangular facets of different tectonic environments.
B. Triangular facets on the north side of the Wheeler Ridge anticline, south edge of the San Joaquin Valley, California. The topographic benches may be the result of mass movement processes (Bielecki and Mueller, 2002).

ment into drainage basins. Each spur ridge terminates at the range front in a characteristic triangular outline. Initial development of faceted spurs is similar, even where uplift is along a reverse fault that dips into the mountains (Fig. 3.5C). An early stage consists of crudely planar 20° to 40° hillslopes. Keller and Pinter (2002, p. 10) nicely depict the key topographic and stratigraphic features for many normal faults (Fig. 3.6). Uplift of range fronts in west-central Nevada proceeds as 1 to 3 m surface ruptures every 5 to 10 ky (Wallace, 1978). These may seem small and infrequent, but the resulting topography is spectacular and the landforms are truly indicative of the relative rates of rock uplift and erosion.

Wallace's block diagrams depict the evolution of a fault-generated mountainous escarpment (Fig. 3.7) that reflects the long-term (>10 My) interactions between uplift and denudation. Initial faulting (stage A) creates a linear scarp crest that migrates away from the range boundary. A range crest is created by merging of scarp crests that migrate from the range-bounding faults on opposite sides of the rising block. Valley floors notched into the rising block (stage B) are zones of most rapid tectonically induced downcutting. The intervening spur ridges and the range crest are gently sloping. These landscape elements have the slowest rates of tectonically induced degradation (stages C and D). The mountain–piedmont junction continues to be straight and the valley floors narrow during continuing rapid uplift, even where rocks are

soft. An aerial view of the Tobin Range (Fig. 3.8) reveals the simplicity of the terrain from which Wallace developed his concepts of landscape evolution.

Rugged faceted spur ridges owe their substantial heights to sustained tectonic base-level fall at the range-front landscape boundary, and to the profound initial difference between valley-floor and ridgecrest rates of denudation. Increase of hillslope steepness and relief also increases the rate of hillslope erosion. Landslide processes become more important as the ever steeper valley side slopes become progressively more unstable (Pain and Bowler, 1973; Pearce and Watson, 1986; Keefer, 1994; Hovius et al., 2000; Dadson et al., 2003, 2004).

The style of landscape evolution reverses after cessation of rapid uplift, but hillslopes never attain a steady-state condition. Hillslope denudation rates exceed uplift rates, so the mountain–piedmont junction becomes sinuous as it retreats from the position of the range-bounding fault to create a pediment (stage E of Fig. 3.7). Non-steady state denudation brings ridgecrests closer to the valley floors, which remain at type 1 dynamic equilibrium with valley floors that become progressively wider.

Dissection of range front triangular facets proceeds independently of the stream-channel downcutting of the adjacent trunk valley floors. Consequent drainage nets on young triangular facets initially consist of parallel rills. Capture of flow from adjacent rills occurs. Small elongate watersheds form on these planar surfaces and become more circu-

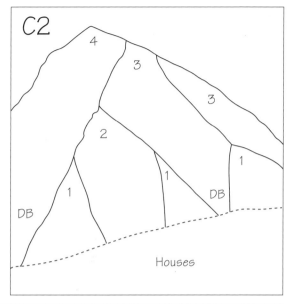

Figure 3.5 Triangular facets of different tectonic environments.
C1. Aerial view of a set of triangular facets that terminates the spur ridge of a thrust-faulted mountain front of the San Gabriel Mountains, Southern California.
C2. Facet-dissection stages as described in Table 3.1. Younger facets 1 and 2 are nested inside older higher stage 3 and stage 4 facets that are deeply incised by small stream channels.
DB, basins to catch debris swept off recently burned steep hillslopes during winter storms.

Figure 3.5 Triangular facets of different tectonic environments.
D. Triangular facets along the edge of a fanhead embayment at Cucamonga Canyon, San Gabriel Mountains, southern California. These nontectonic facets were created by lateral erosion induced baselevel fall caused by streamflow.

lar through the processes first described by Horton (1945). Development of progressively larger drainage nets at the ends of spur ridges concentrates available stream power, promoting efficient erosional destruction of the triangular facets. Planar facets with numerous closely spaced parallel rills (Fig. 2.20A) are eventually transformed into a deeply dissected ridge-and-ravine topography (Fig. 2.20B) in which the triangular shape of the facet is less obvious (Fig. 3.9).

Erosional dissection of faceted spurs can be described as six stages (Table 3.1). The stages are easy to distinguish in the field, on aerial photographs, or on detailed topographic maps. The time needed to achieve a given stage is a function of two compensating processes. Uplift increases facet height, and fluvial erosion deepens valleys. With the passage of time, faceted spurs adjacent to an active fault become higher and more dissected. The lowest, most recently created, part of a facet is less dissected (Fig. 3.5C) because it has been exposed to erosion for time. These qualitative descriptions of facet dissection are used to help define mountain front tectonic activity classes in Chapter 4.

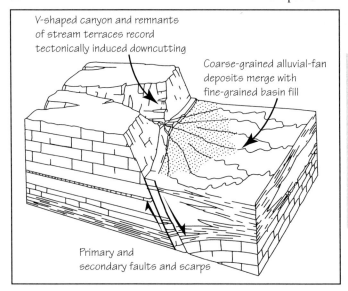

V-shaped canyon and remnants
of stream terraces record
tectonically induced downcutting

Coarse-grained alluvial-fan
deposits merge with
fine-grained basin fill

Primary and
secondary faults and scarps

Figure 3.6 Diagrammatic sketch of the topographic expression of an active normal-fault system. Uplift on a range-bounding normal fault creates a base-level fall that causes deep valleys to be eroded in the mountain block. This tectonic displacement favors accumulation of thick alluvial-fan deposits downstream from the normal-fault zone. From Keller and Pinter (2002, Fig. 1.7C).

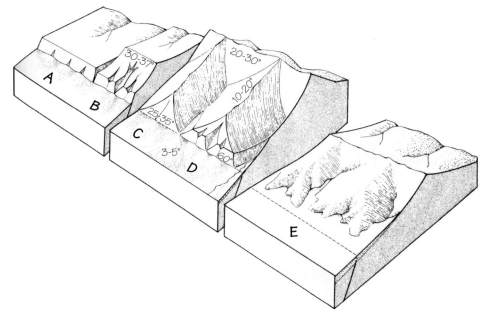

Figure 3.7 Block diagrams showing the sequential development of a fault-generated mountainous escarpment. A. Initial faulting creates a linear scarp. B. Scarp crest migrates away from the rising range boundary to form range crest. C. Valleys are notched into the rising block; their floors are the locations of rapid tectonically induced downcutting by streams. The crests of spur ridges are the locations of slow tectonically induced degradation. D. Episodic displacement along the range-bounding fault maintains a steep, straight mountain–piedmont junction. Main and spur ridge divides continue to rise faster than degradation can lower them. E. The mountain–piedmont junction becomes sinuous and the valley floors become wider after cessation of uplift. Relief becomes less as degradation lowers the ridgecrests. Figure 4 of Wallace (1977).

Figure 3.8 View of the west side of the normal-faulted terrain of the Tobin Range, west-central Nevada showing the basic landscape elements of a tectonically rising landscape as described by Wallace (1977). These include the range crest, spur ridges extending to triangular facets at a straight range front, and a deep valley with a narrow valley-floor width. The irregular dark line at the mountain front is the surface rupture of the 1915 earthquake.

3.2.2 Mountain–piedmont junctions

Transitions between mountainous escarpments and adjacent basins typically are abrupt. Steep hills give way to gentle piedmont slopes in both tectonically active and tectonically inactive landscapes. Piedmonts may consist of either the depositional environment of coalescing alluvial fans, or the erosional environment of pedimented terrain. Fans and pediments may be smooth in the arid realm. They tend to be dissected and less obvious in humid regions where floodplains are a common piedmont landform and forests may cloak subtle features of the landscape.

The planimetric trace of this topographic transition between mountains and piedmont is useful for assessing whether or not the mountain front coincides with an active range-bounding fault zone. The sinuosity of the mountain–piedmont junction represents a balance between 1) the tendency of uplift to maintain a sinuosity as low as that of the range-bounding fault or fold, and 2) the tendency of streams to erode an irregular junction between the mountains and the plains. Straight mountain–piedmont junctions generally indicate the presence of an active fault. Embayed, pedimented mountain–piedmont junctions suggest tectonic quiescence. Downstream

Facet class	Erosional landforms
1	Planar surface with only rills. Includes scarps that have yet to be carved into facets by streams flowing across the scarp.
2	Planar surface with shallow valleys extending a short distance into the facet.
3	Valleys extend more than 0.7 the horizontal distance between the base and top of the facet.
4	Deep valleys extend more than 0.7 the horizontal distance.
5	Greatly dissected but the general form is still obvious.
6	So dissected that the general form of a facet is not obvious.
7	Triangular facets are not present because they have been removed by erosion, or they never existed.

Table 3.1 Stages of dissection of triangular facets.

Figure 3.9 Aerial view of triangular facets near La Canada, San Gabriel Mountains, Southern California. Stage 4 of Table 3.1.

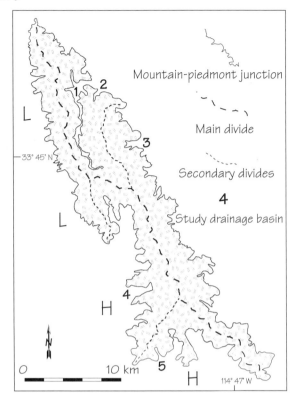

Figure 3.10 Tectonically inactive mountain fronts of the McCoy Mountains in southeastern California.
A. Map of the mountain–piedmont junction and watershed divides. L and H are low and high sinuosity mountain fronts. 1–5 are study watersheds.

increase in stream power maximizes the potential for downcutting and lateral erosion where streams leave the mountains. The result is a highly sinuous mountain–piedmont junction, even in homogenous rocks, but only under tectonically inactive conditions.

Small structures, such as joints, foliation, and bedding planes, also influence sinuosity of mountain–piedmont junctions. Tectonically inactive mountain fronts with structures that parallel the range front may have an anomalously straight mountain–piedmont junction and well defined triangular facets.

Sinuosity of the mountain–piedmont junction also is a function of the width of a mountain range (Parsons and Abrahams, 1984; Mayer, 1986). Wide mountain ranges have large drainage basins that are more likely to have sufficient stream power to quickly attain the base level of erosion and create pediment embayments after uplift has ceased. Range width decreases with erosional retreat of range fronts. Drainage-basin size, and stream power, become less. Mountain–piedmont junction sinuosity may become lower as the mountain landscapes are progressively replaced by the beveled bedrock of pediments.

The constraint of drainage-basin size on sinuosity of the mountain–piedmont junction is illustrated by the McCoy Mountains of southeastern California. Geophysical studies by Rotstein et al. (1976) suggest that the faults initially bounding the McCoy Mountains structural block now are 1 to 2 km from the present range front. The mountains are only half of their original width. Average drainage-basin length

for fronts at L (Fig. 3.10A) is only about 1 km and the sinuosity of the mountain–piedmont junction is moderate. Mountain fronts are highly sinuous at H where drainage basins are twice as large.

Granitic rocks weather slowly and streamflow is ephemeral in this arid, hyperthermic, moderately seasonal climate (Table 2.1). Unit stream power is large during flash floods during infrequent incursions of tropical depressions into the southeastern Mojave Desert or during some wintertime cyclonic storms. The stream channels of the McCoy Mountains degrade by abrasion and plucking during floods, but long-term weathering of the granitic rocks into small particles plays a much larger role than in humid settings such as the Potomac River (Fig. 2.29). McCoy

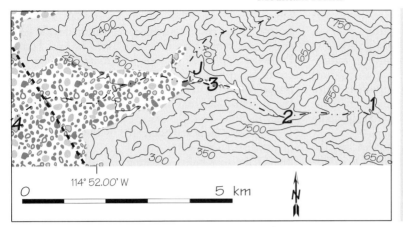

Figure 3.10 Tectonically inactive mountain fronts of the McCoy Mountains, California.
B. Topographic map of watershed 4. Contour interval is 50 m. Gray is mountain bedrock. Pattern is piedmont alluvium whose extent defines the present mountain-piedmont junction. Heavy dashed line is present trace of ends of spur ridges. J (near 3) is stream junction. 1, 2, 3, 4 are defined in Figure 3.10C.

fluvial systems have had >10 My to greatly modify this landscape. There has been ample time for erosion of bedrock stream channels to achieve equilibrium relationships, in marked contrast to a rapidly rising mountain front such as the Charwell River (Figs. 2.16, 2.24–2.28). Planimetric and longitudinal-profile aspects of fluvial-system equilibrium in a tectonically inactive watershed are shown in Figures 3.10 B, C.

The 2 km wide strath at the mountain front is a pediment embayment indicative of ample time and stream power to achieve prolonged type 1 dynamic equilibrium. Diminishing watershed area resulting from 2 km of mountain front retreat to its present position is offset by the prolonged time span for the ephemeral stream to do this work. Pediments are not a special landform when viewed in terms of processes. They form where stream(s) at the base level of erosion bevel straths that coalesce to form the beveled bedrock piedmont landform that we call a pediment. It takes millions of years for this process to remove spur ridges between adjacent drainage basins.

The smooth slightly concave longitudinal profile in the reach between locations 3 and 4 records attainment of type 1 dynamic equilibrium conditions. The upstream narrowing of the pediment embayment (Fig. 3.10B) reflects the importance of concomitant spatial decrease of unit stream power and the relative increase of the importance of rock mass strength of materials beneath the stream channel. Stream power prevails at the mountain front but eventually a threshold is crossed where unit stream power is insufficient to overcome rock mass strength.

So the upstream end of the pediment embayment coincides with the junction (J) of the two largest streams in this drainage basin. Stream power upstream from this junction is insufficient to bevel broad valley floors in this rock type.

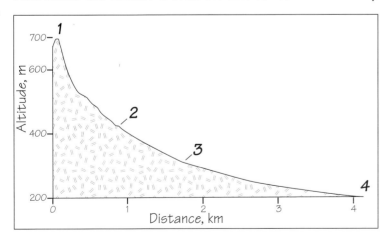

Figure 3.10 Tectonically inactive mountain fronts of the McCoy Mountains.
C. Longitudinal profile of trunk stream channel of watershed 4 of figure 3.10A. 1 is watershed divide. 1–2 is disequilbrium reach. 2–3 is type 2 dynamic equilibrium reach. 3-4 is type 1 dynamic equilibrium reach.
Vertical exaggeration is 4.0.

Valley-floor widening of the strath extends only a short distance further to location 3 (Figs. 3.10B and 3.10C). The longitudinal profile becomes steeper upstream from 3, but still has a form indicative of attainment of equilibrium. But, in contrast with the downstream reach (inclusive gradient-index is 129), the valley floor is narrow (inclusive gradient-index is 200). So, reach 2–3 is best regarded as being type 2 dynamic equilibrium.

Disequilibrium conditions prevail in the headwater's reach, 1–2. Unit stream power is miniscule relative to rock mass strength. Rates of stream-channel downcutting are so slow that disequilibrium prevails even after 10 My of tectonic quiescence.

The watershed area needed to generate sufficient stream power to achieve type 1 dynamic equilibrium in the McCoy Mountains is partly a function of drainage net configuration (Fig. 3.10D). Headwaters reaches, such as 1–2 or 1–3, occupy only a small portion of total watershed area in large drainage basins, so equilibrium conditions are achieved at a basin-position coordinate (Section 2.8) of only 0.25. An example is watershed 1 of Figure 3.10A. The location at which type 1 dynamic equilibrium is attained is farther downstream where headwater's source areas are smaller.

3.2.3 Piedmont Forelands

Normal, reverse, and strike-slip faults associated with tectonically active landscapes may be classed geomor-

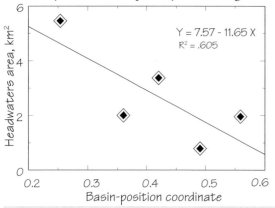

Figure 3.10 Tectonically inactive mountain fronts of the McCoy Mountains.
D. Regression of drainage-basin areas upstream from type 1 dynamic equilibrium stream channels and basin-position coordinates.

phically as internal, bounding, or external. ***Internal faults*** rupture mountain bedrock. Internal strike-slip and thrust faults may have associated normal faults. Prolonged uplift along ***range bounding faults*** can create escarpments rising above adjacent plains. ***External faults*** occur in the depositional reaches of fluvial systems, and may be incipient mountain fronts. They include:
1) Piedmont fault scarps created by the newest splays of a thrust-fault system,
2) Strike-slip faults crossing the basins between mountain ranges, or
3) Inactive former range-bounding faults that no longer coincide with the mountain–piedmont junction because erosion has caused the mountain front to retreat after cessation of uplift.
This section describes landscapes of a compressional tectonic setting that have all three geomorphic fault classes.

The topography associated with active thrust faults (Figs. 3.4, 3.11, 3.12) is much different than that for normal faults (Fig. 3.6). Uplift along a range-bounding thrust fault can create an imposing escarpment (Fig. 3.1), but primary surface-rupture locations shift when new splays of a propagating thrust-fault zone encroach into the adjacent basin (Yielding et al., 1981). The former range-bounding fault becomes an internal mountain front that is less tectonically active, or becomes inactive. The new fault at the edge of rapidly rising hills is the latest in a series of range-bounding faults. Tectonic deformation ruptures and folds this piedmont foreland, which is a bedrock block, capped by remnants of piedmont alluvium. New streams dissect the former depositional slopes as a consequence of piedmont terrain being incorporated into ever-broadening mountain range.

Paleoseismology of piedmont forelands is challenging to the tectonic geomorphologist. Each structural block has a different tectonic history, and recently created external faults pose a deceptively high earthquake hazard that is largely hidden from view. Locations and magnitudes of earthquakes have changed during the late Quaternary and blocks between thrust faults are deformed by folding and antithetic faulting. After an earthquake, the effects of synchronous vertical displacements on several fault zones may overlap where they migrate upstream as knickpoints (Section 2.5.1). Each perturbation diminishes in magnitude as it moves upstream.

A good example of active tectonics of suites of thrust faults is the area of the San Fernando earth-

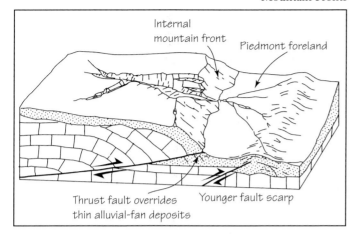

Internal mountain front

Piedmont foreland

Thrust fault overrides thin alluvial-fan deposits

Younger fault scarp

Figure 3.11 Diagrammatic sketch of the topographic expression of an active thrust-fault system. Prolonged uplift on an internal fault has created a high scarp. Uplift may be rapid on the range-bounding fault but insufficient time has elapsed for this scarp to become an impressive landscape element. From Keller and Pinter (2002, Fig. 1.7B).

quake of 1971. Geologists have made extensive and diverse studies in southern California (Crook et al., 1987; Southern California Earthquake Center Group C, 2001). However, the San Fernando fault zone was not included on the detailed geologic map by Oakeshott (1958). This range-bounding fault became obvious after the Mw 6.4 earthquake ruptured 12 km of a range-bounding piedmont foreland (Fig. 3.12). The Mw 6.7 Northridge earthquake of 1994 occurred on a nearby blind (buried and hidden from view) thrust fault where rupture did not reach the surface. Although moderate in size both earthquakes caused major damage and loss of life.

Landscapes associated with thrust-fault migration have strikingly similar diagnostic features. The piedmont foreland bounded by the San Fernando fault zone (Fig. 3.12) resembles those of the Mahiru and Misaka Ranges of Japan (Figs. 3.13A, B). A 2 to 5 km wide piedmont foreland is present between two thrust-faulted mountain fronts. The front created by displacement on an external fault, although low, has been the more active of the two during the late Quaternary. Tectonic landforms include stage 1 and 2 triangular facets (Figure 3.12) and a straight mountain front (not embayed by erosion). Only rapid recent uplift can maintain such tectonic

Figure 3.12 Aerial view of the piedmont foreland near Little Tujunga Canyon, San Gabriel Mountains, southern California. The Sunland fault zone bounds the internal front, and the Lakeview fault zone bounds the range front in the foreground; it ruptured in 1971.

signatures in a landscape underlain by soft sandstone, mudstone, and weakly cemented conglomerate. The smooth surface areas of the piedmont foreland are capped by stream gravel, which was laid down when it was the depositional part of the fluvial system. Active tectonism has converted a former depositional reach of the fluvial system into hilly terrain.

The Mahiru Range of northeastern Japan has excellent examples of piedmont forelands created by thrust-fault migration (Fig. 3.13A). The Kawaguchi fault bounds an impressive internal front and separates Miocene volcanic rocks, mudstone, and sandstone from the Pliocene sandstone and mudstone of the Senya formation (Ikeda, 1983). The younger Senya fault is 3 km basinward of the Kawaguchi fault and bounds the 120 m high Senya Hills. It was the site of the 1896 Mw magnitude 7.2 Rikuu earthquake. Ikeda did not find evidence for an 1896 surface rupture along the internal Kawaguchi fault. Despite the rugged nature of the Mahiru Range, erosional embayments extend far into valleys whose broad floors suggest attainment of the base level of erosion. Bear in mind that tectonic landforms in soft rocks erode quickly in a humid region (Fig. 2.20B). Mahiru Range streams may degrade fast, and/or Holocene uplift rates have been negligible or slow. The strikingly similar piedmont foreland of the Sone Hills

(Figs. 3.13B, C) bordering the Misaka Range is also capped by several remnants of a thin blanket of fluvial deposits.

Ikeda's model (Fig. 3.14) describes encroachment into piedmonts by active thrust faults (Fig. 3.15). Although the older internal front has more topographic relief, the range-bounding fault is presently more active. Many fault splays and folds are present in the hanging wall of either thrust fault. Earth deformation may tilt part of the piedmont foreland back toward the main range. Backtilting, synclinal folding, and antithetic faulting may result in local subsidence or an uplift rate that is slower than that of the tectonic bulge.

Thrust-fault surface ruptures of the 1896 Rikuu event in northeastern Japan and the 1971 San Fernando event in southern California have the same geomorphic and tectonic features. Figure 3.14 describes this common tectonic style by assessing relative rates of uplift for the "basin fill", "piedmont foreland", and "mountain range" structural blocks. Only local components of surface uplift are evaluated here – faulting, folding, erosion, and deposition. Total tectonic uplift (equation 1.2) would include broader wavelength styles of uplift. The sum of local and more regional components of uplift probably would result in all three structural blocks of Figure 3.14 ris-

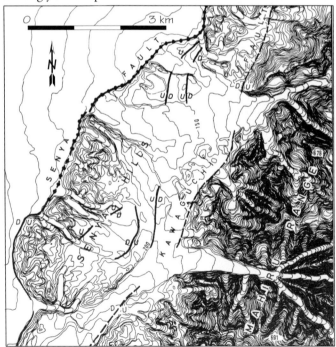

Figure 3.13 Characteristics of thrust faulted mountain fronts in Japan. A. Contour map (interval 10 m) of the Senya Hills piedmont foreland of the Mahiru Range, in northern Honshu Island draining to the Sea of Japan. The Kawaguchi thrust fault (dashed line) bounds the internal front. The dotted line along the range-bounding Senya thrust fault indicates the surface rupture of the 1896 Rikuu earthquake. From Figure 4 of Ikeda, 1983.

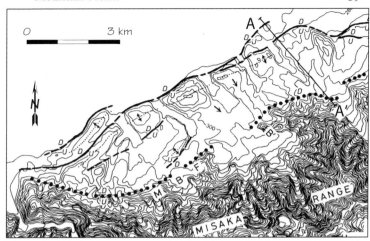

Figure 3.13 Characteristics of thrust faulted mountain fronts in Japan.
B. Contour map (interval 20 m) of the Sone Hills piedmont foreland of the Misaka Range. Active faults are shown by lines, active anticlinal axes with small dotted lines, and inactive internal fault zone with large dotted lines. Arrows show terrace-tread tilt directions. Figure 7 of Ikeda (1983).

ing during the late Pleistocene and at the present. The present internal fault was the range-bounding fault of a rugged, fast-rising mountain range during the late Pleistocene. Fractured ground along the trace of this fault caused by the 1971 San Fernando earthquake suggests that it has not become totally inactive, but much of the slip now occurs on the splay of the fault system that terminates as the new range-bounding fault. The present rate of rock uplift may be less than the surface denudation rate, hence the interpretation of slight net local surface subsidence of the "mountain range" block at present. Erosional processes also reduce the present rate of surface uplift of the tectonic bulge. Note the volume of material eroded below the former piedmont surface during the past ~0.4 My (Figure 3.12). Present basin-fill surface uplift at A is the sum of rock uplift due to tectonic scrunching

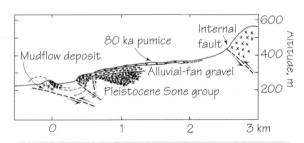

Figure 3.13 Characteristics of thrust faulted mountain fronts in Japan.
C. Geologic cross section A–A' of Figure 3.13B across the Sone Hills showing tectonic setting of typical piedmont foreland. From Figure of Ikeda (1983).

processes and aggradation of alluvial-fan gravels, as was the similar tectonic setting at A' before inception of the range-bounding fault. The model used here has tectonic basin subsidence being a bit more than surface uplift caused by alluvial-fan deposition.

Ikeda's model nicely explains the general topography of the Little Tujunga Canyon landscape. Many structural complications surely are present. Sharp (1975) notes that the fault dip is steeper than 20°. Yeats et al. (1997) summarize evidence for lateral components of slip. Correspondence with Bob Yeats points out the likelihood of bedding-plane faults, and Hiro Tsutsumi of the Kyoto Institute of Geophysics considers that the 1971 fault may be a well-expressed flexural-slip fault on the side of the Merrick syncline (Tsutsumi and Yeats, 1999; Tsutsumi et al., 2001). Structural geologists have yet to connect the 1971 fault with a large-displacement fault at depth. I proceed with a geomorphic appraisal using Ikeda's modeling as one example, realizing that improved structural models will be forthcoming.

Many earth scientists have studied the geometry of the faults and the displacements associated with the 1971 San Fernando earthquake and the more recent adjacent Northridge earthquake of 1994. The Figure 3.14 model can be evaluated with geophysical data (Allen et al., 1971; Sharp, 1975; Ikeda and Yonekura, 1979; Ikeda, 1983), by crustal-movement surveys (Castle et al., 1975; Savage et al., 1986), and by geologic information (Kamb et al., 1971; Proctor et al., 1972; U.S. Geological Survey, 1996).

Ikeda's model illustrates some likely complexities of Quaternary landscape change associated with multiple thrust faults, each having a different

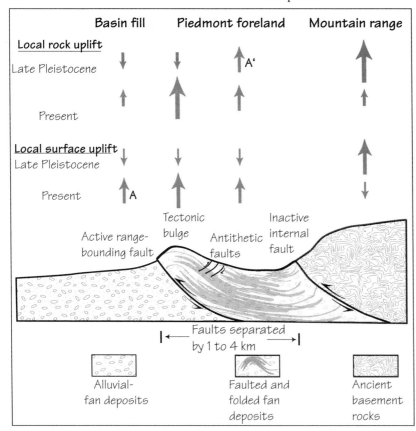

Figure 3.14 Tectonic elements of a piedmont foreland created by Quaternary migration of thrust faults. Adapted from Figure 1 of Ikeda (1983). Spatial and temporal variations of relative magnitudes of local uplift or subsidence are indicated by the sizes of the gray arrows. A and A' are where aggradation is part of the surface uplift.

displacement history. The location of the 1971 hypocenter and its 52° dip dictate a decrease in the angle of faulting toward the surface. In Ikeda's model the dip abruptly decreases to about 13° at a depth of about 2.5 km (Fig. 3.15A). More than one steeply dipping fault may have been active in 1971. Heaton and Helmberger 's model (1979) has two hypocenters: one at 13 km depth on a steeply dipping fault, and a second hypocenter on a more gently dipping range-bounding fault. Mori et al. (1995) show that the 1971 and 1994 events overlap. Figure 3.15B suggests that the internal, as well as the range-bounding front, was uplifted in 1971. In the Figure 3.15A model, a short 36° dipping fault segment was introduced to improve the fit between observed and modeled surface displacements. In the model shown, Ikeda assigns a uniform slip of 5.5 m over the two deeper fault segments. A variety of surface displacement patterns are depicted in Figure 3.15B, where θ is the dip of the more gently dipping fault segment as a thrust fault approaches the surface. Magnitudes

of modeled local subsidence increase with decreasing dip of the shallow fault segment. Ikeda concludes that an abrupt decrease in thrust-fault dip results in backtilting of the range-bounding front, and in continued uplift of the internal front where surface rupture is no longer obvious. The pronounced tectonic buckling associated with the range-bounding fault requires a large horizontal component of slip on a gently dipping fault that abruptly terminates in a set of secondary fault splays (36° in Fig. 3.15A). Once again, tectonic scrunching is an important component of rock uplift.

Such spatial variations in tectonic deformation should profoundly affect consequences on the behavior of streams flowing across an active thrust fault and fold belt. For example, subsidence of the Merrick Syncline is only relative between distances of 1 and 4 km in Figure 3.15B. Streams in this reach should reflect the effects of slower uplift (Fig. 3.14) if Ikeda's model is correct. Landscape assemblages test such models.

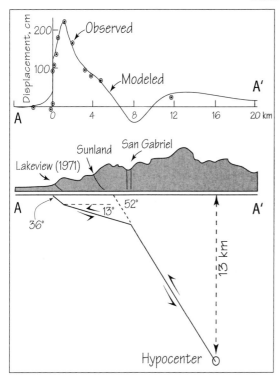

A broad alluvial valley floor where a stream is at the base level of erosion (Figure 2.26) might look much the same as an aggrading valley floor to most casual observers. The tectonic implications of these equilibrium and aggrading modes of stream operation are quite different. Seismic refraction surveys were used to assess stream-channel alluvial thicknesses to determine which thrust faults are active, and if the Merrick Syncline reach is truly subsiding relative to the base level of erosion reference surface.

Streams that cross a belt of active thrust faults flow through alternating *embayment reaches* and *gorge reaches*. These diagnostic landform assemblages are the result of spatial variations in the rate of rock uplift. Rapid local tectonically induced downcutting creates a rugged gorge that ends abruptly after a stream crosses splays of an active zone of thrust faulting. Stream terraces may not form in gorge reaches

Figure 3.15 Vertical displacements and fault geometry of the San Gabriel Mountains. Line of section A–A' parallels and is 2 km west of Little Tujunga Canyon.
A. Cross section and calculations of theoretical vertical displacements compared with the observed 1971 displacements. Fault geometry is constrained by the 13 km depth and 52° dip of the focal mechanism of the main shock of the 1971 San Fernando earthquake. The curve is the best of three models but suggests true net subsidence between distances of 6.5 and 10 km. Note the large vertical displacement but minimal subsidence at the front that ruptured in 1971. From Ikeda (1983, Fig. 14).

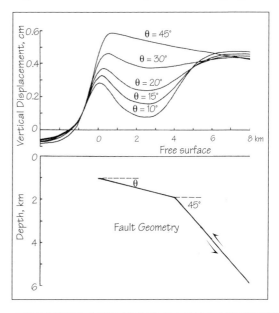

Figure 3.15 Vertical displacements and fault geometry of the San Gabriel Mountains.
B. The effects of bending of a thrust fault plane on rock uplift, Δu. Uplift of the internal front, formation of a narrow tectonic bulge at the range-bounding fault, and backtilting and relative local subsidence of the piedmont foreland occur simultaneously where the dip of the shallow fault plane is 20° or less. From Ikeda (1983, Fig. 17).

A tectonic geomorphologist would predict episodic attainment of the base level of erosion by looking for the presence of strath terraces. These reference surfaces would be parallel where passing through a reach with an inactive thrust fault. Strath terraces would diverge as a stream approaches an active thrust fault and then converge downstream from the tectonic perturbation, ending abruptly at the range-bounding fault zone.

if uplift rates are sufficiently rapid, relative to stream-channel downcutting rates, to preclude attainment of type 1 dynamic equilibrium (Section 2.4.3).

Abrupt widening of the active channel and the valley floor characterizes embayment reaches, which begin immediately downstream from the most downstream splay of a thrust-fault zone. Embayment reaches gradually narrow downstream and assume the characteristics of gorge reaches. The wide valley floors of embayment reaches may appear to be sites of thick alluvium. The presence of late Pleistocene and Holocene gravel strata would confirm a model of backtilting, synclinal folding, and local downwarping associated with antithetic faults that characterize part of the piedmont foreland tectonic domain. Conversely, the presence of strath terraces would attest to long-term net uplift (Section 2.6.1) of embayment reaches (Fig. 3.16). Amounts of net uplift relative to the reference datum (base level of erosion) are reduced by local tectonic downwarping in embayment reaches (Fig. 3.14).

Thicknesses of stream-channel alluvium are functions of rates of stream-channel downcutting and magnitudes of local base-level change created by active thrust faulting. Local faulting of stream-beds crossing fold-and-thrust belts results in variable thicknesses of valley-floor alluvium.

Streams don't remain at their base level of erosion after a surface rupture event on an active thrust fault. Each event induces aggradation downstream and degradation upstream from the fault (Fig. 3.17). A pocket of streambed gravel downstream from a thrust fault is anomalous and indicates a recent surface-rupture event. Tectonically induced downcutting will eventually eliminate anomalous thicknesses of gravel. Duration of such local deposition in a reach characterized by long-term degradation is a function of the magnitude and frequency of tectonic displacements, resistance of streambed bedrock to abrasion and plucking, and the annual tractive force of the stream.

Variations in thicknesses of streambed gravel were assessed by shallow seismic refraction surveys for two streams. One survey was along Little Tujunga Canyon (Fig. 3.18). Gneissic and plutonic rocks underlie the headwater's half of the 49 km² watershed and soft marine sediments are present in the downstream half.

The headwater's reaches have numerous exposures of bedrock with intervening patches of thin gravel. Such characteristics are typical of disequilib-

Figure 3.16 Late Pleistocene strath terrace in the embayment reach of Little Tujunga Canyon. The internal mountain front rises in the background as a series of structural and fluvial benches.

rium reaches that prevail in mountainous streams and provide little information about recent surface ruptures. Our basic reference surface – the base level of erosion is missing in such reaches. Non – geomorphic studies indicate that the oblique-dextral San Gabriel fault zone may be still active (Lee and Schwarcz, 1996).

Streambed gravels are only 1 m thick in the channel just upstream from the internal Sunland fault and increase almost threefold to about 2.7 m immediately downstream from the fault trace. Gravel thicknesses remain the same through the Merrick syncline reach but thin to 0.8 to 1.5 m as the stream approaches the gorge reach upstream from the range-bounding Lakeview fault. Gravel thickness then increases abruptly to 19 m only 0.1 km downstream from the fault zone (Fig. 3.18). Thus the most obvious increase in thickness of streambed gravels along Little Tujunga Canyon is associated with the most active fault zone. Sinestral-reverse oblique slip of the land surface of as much as 2 m occurred during the Mw magnitude 6.4 San Fernando earthquake of 1971 (Barrows et al., 1973), only 6 years before our survey of streambed gravel thicknesses. This historic

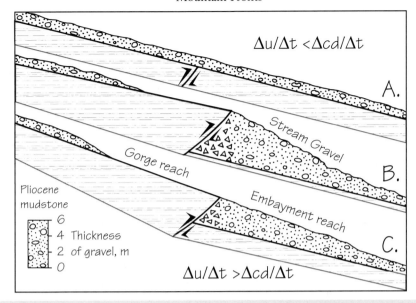

Figure 3.17 Hypothetical distribution of deposits beneath a stream flowing across an active thrust fault. A. Longitudinal profile of stream that approximates a condition of type 2 dynamic equilibrium. The thin blanket of gravel is easily entrained by large flow events, which have beveled the soft bedrock substrate and created a strath since the most recent surface rupture.
B. Thrust faulting raises the reach upstream from the fault and tilts it upstream. Scour removes the gravel. Collapse of the scarp creates colluvium downstream from the fault and the tectonic base-level fall creates space for deposition of streambed gravels.
C. Valley-floor degradation continues during an interval of no additional faulting. Part of the gravel adjacent to the fault is removed and deposited further downstream. More bedrock is exposed upstream from the fault as the stream incises to deepen a bedrock narrows reach. u is uplift, cd is stream-channel downcutting, and t is time. The uplift rate–channel downcutting rate relation is the reverse of A.

event may be partly responsible for the small stream-bed gravel anomaly downstream from the internal Sunland fault.

These several variations in streambed gravel thickness might be only random variations about mean gravel thickness of covering a strath in a tectonically inactive fluvial system. They can also be interpreted as varying systematically with each geologic structure. If so, they indicate continuing uplift along the Sunland fault but at an order of magnitude slower rate than for the range-bounding Lakeview fault. Little deformation appears to have occurred recently in the Merrick syncline reach. Sustained synclinal folding would have depressed the streambed longitudinal profile below the base level of erosion and thick gravels would then characterize this reach. Instead, rates of long-term stream-channel downcutting exceed rates of tectonic downwarping.

Total watershed relief upstream from the Lakeview fault zone has increased about 102 m. The stream of Little Tujunga Canyon upstream from the mountain front is large enough to be able to repeatedly downcut to its base level of erosion. Rapid vertical tectonic displacement has created an anomalously steep reach, especially considering the soft nature of the underlying Pliocene marine mudstones.

Strath heights above the active channel record tectonic uplift. As much as 30 m separates some straths. This suggests that: 1) surface ruptures were so frequent that the base level of erosion was rarely attained, 2) the strath record is incomplete because erosion has destroyed some straths or has made them difficult to locate, or 3) modulation of times of stream-channel downcutting by climatic conditions (Section 2.6.2) may have permitted only occasional return to the base level of erosion.

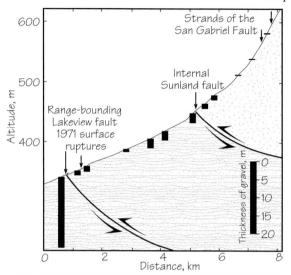

Figure 3.18 Variations in thickness of streambed gravel along Little Tujunga Canyon where it crosses active and inactive thrust faults. Subsurface faults and geology are diagrammatic.

The alfisols soils chronosequence for the Transverse Ranges (see Tables 4.2 and 4.3 in Bull, 1991), together with a flight of terraces permits evaluation of changes in late Quaternary uplift rates on the range-bounding Lakeview fault zone. The present interglacial climate is associated with the process of major strath formation rather than aggradation or degradation. A similar climate control of processes was assumed in estimating the ages of older straths. Soils chronosequence age estimates have poor precision and accuracy when compared to other dating methods (see Figure 6.1 and the comparison with eight other methods discussed in Section 6.1.1). Ages based on estimates of the time needed to form soils on terrace treads have larger uncertainties with increase of terrace age. Even so, Figure 3.19 can be used to illustrate how to use stream terraces, and it provides an obvious tectonic conclusion.

Mean rates of valley-floor deepening seem to have increased from ~0.15 m/ky between 330 and 125 ka to ~3.17 m/ky since 6 ka. This apparently rapid rate of Holocene faulting is similar to that documented by Morton and Matti (1987) for the Cucamonga fault zone at the east end of the San Gabriel Mountains. The most straightforward interpretation of the large increase in tectonically induced downcutting rates is that thrust-fault propagation shifted the

location of maximum tectonic deformation gradually, but in an accelerating manner, from the internal fault zone to the range-bounding fault over a time span of 0.3 My.

The exceptionally rapid Holocene uplift rates may not represent the true long-term increase of uplift rates. First, recent stream-channel downcutting might be anomalous if only a response to a recent temporal cluster of surface ruptures. It helps to know how the most recent event fits in with the characteristic earthquake recurrence interval for a particular fault. Second, Little Tujunga Canyon stream-channel downcutting since the mid Holocene may exceed the long-term rate as the stream catches up after an episode of climatically induced aggradation induced by the Pleistocene–Holocene climatic change. The youngest strath terrace might be an internal adjustment terrace instead of representing a return to the base level of erosion datum. Two examples of such anomalously rapid stream-channel downcutting rates are shown in Figure 2.16. The average tectonically

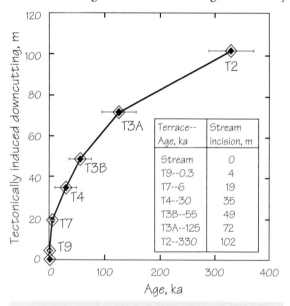

Terrace-- Age, ka	Stream incision, m
Stream	0
T9--0.3	4
T7--6	19
T4--30	35
T3B--55	49
T3A--125	72
T2--330	102

Figure 3.19 Uplift-rate trend for the range-bounding fault where it truncates the terraces of Little Tujunga Creek, based on heights of stream-terrace straths above active-channel strath. Progressive increase in tectonically induced downcutting accompanied the transfer of active thrust faulting from the internal Sunland to the range-bounding Lakeview fault.

induced downcutting rate of ~1.2 m/ky since 30 ka is a more conservative estimate than the short-term rates for estimating recent increases in uplift rate along the Lakeview fault zone.

Tectonic offsets of the stream channel result in deposition downstream from the fault zone as well as incision in the upstream reach, so an apparent slip rate of 1.2 m/ky is a minimum. Total tectonic displacement needs to be increased by at least 19 m. But unlike the tectonic landform of strath terraces, we don't know when episodes of tectonically induced deposition of basin fill occurred unless many layers are dated.

The late Quaternary shift in tectonic displacement from the internal to the range-bounding fault is confirmed by general trends of stream-terrace longitudinal profiles. All terraces in the reach upstream from the range-bounding Lakeview fault have straths that parallel each other and parallel the strath beneath the active channel. Unit stream power is ample to keep pace with uplift because of the quite low rock mass strength and the abrasive tools provided by cobbles and boulders derived from resistant rock types upstream from the internal fault. The Holocene strath terrace and active-channel strath are parallel in the reach that passes through the internal Sunland fault zone, implying lack of significant tectonic deformation. In contrast the 30 ka T4 terrace diverges upstream and converges with the active channel downstream from the internal fault zone, which suggests distributed tectonic deformation between 30 ka and 7 ka.

The 125 and 30 ka stream terraces are separated by more than 20 m in the downstream reaches of Little Tujunga Canyon. In the North Fork of the San Gabriel River (Fig. 3.3), the 55 ka aggradation event was strong enough to bury the strongly developed soil profile on the 125 ka terrace tread. Such composite alluvial fills did not form in Little Tujunga Canyon because rapid tectonically induced downcutting preserved each aggradation event as a separate stream terrace.

Stream gradient (*SL*) indices (Section 2.6.1) have been used to evaluate tectonic and lithologic controls on streams of the San Gabriel Mountains (Keller and Rockwell, 1984). Maps of regional trends are labor intensive. Hack calculated values of the *SL* index, both narrow and inclusive, for 400 reaches in a single topographic quadrangle, and Keller undertook the ambitious task of mapping the *SL* indices for the entire San Gabriel Mountains (Figure 4.9 of Keller

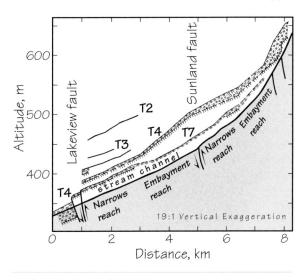

Figure 3.20 Longitudinal profiles of late Quaternary strath terraces of Little Tujunga Canyon.

and Pinter, 2002). Their focus is on longer reaches (inclusive stream-gradient index), and they assume that the larger streams of the San Gabriel Mountains adjust quickly to lithologic and tectonic controlling variables. Anomalously high values of the inclusive gradient-index should be indicative of resistant rocks or long reaches that have been affected by recent tectonic deformation.

The magnitude and extent of an inclusive gradient-index anomaly in the drainage basin of Little Tujunga Canyon (Fig. 3.21) records rapid uplift on the range-bounding fault. The inclusive stream-gradient index is in the highest 5% category (for the entire San Gabriel Mountains) near the mouth of Little Tujunga Canyon, despite the presence of soft Pliocene mudstones. This indicates rapid uplift rates as suggested by Figure 3.19. This inclusive stream-gradient index anomaly has been present since about 125 ka as shown by the flight of parallel strath terraces. The maximum *SL* anomaly extends upstream from the internal Sunland fault. Presumably this extent represents the effects of rapid uplift on the range-bounding fault with only a minor base-level fall contribution from the internal fault.

The second seismic refraction survey was in rugged Cucamonga Canyon (Fig. 3.22), whose 34 km² drainage basin is underlain by gneissic and metasedimentary rocks. Morton and Matti's (1987) studies of thrust-fault scarps on late Quaternary pied-

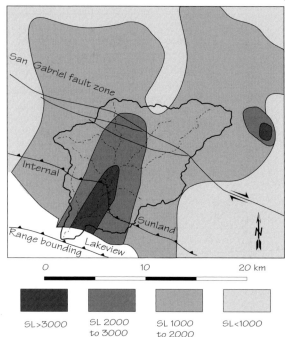

SL>3000   SL 2000   SL 1000   SL<1000
          to 3000   to 2000

Figure 3.21 Relation of stream-gradient (SL) indices to fault zones in the watershed of Little Tujunga Canyon. Holocene displacements have been large and frequent on the range-bounding fault, and minimal to rare on the internal and San Gabriel fault zones. SL mapping is from Figure 4.9 of Keller and Pinter, 2002.

mont alluvial surfaces show that this part of the San Gabriel Mountains also is rising rapidly, as does the *SL*-index map. They estimate an earthquake recurrence interval of roughly 0.6 ky, fault displacements of 2 m, and a slip rate of 5 m/ky for the past 13 ky.

Three escarpments are present in a 4 km long study reach. The Demens and Cucamonga faults cross the Cucamonga Canyon at fronts 23A and 23C, respectively, and a dissected topographic escarpment suggests the presence of a third unmapped thrust fault at front 23B. Bedrock is not exposed in the stream channel upstream from front 23A, and thickness of streambed gravel above a planar strath does not change across the fault zone. Either the Demens fault is no longer active, or stream-channel downcutting rates exceed fault slip rates. A thin blanket of gravel would represent the cutting tools left after the recent large streamflow events scoured the strath.

Front 23B is a nice example of how geomorphic studies can identify a previously unmapped ac-

tive thrust fault. A narrow gorge, only 8 to 20 m wide and flanked by vertical cliffs, is present in the reach upstream from where an escarpment, that can be seen on nearby hillslopes, crosses the valley floor of Cucamonga Canyon. This tendency for stream-channel degradation reverses abruptly at the conjectural fault zone. About 10 m of gravel are present below the stream as a wedge of alluvium that thins to 4 m at 200 m downstream from the active thrust fault. At least 10 m of throw has occurred recently on this thrust fault. Total tectonic displacement would be the sum of stream-channel changes in the two adjacent reaches. The thickness of the streambed gravel anomaly should be added to the amount of stream-channel incision into the bedrock.

Streamflow diverges and becomes braided upon entering the broad fanhead trench of Cucamonga Canyon, where about 1.8 to 2.6 m of gravel overlie bedrock. Channel gravels thicken abruptly to 10 m just downstream from a streambank exposure of the range-bounding Cucamonga fault zone (same size of streambed gravel anomaly as for the postulated Front 23B active fault zone), and increase further in thickness to 19 m just downstream from a nearby piedmont fault scarp (Front 23D). I conclude that the reach at front 23A may approximate equilibrium conditions. Pulses of uplift appear to exceed channel downcutting rates at fronts 23B and 23C because tectonically induced downcutting has yet to eliminate the most recent surface rupture. Tectonic base-level falls downstream from Fronts 23C and 23D have created space for accumulation of basin fill.

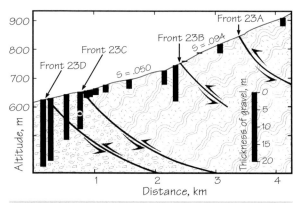

Figure 3.22 Variations in thickness of streambed gravel along Cucamonga Canyon where it crosses active and inactive thrust faults. Subsurface faults and geology are diagrammatic.

In summary, tectonic geomorphology studies of thrust-faulted terrains focus on the distinctive landscape assemblage of piedmont forelands. Recognition of piedmont forelands helps define the tectonic framework of a study area. Studies of the responses of streams to tectonic deformation can also help locate those active thrust faults whose surface ruptures do not extend to the surface – the "blind" thrusts. Useful fluvial landforms include stream terraces, alternating gorge and embayment reaches, streambed longitudinal profiles and stream-gradient indices, and fault offset of bedrock floored stream channels that preserve anomalous pockets of gravel. Studies using these geomorphic tools can identify which thrust faults are most likely to generate future earthquakes.

3.3 Fault Segmentation of Mountain Fronts

3.3.1 Different Ways to Study Active Faults

Spectacular mountain ranges commonly are bounded by active faults that pose a hazard to cities in the adjacent basins (Fig. 3.1). Paleoseismologists use diverse information in their quest to learn more about the potentially hazardous fault zones. We date prehistorical earthquakes and estimate their magnitudes, ascertain surface-rupture heights and lengths, make maps depicting patterns of seismic shaking, and determine if these characteristics are similar or different for consecutive earthquakes. This information in needed not only for range-bounding faults, but also for faults that terminate below the ground surface, and for submarine fault zones. Our quest has taken us to diverse tectonic settings and has resulted in new ways to study earthquakes.

Paleoseismologists subdivide long fault zones into sections. Many studies suggest that surface-rupturing earthquakes occur on sections of fault zones with distinctly different structural, lithologic, and topographic characteristics (Allen, 1968; Wallace, 1970; Bull, 1978; Matsuda, 1978; Sieh and Jahns, 1984; Schwartz and Coppersmith, 1986; Schwartz and Crone, 1985; Bull and Pearthree, 1988; and Crone and Haller, 1989; Wells and Coppersmith, 1994). For some faults, recent vertical or horizontal displacements appear to be similar for consecutive earthquakes, which promoted the concept of a characteristic (typical) earthquake as part of the fault segmentation model.

We realize the importance of determining if earthquake behavior is persistent for a particular fault.

Attempts to simplify repetitive earthquake behavior take rupture length into account. Rupture length is crucial because moment magnitudes of earthquakes, Mw, are a function of the area of a fault plane that is ruptured. A fault zone continues to propagate and evolve with each event, so past behavior may provide only a general guide to future style, timing, and magnitude of earthquakes.

Even the best models may not be applicable in all tectonic settings. The Mw 7.3 Landers earthquake of 1992 does not fit simple earthquake models. Dextral slip side-stepped through consecutive "segments" on five "separate" fault zones in the Mojave Desert of southern California (Hart et al., 1993; Sieh et al., 1993). Diverse approaches to paleoseismology are needed when hypotheses of persistent earthquake behavior don't match subsequent events.

Fault segmentation studies during the past 20 years have emphasized dating of colluvial, alluvial, and swamp deposits with their intercalated paleosols, that are exposed by excavations across fault zones (Swan et al., 1981; Schwartz and Coppersmith, 1984; Sieh 1984). Trench studies have utilized new dating techniques such as optical and thermoluminescence (McCalpin and Forman, 1991; Harrison et al., 1997), electron spin resonance (Lee and Schwarcz, 1996), and have re-defined the state of knowledge for specific active faults (Hancock et al., 1991). Mathematical refinements of radiocarbon dates and their stratigraphic layers (Biasi and Weldon, 1994; Biasi et al., 2002) are now used by many workers (Grant and Lettis, 2002).

These worthwhile efforts have opened interesting new pages about recent earth history. The applied spin-off is much improved knowledge about earthquake hazards posed by active faults such as the San Andreas transform and the Wasatch fault zones in California and Utah.

Let us consider the merits and shortcomings of the trench-and-date stratigraphic approach to paleoseismology. This sets the stage for geomorphic studies of active range-bounding faults.

Major advantages of trenching fault scarps include:
1) Identification of the fault responsible for a specific surface-rupture event.
2) Presence of distorted strata and liquefaction features that unequivocally demonstrate that slip on the fault plane was seismic in nature.
3) Detailed local information about style and magnitude of surficial faulting.

4) Potential to estimate the times of earthquakes of the past 40 ka, using the widely preferred radiocarbon dating method. Organic materials obtained from faulted and unfaulted strata, deposited before and after the event, constrain the interval during which a prehistorical earthquake occurred.

5) Opportunities to measure offsets of bedding and buried stream channels in order to separate the events responsible for a multiple-rupture event fault scarp. Displacement amounts can be used to estimate magnitude ranges for prehistorical earthquakes.

6) Identification of which scarps contain faults and thus are potentially dangerous building sites, and which were the result of nontectonic processes.

Stratigraphic studies made in trenches excavated across fault scarps continue to be important for paleoseismology (Machette, 1978; Weldon and Sieh, 1985; Rockwell and Pinnault, 1986; Sieh et al., 1989; Sieh and Williams, 1990; McGill and Sieh, 1991; Machette et al., 1992; Grant and Sieh, 1993; Lindvall and Rockwell, 1995; and in the book edited by McCalpin, 1996). The November 2002 issue of the Bulletin of the Seismological Society of America is devoted to San Andreas fault earthquakes. It is a treatise about trench-and-date paleoseismology. Age estimates of organic materials created before or after a surface-rupture event have greatly improved our perception of earthquake recurrence intervals and surface rupture characteristics (Sieh et al., 1989: Fumal et al., 1993; McCalpin, 1996; Sietz et al., 1997; Biasi et al., 2002). Fewer hazards assessments have been made for thrust faults hidden in folds (Cifuentes, 1989; Scientists of the U.S. Geological Survey and the Southern California Earthquake Center, 1994; Keller et al., 1998), or for submarine subduction zone thrust faults (Atwater et al., 1991, Atwater and Yamaguchi, 1991; Atwater et al., 1995).

Three examples of the marvelous insights that trench-and-date stratigraphic studies may provide are included in this book. Critical details of stratigraphy ruptured by a thrust fault at the Loma Alta site in the San Gabriel Mountains are presented next. Using soil profiles as stratigraphic time lines is illustrated in Chapter 4. Figure 6.50 is an analysis of the Honey Lake fault zone, a major right-lateral strike-slip fault in the Walker Lane tectonic belt.

Site selection, careful stratigraphic description and sampling, and making the sensible analyses indeed required the combined talents of Charlie Rubin, Scott Lindvall, and Tom Rockwell (1998) at the Loma Alta site. I summarize their project here because of their diverse trenching experience and flexibility in making interpretations.

Their goal was to answer an important hazards question. Were recent earthquakes indicative of the hazard level for the Los Angeles metropolitan area? These include the Mw magnitude 6.7 San Fernando earthquake of 1971, Mw 5.9 Whittier Narrows event of 1987, and the most damaging earthquake in the history of the United States, the 1994 Northridge Mw 6.7 event. Even moderate-size surface ruptures of the Sierra Madre range-bounding thrust fault would direct enormous seismic energy southward into the adjacent densely populated metropolis. Crook et al. (1987) were unable to identify fault scarps or ruptured strata younger than late Pleistocene, and concluded that this fault segment was not as tectonically active as segments to the east and west. It may have not produced a Holocene earthquake.

A thorough geomorphic reconnaissance was made before spending >$100,000 on a trench. Which topographic scarps might record a prehistorical surface rupture? The ancient plane-table surveys used to make the first topographic maps of the American southwest provide details not available in most modern maps. Rubin's assessment of tectonic and nontectonic scarps included re-mapping the entire mountain front using pre-urbanization aerial photographs taken between 1928 and 1935 and early versions of topographic maps, which have 1.5 m (5 foot contours). The emphasis was on locating and then field checking low steps in alluvial surfaces that might be late Quaternary fault scarps.

This geomorphic approach was an exercise in understanding the behavior of fluvial systems. They needed to predict the likelihood of having discrete alluvial strata needed to define the locations and magnitudes of fault ruptures, and which sites might allow identification of multiple colluvial wedges in a thrust-fault setting. The deposits had to be young and fine grained enough to contain detrital charcoal. Small alluvial fans fed by minor canyons fit these guidelines better than the bouldery floodplains of large rivers. Loma Alta made the list of finalists.

The 5 m deep and 25 m long Loma Alta trench was excavated through a 2-m high fault scarp in a Pasadena city park. It is on a fill terrace that crosses the Sierra Madre fault zone 1.2 km east of Caltech's Jet Propulsion Laboratory (which is next to Arroyo Seco on Figure 3.3). Rubin and Lindvall pegged a 50 cm grid on the trench wall and spent 2 to 3 weeks mapping in fine detail. The location of each cobble was noted (Figure 3.23 shows diagrammatic cobble and boulders) and the rationale for collecting each radiocarbon sample was evaluated. Tom Rockwell did the soil profile sampling and description. Soil profiles are essential to frame this small stratigraphic section and correlate it to the San Gabriel Mountains soils chronosequence. Soils of the type sections are dated with more than detrital charcoal and provide a crosscheck of the conclusions reached by the Loma Alta batch of radiocarbon age estimates.

The Loma Alta stratigraphic section contains obvious climatic and tectonic signatures. Brief pulses of regional late Quaternary climate change induced aggradation that temporarily reversed the overall trend for tectonically induced downcutting of valley floors in the San Gabriel Mountains. Climate change caused partial stripping of hillslope sediment reservoirs. The resulting fill-terrace treads became sites of soil-profiles that became more strongly developed with the passage of time. Terrace-tread soils may be buried at mountain fronts with active thrust faults. Burial stops surficial weathering and pedogenic processes, but key soil-profile characteristics are preserved

in the stratigraphy. The alfisols soils chronosequence for the Transverse Ranges (Table 4.11, Bull, 1991) provides likely matches for the Loma Alta site soil profiles. The young soil above unit 4b is an A horizon that dates to the mid-to-late Holocene. The 46 cm thick buried argillic soil-profile horizon of unit 2 (Fig. 3.23) is overlain by a 36 cm A horizon. This suggests a latest Pleistocene to early Holocene age with the older age estimate preferred because the soil profile would be more strongly developed if it had not been buried.

Detrital organic matter for radiocarbon dating is usually difficult to find in gravelly alluvium, even in humid regions. The latest Pleistocene aggradation event gravels in humid New Zealand have yielded few samples from hundreds of exposures. Wood floats downstream or rots. Bouldery braided streams destroy charcoal fragments.

Streambank exposures and fault-scarp trenches in the arid Gobi Altai of Mongolia did not provide detrital organic matter. Paleoseismologists turned to other ways of dating displacements. Stratigraphers collected samples for luminescence dating (thermo-

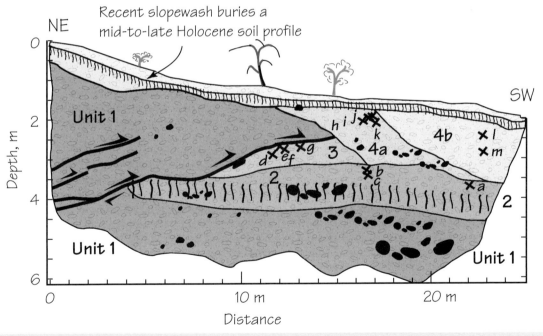

Figure 3.23. Stratigraphic section exposed in the trench wall at Loma Alta, southern front of the San Gabriel Mountains of southern California showing fault traces and stratigraphic units 1 through 4b. Lettered x's show the locations of charcoal fragments collected for radiocarbon dating described in Table 3.2. Redrafted from Figure 3 of Rubin et al. (1998).

luminescence, TL, and optically stimulated lumines-cence, OSL of silt fragments). See the work of Pren-tice et al., (2002) for applications of luminescence dating, and for splendid examples of trench logs that efficiently focus the reader on key aspects of the data and interpretations. Geomorphologists used terres-trial cosmogenic isotopes such as [10]Be to date alluvial fan surfaces that had been tectonically translocated by strike-slip faulting or have been incised by tectoni-cally induced downcutting. See the work of Vassallo et al. (2005) for clear-cut examples of results and de-scription of the method.

The San Gabriel Mountains provide abun-dant charcoal fragments to stream channels. Hill-slopes covered with chaparral consisting of waxy and highly flammable plants favor frequent fires and are the source of abundant charcoal that is flushed downstream and deposited with alluvium. The key to dating success involves recognizing three classes of charcoal fragments.

1) Fragments eroded from old alluvial and colluvial deposits, transported a short distance downstream and redeposited with new alluvium. These charcoal samples look nice in the field but yield dismaying re-sults (wide range of ages, with nearly all being much too old).

2) Charcoal created just before deposition with al-luvium. For example, a late summer firestorm leaves hillslopes barren and creates much new charcoal, which is transported and deposited by debris-flow slurries during the intense rainstorms of the next win-ter. Even these preferred materials for dating a single bed can't tell us the length of time between growth of a plant stem and the time of the fire.

3) Charcoal fragments created by brush fires on a san-dy stream-terrace tread and then taken underground by subsequent bioturbation processes such as bur-rowing rodents. The radiocarbon age estimate tends to be much too young, sometimes even modern.

One needs many radiocarbon dates to make appropriate age interpretations. A single date for a trench site, such as in the Figure 6.50 example, may be better than none. Having the range of possible ages suggested by the 13 radiocarbon samples at the Loma Alta (Table 3.2) site allowed selection of the most reasonable age estimates for each of the strati-graphic units.

Interpretation of radiocarbon and soils in-formation varies depending on the biases and back-ground of paleoseismologists. The nice Rubin et al. scenario is presented first, then my thoughts that use the buried soils to a greater extent. You, as readers, can be expected to further develop alternative mod-els. Welcome to the intricate world of "trench-and-date" paleoseismology.

Unit 1 is crudely stratified sandy boulder- to pebble-sized gravel. Moderate diagenetic weathering of diorite cobbles suggests that it is significantly older than unit 2. Unit 2 is a fine sandy gravelly loam. A buried soil (A, Bt, and Cox horizon), with a 50-cm thick Bt horizon extends into the top of unit 1. This soil-forming interval of landscape stability ended with the first of two thrust-fault events, which raised part of unit 2 sufficiently to strip it from the hanging wall and redeposit it as the colluvial wedge of unit 3. A second rupture event resulted in deposition of unit 4a, which is a massive colluvial wedge of boulders and gravel in an organic-rich silt and sand matrix. The bouldery unit 4a grades laterally into the exten-sively bioturbated, organic-rich silty sand of unit 4b of Figure 3.23.

Ideally, radiocarbon ages are progressively younger in overlying strata. Charcoal from each of the Loma Alta stratigraphic units provided mixed re-sults. Charcoal fragments can also be anomalously young if introduced by bioturbation after deposition of the unit. This left the research team guessing as to whether the dated fragment was detrital or bio-turbated, reworked from older deposits, or created by fire just before deposition. All detrital charcoal provides a maximum age estimate of the stratum that contains it, so the youngest age in a unit is considered as the maximum possible age for a unit. Three lines of evidence indicate the presence of older reworked charcoal in younger stratigraphic units:

1) A radiocarbon age of 29 ± 4 ka was obtained for charcoal from an older stream terrace located about 1 km uphill from the excavation site. Detrital charcoal derived from such older units can be incorporated into any younger alluvium.

2) Angular and rounded charcoal fragments were analyzed from the same sample locality. Sample j of Table 3.2 was angular fragment that provided a radio-carbon age of 14.2 ± 1 ka, and the sample i rounded fragment an older radiocarbon age of 18.2 ± 1 ka.

3) Radiocarbon ages vary greatly for the same strati-graphic horizon, which suggests recycling of older detrital charcoal.

An opposite problem may affect charcoal in extensively bioturbated deposits, such as Unit 4b, where young carbon ages may reflect incorporation of charcoal that postdates the time of deposition.

Radiocarbon sample	Radiocarbon age (years B.P. ±1 σ)	Calendric age range (years B.P. ±2 σ)	Comments and interpretations
Unit 2			
a	16,330 ± 110	18,912–19,604	All three samples are charcoal fragments in a thick buried A soil-profile horizon
b	24,360 ± 300		
c	16,175 ± 80	18,804–19,349	
Unit 3			
d	28,760 ± 370		Recycled charcoal fragments
e	29,900 ± 390		Recycled charcoal fragments
f	15,600 ± 140	18,191–18,816	Maximum age for deposition of 3
g	34,300 ± 800		Recycled charcoal fragments
Unit 4a			
h	14,360 ± 320	16,445–17,924	Recycled charcoal fragments
i	15,235 ± 95	17,901–18,405	Recycled rounded fragment
j	12.160 ± 75	13,897–14,530	Recycled angular fragment
k	9,495 ± 75	10,349–10,902	Maximum age for deposition of 4
Unit 4b			
l	1,185 ± 95	937–1,277	Both may be minimum ages if the charcoal dates bioturbation
m	3,335 ± 35	3,471–3,360	

Table 3.2. Radiocarbon analyses of charcoal samples. Samples are listed in stratigraphic order; unit 2 is the oldest, 4b the youngest. See Figure 3.23 for sample locations of a–m. Analytical uncertainties are only for the laboratory measurements. From Table 1 of Rubin et al. (1998).

The buried colluvial wedges (units 3 and 4a) are well stratified and show little evidence of bioturbation; thus incorporation of young carbon is unlikely.

Rubin et al. conclude that units 2 through 4a are latest Pleistocene to early Holocene in age and that unit 4b is younger. Four fragments of detrital charcoal from unit 3 yielded ages of between 18 and 34 ka. Three appear to be recycled charcoal from older alluvial deposits and even 18 ka is considered as the maximum age for deposition of unit 3 colluvial deposits. Four fragments of detrital charcoal from unit 4a yielded ages ranging from 11 to 18 ka. The 11 ka estimate is considered the maximum age. Two detrital charcoal fragments from unit 4b yielded radiocarbon ages of 1 and 3 ka, and are regarded as minimum ages because it is likely that the charcoal was emplaced by bioturbation after deposition of unit 4b.

Two thrust faulting events have occurred since the formation of the soil profile on the unit 2 former land surface (Fig. 3.23). The first surface rupture event is the obvious truncation of unit 2 as slip along the fault plane shoved unit 1 over unit 2. This local landscape instability created the space needed for deposition of the unit 3 colluvial wedge. It was preceded by an interval of landscape stability long enough to form the soil profile. The second event cut through the unit 3 deposits and resulted in deposition of the unit 4a colluvial wedge. Erosion has removed the unit 3 deposits above the fault and has removed part of unit 4a.

The magnitudes and importance of the two recent thrust-fault rupture events are clear-cut. Total slip for the two events was 10.5 m, assuming that unit 2 was continuous across the fault zone. Restoration of the upper plate for the most recent earthquake yields a minimum slip of 3.9 ± 0.1 m.

Prehistorical earthquakes at the Loma Alta site were substantially larger than the 1971 to 1994 earthquakes along reverse faults in the region. Large displacements of 4 m or greater are inconsistent with short (15 to 20 km) ruptures of the Sierra Madre fault and imply that past earthquakes ruptured much of the fault zone. The component of oblique thrust slip at depth for the 1971 Mw 6.7 San Fernando earthquake was as much as 3 m (Heaton, 1982), but the surface rupture was generally < 1 m (Kamb et al., 1971; U.S. Geological Survey, 1971).

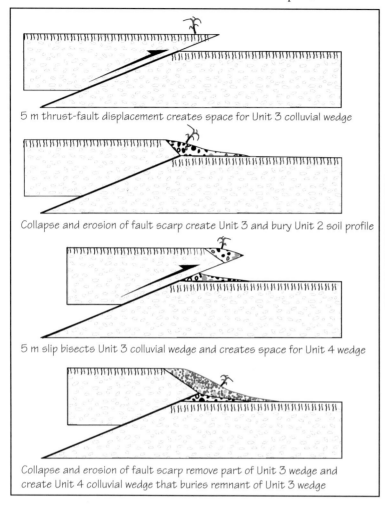

5 m thrust-fault displacement creates space for Unit 3 colluvial wedge

Collapse and erosion of fault scarp create Unit 3 and bury Unit 2 soil profile

5 m slip bisects Unit 3 colluvial wedge and creates space for Unit 4 wedge

Collapse and erosion of fault scarp remove part of Unit 3 wedge and create Unit 4 colluvial wedge that buries remnant of Unit 3 wedge

Figure 3.24. Cartoons showing development of colluvial wedges from two successive earthquakes at the Loma Alta trench site on the range-bounding Sierra Madre thrust fault, Southern California. Redrafted from Figure 5 of Rubin et al. (1998).

If we assume that 5.25 m represents the average surface displacement, a regression analysis predicts a Mw magnitude earthquake of 7.5 ± 0.5. With an average slip of 5.25 m, a strike length of 65 km, and a seismogenic depth of 18 km, the seismic moment for the most recent earthquake is 2.65 x 1027 dyne-cm. Seismic moment (Aki, 1966) is the product of fault-surface area, mean displacement on the fault, and rock rigidity. Converting seismic moment to moment magnitude yields Mw magnitudes (Hanks and Kanamori, 1979) of ~7.5 for both prehistorical earthquakes recorded at the Loma Alta site.

Tucker and Dolan (2001) excavated a trench on the Sierra Madre fault 34 km ESE of Loma Alta. They found evidence for 14 m of slip between 24 and 8 ka. Infrequent, but large (Mw > 7) earthquakes at

their site agree with the Rubin et al. (1998) conclusion of recent large earthquake(s).

Both recent earthquakes at the Loma Alta site were much larger than recent historical earthquakes. The results of the trench study support Hough's (1995) fractal model that most of the seismic moment release in the greater Los Angeles area is by infrequent but large events (Mw = 7.4 to 7.5). Damage from large magnitude earthquakes along the Sierra Madre fault would be substantially larger than that caused by the 1994 Mw magnitude 6.7 Northridge earthquake. Near-field, high-amplitude ground motions would ripple through a much larger area and for a longer duration. Unlike the Northridge earthquake that ruptured northward away from the metropolitan region, a Mw magnitude >7 earthquake on the Sierra

Madre fault would rupture southward, directing its energy into the adjacent densely populated basin.

The Loma Alta trench project resulted in re-assessment of seismic hazard and underscores the value of the stratigraphic approach to paleoseismology. The research team was careful and conservative in evaluating the times of the surface rupture events. Of course one would prefer several independent ways to cross-check the conclusions based on radiocarbon dating of charcoal fragments. Lacking these, one can use soil-profile characteristics to see if they agree with the overall conclusions. Each soil profile records a time span postdating the alluvium it is formed in. So times of faulting postdate both the stratigraphic and soil-profile ages.

Unit 2 is central to any temporal analysis because it has a well-developed soil that was overridden by the first thrust faulting event. So let us consider two alternative times of faulting. Charcoal fragments constrain when the alluvium was deposited, but not the time span required for subsequent soil-profile formation. Radiocarbon samples a and c date to 19 ka and like recycled fragment b are in the A soil-profile horizon. Assume that both are contemporaneous with the deposition of unit 2, and indeed represent the maximum possible age. Deposition was followed by pedogenesis long enough to create a strongly developed soil before scrunch tectonics offset and buried the surface of the fill terrace. The youngest soil with an argillic horizon in the soils chronosequence for the San Gabriel Mountains has an age of 4 to 7 ka (Bull, 1991, Chapter 4). It might take 5 to 10 ky to form the unit 2 soil. If so, the penultimate event may have occurred in the early Holocene.

A second scenario to consider. The sandy, organic, A horizon of the unit 2 buried soil would have been favorable for bioturbation. If so, samples a and c could be younger than when unit 2 deposition ceased. This leaves sample b, with a conventional radiocarbon age of 24 ky, as being more representative of the maximum age of unit 2. The likely timing of the first surface rupture event becomes much older, but still is offset by time needed to create the now buried soil profile (~24 ka minus ~8 ka =16 ± 5 ka).

Both unit 4b charcoal dates were considered minima. But note that the incipient soil profile capping this unit has characteristics requiring only 1 to 3 ky to form elsewhere in the San Gabriel Mountains. Both samples may have been moved about by bioturbation, but could well have been initially deposited with the recent sandy alluvium.

None of my second thoughts about the timing of depositional, pedogenic, and tectonic events change the important conclusions of Rubin et al. (1998). Two large recent prehistorical earthquakes ruptured much of the Sierra Madre fault zone with important implications for seismic hazard planning.

Trench-and-date paleoseismology has several drawbacks.
1) Not all faults have accessible scarps including blind thrust faults hidden in anticlines (Unruh and Moores, 1992; Bullard and Lettis, 1993; Unruh, 2001), submarine fault scarps, and faults beneath large cities. California earthquakes between 1983 and 1994 at Coalinga (Atwater et al., 1990), Kettleman Hills (Ekström et al., 1992), Whittier Narrows (Hauksson et al., 1988), Loma Prieta (McNutt and Sydnor, 1990), and Northridge (Hudnut et al., 1996) call attention to severe seismic hazards posed by faults that splay into concealed branches, or that terminate in folds, before reaching the surface. Earthquakes on these blind thrust faults caused huge financial loss and cost many lives. Surface ruptures of great subduction zone earthquakes are concealed beneath the sea (Heaton and Kanamori, 1984). Trench-and-date stratigraphic work is not possible in such settings.
2) Organic matter dated by radiocarbon analysis either predates or postdates times of disruption of seismic stratigraphy. Using multiple samples the time of an earthquake may be bracketed by analyses of organic matter formed before and after the event, but we cannot radiocarbon date the time of an earthquake. Atmospheric radiocarbon production rates may vary sufficiently to result in multiple possible calendric radiocarbon ages for a single sample (Stuiver et al., 1998). This problem is especially acute for the crucial post-1700 A.D. interval. When combined with the usual laboratory analytical error of ±40 years, the uncertainties for many radiocarbon estimates of times of earthquakes are larger than generally acknowledged. These complications make it difficult to use radiocarbon dating to separate events on faults where the earthquake recurrence interval is 100–300 years.

Detrital organic matter, such as charcoal and wood deposited by streamflows, presents additional challenges for the paleoseismologist because such materials are older than their time of deposition by an unknown amount. The material that we date in a layer that is below a faulted former land surface at the time of a prehistorical earthquake predates both its time of deposition and the time of the earthquake. Detrital organic matter in a layer above the former faulted land surface might have an age that is younger, older, or the same as the time of the earthquake.

Fortunately, new stratigraphic and surface-exposure dating methods, such as optical stimulation luminescence (OSL) and terrestrial cosmogenic nuclides (TCN), extend the range of dating and provide cross-checks. See the excellent book by Noller et al. (2000). Precision and accuracy of nine dating methods are compared in Figure 6.1 of this book.

3) Earthquakes occurring during times of nondeposition at a trench site are not likely to be recognized. Omission of such earthquakes creates a false impression of irregular (clustered earthquake) behavior (Bull, 1996a).

4) Trench studies of fault segmentation are time consuming and expensive. Decades of work in two dozen trenches were required to better understand the behavior of the Wasatch Range bounding fault zone in Utah. Little is known about overlap between adjacent segments because few excavations are made in segment boundaries.

5) Trench-and-date stratigraphic studies cannot evaluate the extent and intensity of seismic shaking and are a cumbersome way to describe extent of a surface rupture.

New approaches are needed in paleoseismology. Tectonic geomorphology is an important but under utilized tool with potential to complement important trench-and-date studies of earthquakes. Mapping and soils dating of alluvial surfaces allows optimal selection of trench sites by defining the extents and approximate times of prehistorical surface ruptures. Tectonic geomorphology studies use diverse data from topographic, pedogenic, stratigraphic, hydrologic, botanical, and structural geology sources. The geomorphic approach compliments geophysical studies and may not require expensive and time-consuming trenching projects. Multidisciplinary approaches can solve more problems.

Several new approaches to paleoseismology have been selected for this book. These include the work of Kirk Vincent (1995) who made detailed studies of longitudinal profiles of faulted alluvial fans. His analysis of segment-boundary behavior of tectonically active mountain fronts provides the most rigorous field test to date of the fault segmentation model of Schwartz and Coppersmith (1984). Regional appraisal of the relative tectonic activity of mountain fronts (Chapter 4) is important in selecting safe locations for nuclear power electrical generating stations. Seismic risk is minimal for range-bounding

faults that have been inactive for 1 My and are distant from active fault zones. Geomorphic processes that are sensitive to seismic shaking can be used to date and locate earthquakes, estimate their magnitudes, and describe regional patterns of seismic shaking caused by prehistorical earthquakes (Chapter 6).

The best approach to paleoseismology is to make both stratigraphic and geomorphic studies. The need to make geomorphic maps before trenching is stressed by McCalpin (1996, p. 34). Traditional dating of faulted stratigraphic sections in trenches is an essential next step. Geomorphic studies that define rupture lengths and displacement amounts for each event allow evaluation of how slip changes where a fault scarp extends into a segment boundary. Surface-exposure dating of coseismic landslides can recognize earthquakes that occur during times of nondeposition at a trench site (Bull, 1996a), and can be used to make maps of seismic shaking comparable to Mercalli Intensity maps (Bull and Brandon, 1998). Then, with both stratigraphic and geomorphic data in hand, we can better appraise the persistence of fault-zone behavior and potential surface rupture and seismic-shaking hazards.

3.3.2 Segmentation Concepts and Classification

Even large earthquakes do not rupture the entire length of long fault zones. The world's longest known historical rupture of a normal fault may be only 70–75 km (Section 5.4.2). The strike-slip San Andreas fault of California does not rupture its entire length; instead surface ruptures occur along one of four segments that are 150 to 500 km long (Allen, 1968). The fault-segment model is used to estimate lengths of potential surface ruptures, earthquake magnitudes, and to calculate seismic moment (Aki, 1969; Tsai and Aki, 1969; and seismic moment rate, Brune, 1968). A major endeavor in active tectonics is to decipher the spatial and temporal distribution of consecutive surface ruptures on normal, reverse, and strike-slip faults. Such work provides insight about the frequency, magnitude, and style of stress release along plate boundary and intra-plate fault zones. This research has significant implications for studies of crustal mechanics, origins of mountainous landscapes, and earthquake hazards.

Defining fault segments is a paleoseismic endeavor that strives for better understanding of the behavior of active faults and of the seismic risk they pose. **Fault segmentation** is a conceptual model that provides constraints for estimating earthquake size by recognizing adjacent sections of faults, each with their own style, magnitude, and timing of surface ruptures. The theme of this model of fault behavior is that a fault segment tends to rupture repeatedly in similar-size, or characteristic, events that approximate the maximum possible magnitude earthquake. The model implies temporally constant length and displacement amounts for each of a sequence of fault segments.

A basic premise is that future large earthquakes – the maximum magnitude earthquake model of Wesnousky (1986) – are most likely to occur on those faults with the largest cumulative displacements and most rapid slip rates during the late Quaternary (Allen, 1975; Matsuda, 1978). Identification of segments of a fault zone that repeatedly rupture independently of one another is not easy. Ideally, one would use repetitions of historical surface ruptures. Such data are rarely available, so paleoseismologists are forced to use less reliable approaches.

Knuepfer (1989, Table 3) ranked criteria for classification of earthquake segments in terms of the likelihood of being future locations of earthquake surface ruptures. Rupture limits for historical earthquakes are ideal (100%). Rupture limits for well-dated prehistorical earthquakes could be almost as good, but even multiple trench sites usually provide information about the most recent one or two earthquakes. Topographic, lithologic, and geophysical changes along a fault zone have a reliability of less than 50% (39%), as do geologic characteristics such as structural branching and intersections with other faults (31%). Spatial changes in style of faulting or slip rates and in earthquake recurrence intervals are of little use (26%), as are changes in fault orientation such as bends and stepovers (18%). It seems that most attempts to postulate fault-segment boundaries are merely models to be tested by innovative studies and future earthquakes. We should also always allow for the possibility that some fault zones may alternate between different styles of behavior. Topographic changes are used in this chapter because they record persistent tectonic deformation quite well. Whether or not they indicate the extent of the next surface rupture depends on how many segments are involved in the event.

Landscape analyses can decipher spatial variations of vertical tectonic displacements of mountain fronts that range from high escarpments to low fault scarps, using the landforms of hillslopes, streams, and alluvial fans. Mountain-front topography may not tell us much about the magnitude and extent of future slip events. Its primary value is the way it integrates late Quaternary fault displacements and erosion to create a record of the long-term rate, magnitude, and style of tectonic deformation. Having a landscape persistence perspective for the past 100 ky provides a useful counterpoint for radiocarbon dating of earthquakes for the past 40 ky.

3.3.3 Fault-Segment Boundaries

Barriers to rupture propagation define the ends of fault segments in a variety of settings ranging from subduction zones (Aki, 1984) to Basin and Range Province normal faults (Wheeler et al., 1987; Fonseca, 1988). Boundaries between segments typically are structurally complex and range in width from less than 1 km to more than 10 km. We seek information regarding which discontinuities consistently act as barriers to propagation of strain release during earthquakes.

A 370-km long zone of normal faulting bounds the Wasatch Range of Utah, which is the structural boundary between the Basin and Range Province and the Colorado Plateau. Spatial patterns of cumulative displacement were used to define a *characteristic earthquake model* (Schwartz and Coppersmith, 1984; Schwartz and Crone, 1985). Historical earthquakes have not occurred but three decades of trench studies reveal many Quaternary surface ruptures (Swan et al., 1981; Machette et al., 1989, 1991, 1992; Machette and Brown, 1995; McCalpin, 1996). Magnitudes of vertical displacement recorded at a trench site tend to be similar, and individual surface ruptures are largest near centers of segments and become less near terminations.

Of course surface ruptures cannot always be zero at *segment boundaries* (the transition zone between adjacent segments of a fault zone), because those parts of a mountain range also have been raised relative to the valley. So surface ruptures of adjacent segments must overlap (Fig. 3.25). Assume that:
1) the center half of each segment is not affected by ruptures in adjacent segments,
2) ruptures do not stop abruptly upon entering a boundary between fault segments, but continue on for a distance of one-fourth of the length of the next segment, and
3) repetitive displacements occur with similar frequencies and magnitudes for all segments.

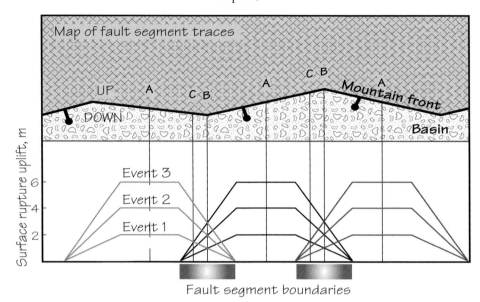

Figure 3.25 Spatial variations of sizes of characteristic surface ruptures in adjacent fault segments assuming 1) rupture events of similar magnitude and size, 2) that the middle half of each segment is not affected by displacements in adjacent segments, and 3) that the surface rupture in a given segment dies out one-fourth of the distance into the adjacent segments. Frequency and magnitude of surface ruptures are different at locations A, B, and C of this model.

Temporal patterns of surface ruptures will vary along the hypothetical segmented fault zone (Fig. 3.25). At location A, the earthquake recurrence interval is assigned a value of 10 ky and amount of displacement is assigned a value of 2 m; each displacement is similar and large. At location B the earthquake recurrence interval would have a value of 5 ky and displacements would have values of 0.5 m; each displacement is similar and small. At location C the earthquake recurrence interval would also have a value of 5 ky, but the displacements usually would alternate between large (0.8 m) and small (0.2 m). Real-world fault zones are more complicated than this model, particularly where consecutive surface ruptures have different lengths. Although diagrammatic, Figure 3.25 illustrates the difficulty of using single trench sites to estimate the frequency and magnitude of surface ruptures on range-bounding strike-slip, normal, and thrust faults.

Many variations of the characteristic earthquake model have been proposed and most can be tested in the field. Using tectonic geomorphology, Kirk Vincent devised tests of the model which are described in the next section.

3.3.4 Normal Fault Surface Ruptures

The fault segmentation model needs validation, particularly in regard to the style and pattern of consecutive surface-rupture events where they enter a boundary between two fault segments. This section addresses that topic by summarizing the work of Kirk Vincent on an active left-lateral oblique normal fault in Idaho (Crone et al., 1987). Discussion begins with how to use faulted fluvial landforms to estimate apparent vertical displacements. Then we correct apparent throw to obtain true values of vertical displacements. These geomorphic procedures provide insight into four consecutive late Quaternary earthquakes on the Thousand Springs and Mackay segments of the Lost River Fault, and in the 10 km long Elkhorn Creek segment boundary between the two segments (Fig. 3.26). This essay is about how to measure true tectonic displacement and what happens to surface ruptures in the boundary transition between two distinctive fault segments.

Estimated ages and correlation of fluvial-system aggradation events were based on radiocarbon analyses, a soils chronosequence emphasizing soil-carbonate pebble coatings and stage of carbonate pedogenesis, and the 6.8 ka Mazama volcanic ash (Pierce and Scott, 1982; Scott et al., 1985; Vincent et al., 1994; and Vincent, 1995, Chapter 2). Approximate calendric ages of the prehistorical earthquakes are 5 ± 1 ka, 10.5 ± 1 ka, and 14 ± 2 ka. A Ms magnitude 7.3 earthquake in 1983 provided an opportunity to collect data not readily available for the prehistorical earthquakes.

Kirk Vincent (1995) surveyed topographic profiles of faulted alluvial surfaces. Longitudinal profiles of alluvial fans and stream terraces typically are straight, or have a systematic curvature. This allows estimation of the vertical component of fault displacement for the entire width of a fault zone instead of just the main strand of the fault. It is easy to mistake the height or throw (referred to as **apparent throw** in this discussion) of a fault scarp as being representative of true displacement, especially where tectonic deformation along secondary faults is not obvious. See the disparities illustrated by Figure 3.27. Net throw is best measured as the vertical separation of the topographic profile for the landform that was faulted, but that too is an "apparent" vertical tectonic displacement whose correction is discussed in this section.

Surface ruptures associated with normal faults of the Basin and Range Province typically consist of a zone of deformation that is 10 to 500 m wide. Secondary features such as antithetic and synthetic scarps, tilted blocks, sagged or bulged ground, and rifts become difficult to see with the passage of time (Gilbert, 1890; Slemmons, 1957; Vita-Finzi and King, 1985; and Xiao and Suppe, 1992). Visual prominence of the primary fault scarp can be a misleading indicator of the magnitude of total vertical displacement because tectonic dislocation is distributed over the entire width of the rupture zone in a style that varies along a given fault zone.

Apparent net throw can be determined by surveying the land surface over a distance that exceeds three times the width of the rupture zone. Prefaulting shapes of undissected flood plains, stream terraces, and alluvial fans can be reconstructed with confidence by making surveys with an electronic distance meter.

Vincent surveyed streamflow paths over distances of 200 m to 800 m for many alluvial landforms in his Idaho study area and found that most longitudinal profiles are straight over distances of 100 to 800 m. Straight-line profiles downslope from the Lost River fault have a gradient that generally is slightly less than the gradient upslope from the fault.

First, let us describe the reference surface of a fluvial landform using a graph of surface altitude plotted against horizontal distance along the general path that streams flowed to construct or maintain the landform. The survey transect may be broadly curved, straight, or meandering and the longitudinal profile typically appears straight or changes gradient gradually.

The systematic curvature of a longitudinal profile of a stream that is entraining and depositing bedload reflects orderly adjustments between many variables. The stream may be aggrading, degrading, or at equilibrium. The resulting stream channel, floodplain, or braided depositional surface has a longitudinal profile that may be described by an arithmetic, exponential, or power-function equation.

Consider a reach of a stream at flood stage (when much of the work of streams is done) where

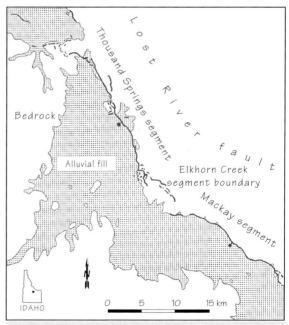

Figure 3.26 Latest Pleistocene, Holocene, and 1983 surface ruptures of the Lost River fault, Idaho. From Figure 6.3 of Vincent (1995).

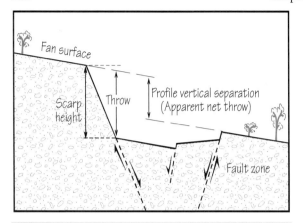

Figure 3.27 Cross section of a hypothetical normal-fault rupture zone. Scarp height is larger than throw because the fan surface is inclined. Most range-bounding faults are fault zones, so neither scarp height nor throw at the main scarp is equivalent to net throw over the entire fault zone. Figure 5.1 of Vincent (1995)

the discharge of water and sediment, channel geometry, and hydraulic roughness are uniform. In this ideal situation, stream power and resisting power are constant but may not be equal. This condition defines a straight **energy grade line**. A landform that evolved during uniform flow conditions will parallel the energy grade line and thus will have a straight longitudinal profile. The longitudinal profile of an alluvial surface created during equilibrium or aggradation conditions is preserved when the mode of operation switches to degradation.

Vincent surveyed such stream terraces and surfaces of incised alluvial fans. Subsequent vertical displacement of the abandoned fluvial surface by a normal fault results in two sections of the longitudinal profile above and below the rupture zone that originally would have had virtually the same gradient. By projecting both undeformed profile sections into the rupture zone, one can estimate vertical offset as the vertical separation of parallel lines. Field examples of faulted longitudinal profiles are used here to illustrate ideal, and not so ideal, situations and to clarify methods of measuring vertical displacement.

The straight narrow grassy floodplain of perennial Willow Creek (Fig. 3.28) has a longitudinal profile that is almost ideal for a tectonic geomorphology analysis. The surface-rupture zone for the normal fault is about 100 m wide, and is bounded by a main

scarp and an antithetic thrust fault. The floodplain slopes 0.035 m/m upstream from the fault zone, and 0.034 m/m in the downstream reach. The tectonically deformed reach within the rupture zone is steep because a block of alluvium was tilted downstream during the 1983 earthquake. The mean gradient of the tilted block, including the scarp height, is 0.052. Three interpretations are possible for the 700-m long survey transect.
1) The two segments of the pre-1983 Borah Peak earthquake stream profile are statistically the same, and the essentially constant gradient reflects uniformly interacting streamflow parameters for the 700 m reach prior to the 1983 surface rupture.
2) The apparent decrease in gradient downstream from 0.035 to 0.034 is real at the 95% confidence level and reflects the pre-earthquake downstream decrease in gradient typical of most streams.
3) Minor tectonic tilting of the blocks upstream or downstream from the fault zone occurred as a result of the 1983 earthquake.

Vincent assumed that either the first or second interpretation applies, which allows us to assess the magnitude of the apparent throw in the 0 to 100 m reach. Only 1 m is expressed as a fault scarp, with the remainder masked by complex deformation in the rupture zone. The straight profile segments above and below the fault zone are projected through the fault zone in order to estimate the total vertical displacement. Vertical separation of profile projections is 2.4 m at the main scarp, 2.3 m in the middle of the fault zone, and 2.2 m at the antithetic thrust. The estimate of total vertical displacement can be defined in one of two ways: as the average of the two values on the upslope and downslope sides of the fault zone, or as measured at the center of the fault zone. The extreme values are used to assign an uncertainty for the precision of the estimate. The total 1983 vertical displacement by the Thousand Springs fault at the Willow Creek site was 2.3 m ± 0.1 m.

The longitudinal profile of a latest Pleistocene incised fan surface is not as ideal as the Willow Creek floodplain, but it is good enough to estimate apparent total vertical displacement by two surface-rupture events. This late Pleistocene aggradation surface is eleven times steeper than the nearby floodplain, and changes in gradient occur even within a distance of 100 m. The rupture zone consists of a graben bounded by a main scarp and an antithetic scarp, both with normal fault displacements. The situation is favorable for three reasons. The fault zone is narrow, be-

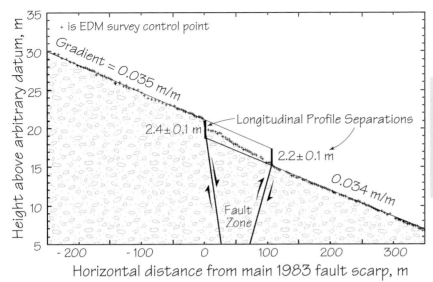

Figure 3.28 Longitudinal profile of the floodplain of Willow Creek, which was faulted during the 1983 Borah Peak earthquake. Figure 5.2 of Vincent (1995)

ing only 30 m wide. Topographic profiles on both sides of the rupture zone are well-constrained straight lines. Fan gradients upstream and downstream from the fault are similar: the footwall-block gradient is 0.41 m/m (22.3° slope), and the downstream hanging-wall block gradient is 0.39 m/m (21.3° slope). The profiles of the two straight fan segments are shown projected through the rupture zone in Figure 3.29. Apparent total vertical displacement resulting from the two surface rupture events is 2.4 m ± 0.2 m at the main scarp and 1.8 m ± 0.2 m at the antithetic scarp. The ± 0.2 m uncertainty is only 12%, which is good for a fan surface that has been faulted twice and has undergone dissection by consequent streams since ~15 ka. Vertical offset is about 2.1 m ± 0.5 m. Unfortunately the apparent throw is much less than true vertical displacement, because of the steep slope of this landform (discussed below).

In summary, Kirk Vincent's method of estimating apparent vertical displacements seems to provide maximum values at the main scarp and minimum values at the antithetic scarp. These extremes provide relative estimates of uncertainty as well as mean values. Measuring the vertical separation of straight-profile projections at the center of a rupture zone is the best measure of apparent throw. Vincent's method can be applied to the occasional case of a curving longitudinal profile, where the only valid estimate of apparent throw is at the center of the fault zone.

In order to complete our discussion of his approach we need to discuss why the estimates of throw discussed so far are only apparent. Let us examine a correction procedure and its application.

The vertical separation of the faulted topographic profiles of the two alluvial geomorphic surfaces (Figs. 3. 28 and 3.29) is an apparent throw because normal faulting of these fans causes both vertical and horizontal tectonic displacements of the land surface (Wallace, 1980). Apparent throw is less than true vertical displacement where a topographic profile is inclined in the direction of dip. Corrections generally can be made to obtain true vertical components of fault motion where longitudinal profile azimuths are within 15° of being perpendicular to fault strikes. Corrections are minor to moderate, but are large where fan slope approaches the fault-plane dip.

Two end members are unlikely, but approximations are sufficiently common to warrant consideration. First, imagine a 20° planar hillslope with a rock slab that is resting on a slide plane with a 20° dip (hillslope and failure plane are parallel). A vertical rift develops and the rock slab slides downslope. Projection of the longitudinal profiles of the two straight hillslope profile segments would suggest that no vertical displacement had occurred, even though the slab moved to a lower altitude. The slab would be extended horizontally and lowered in altitude by the same amount. The second case is the opposite extreme. If either the fault-plane dip is 90° or the land-surface

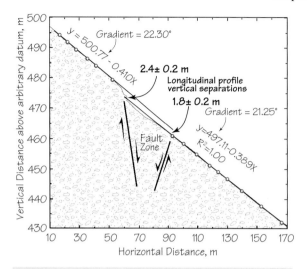

Figure 3.29 Longitudinal profile of a steep alluvial fan that has been faulted twice: during the 1983 Borah Peak earthquake and by an ~ 11 ka earthquake. Figure 5.3 of Vincent (1995).

slope is 0°, apparent throw of topographic profiles would be the same as the true vertical displacements. Both cases are rare.

The general case may be described by an equation and by a field of percentage correction factors illustrated by Figure 3.30, or an equation from Vincent (1995, p. 99). If V_t is true vertical displacement, V_a the apparent vertical displacement, α the dip of the fault plane in degrees, and β the land-surface slope, then

$$V_t = \frac{V_a [(sin\,\alpha)\,(sin(90°+\beta)]}{sin\,(\alpha–\beta)} \qquad (3.1)$$

Figure 3.30 shows graphical percentage correction factors to be added to apparent throw values. Typical Basin and Range Province settings have fault dips of 30° to 60° and fan slopes of <10° that result in minimal, or at least manageable, corrections to obtain true vertical displacements. Fault dip is rarely known with certainty, so fluvial landforms with gradients of < 3° (< 0.05 m/m) offer the best chance for accurate estimates of vertical displacements. Situations where corrections exceed 100% should be avoided.

Consider three faulted fans. If a 3° stream profile was apparently displaced vertically 1.00 m by a 50° fault, the Figure 3.30 correction for profile displacement would be less than 5%. The estimate for true vertical displacement would be 1.05 m. For an alluvial fan sloping less than 5°, apparent throw would be increased to 1.10 m. The steep alluvial fan of Figure 3.29 had a pre-surface rupture slope of about 21.6° and at that location the Lost River Fault dips about 50°. The apparent throw of 2.1 m should be increased by 50% to 3.05 m. The vertical displacements at the edges of the surface-rupture zone that were used to calculate the uncertainty should be increased proportionately. The true vertical displacement of the faulted alluvial fan of Figure 3.29 is 3.0 m ± 0.4 m.

A much clearer picture of the segmentation characteristics of the Lost River fault zone emerges when corrections are made to obtain true vertical displacements of the faulted alluvial surfaces surveyed by Kirk Vincent (Fig. 3.31). Although the 50° dip

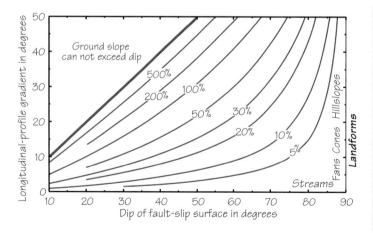

Figure 3.30 Percentage corrections of apparent vertical tectonic displacements (throw) of topographic profiles of alluvial landforms based on equation 3.1, knowing longitudinal profile gradient and fault dip. This chart provides estimates of true vertical components of slip on normal faults. The ground surface gradient has a strong influence on the correction factor where the dip of the slip surface is 30° to 70°, but is < 50% for fluvial landforms with < 10° gradients. From Wallace (1980).

of the range-bounding fault is uniform, fan gradients increase progressively from Willow Creek to Elkhorn Creek. So progressively larger corrections to apparent throw are necessary. Part of the Thousand Springs segment is shown from about 14 to about 5 km on the horizontal distance scale. The Elkhorn Creek segment boundary extends from about 5 to 4 km. The Willow Creek site (Fig. 3.28) is located at about 14 km, and Elkhorn Creek is located at about 3 km. All the surveyed piedmont landforms have been faulted twice, during the 1983 Borah Peak earthquake and during a prehistorical earthquake. Apparent throw (Fig. 3.31A) appears to decrease gradually to the southeast along the entire 16 km of the fault under consideration. In contrast, the spatial pattern of true vertical displacements of longitudinal profiles remains constant at about 4.5 m for the 7 km and then decreases abruptly to less than 1 m in the segment boundary (Fig. 3.31B). A similar procedure for all estimates of throw on the Thousand Springs segment (Fig. 3.32) shows the magnitudes of the corrections needed and reveals highly variable amounts of true displacements.

In summary, apparent throws of longitudinal profiles of alluvial landforms provide nicely constrained estimates of fault displacement. Corrections for the effects of land-surface slope relative to fault-plane dip on true values of vertical displacements are worthwhile if slopes are steep and/or if fault dips are shallow. Both of the postglacial earthquakes on the Thousand Springs segment were large in magnitude and had nearly identical displacement patterns. Cumulative vertical displacement is more than 4 m at the center of the segment, but is only 1 m in the Elkhorn Creek segment boundary. Ruptures entering the northern margin of the segment boundary divide and displacement magnitude decreases dramatically.

Up to this point our discussion has concentrated on field methods and analytical techniques. Care in selecting appropriate and diverse field sites, precise surveying of longitudinal profiles of alluvial landforms, and correction of apparent throw values are all needed to obtain reliable values of true vertical displacement.

Vincent's accurate dataset allows us to better understand fault behavior in the Elkhorn Creek

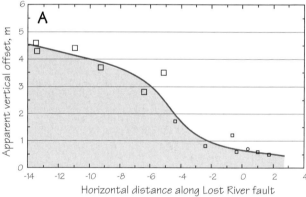

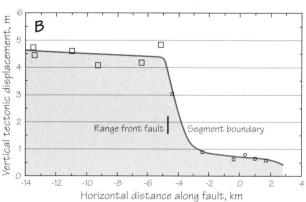

Figure 3.31 Cumulative displacements caused by two earthquakes, the 1983 Borah Peak earthquake and the ~ 11 ka earthquake, in the transition zone between the Thousand Springs segment and Elkhorn Creek segment boundary.

A. Throw (apparent vertical displacements of longitudinal profiles along fault trace).

B. True vertical displacements where the apparent throw data have been corrected for the dip-slope effects using the Figure 3.30 chart. Figure 6 of Vincent and Bull (1990).

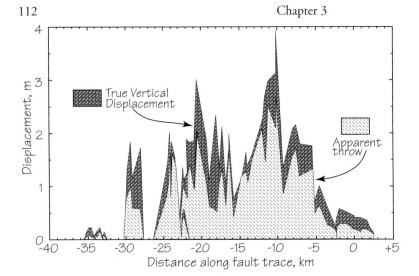

Figure 3.32 Comparison of throw and true vertical displacement for the 1983 Borah Peak, Idaho, earthquake. Figure 5.9 of Vincent (1995).

segment boundary, and to test the fault segmentation models as proposed by Fraser et al. (1964), Sieh (1981), and Schwartz and Coppersmith (1984).

Deciphering sequences of small and large surface ruptures is the basis for evaluating earthquake hazards, rupture initiation and termination, and evolution of mountain ranges. Several segmentation models have been proposed. Surface ruptures of the Mw magnitude 7.5 Hebgen Lake earthquake of 1959 in Montana provided Fraser et al. (1964) data for a model that assumes 50% overlap. Ruptures generated on adjacent fault segments overlap completely. The result over the long term is uniform cumulative slip. Their model accommodates discrete surface-rupture events on range-bounding fault segments, and uniform mountain-range uplift over geologic time spans (uniform range-crest altitudes). The model of Sieh (1981) also accommodates uniform slip that results in uniform range-crest altitudes. It uses major earthquakes close to the maximal size for a given fault segment. Schwartz and Coppersmith (1984) proposed a model of segmentation where similar size maximal earthquakes may recur on a segment, and that smaller earthquakes are not the norm. Rupture overlap between adjacent segments is minor. The result over the long term is nonuniform cumulative slip.

A key input to fault-segmentation models is "how much do ruptures on adjacent segments overlap?" Vincent was able to answer this question in Idaho because surface ruptures are well preserved for four earthquakes on two adjacent fault segments (Fig. 3.33). Both segments have ruptured independently twice since the latest Pleistocene aggradation event,

thus the interactions of four surface-rupture events can be investigated. The Thousand Springs fault segment in the northern portion of his study area ruptured in 1983 and in the early to middle Holocene. The southern portion of his study area is the Mackay fault segment, which ruptured once just after the end of the latest glacial interval and once after 7 ka (Vincent and Bull, 1990).

The central portions of both fault segments are characterized by steep, straight, mountain fronts and recent faulting has occurred near the mountain–piedmont junction. The region separating the two segments is a rupture asperity referred to as the Elkhorn Creek fault segment boundary. The strike of the range-front bends within the 10-km long segment boundary forming a spur of hills, which rise in steps up to the highest peak in Idaho, Borah Peak.

Both post-glacial earthquakes on the Mackay segment were large in magnitude with nearly identical displacement patterns. Ruptures on the Mackay segment penetrated the segment boundary, but details are lacking. Surface ruptures from the Thousand Springs and Mackay segments appear to overlap as much as 6 km in the segment boundary.

The conspicuous cumulative slip low in the Elkhorn Creek segment boundary, which contains Borah Peak, suggests difficulty in locating late Quaternary displacements. Careful mapping by Vincent shows that the trace of the 1983 surface rupture is 14% longer than thought by Crone et al. (1987). He found 10 km of fault scarps and ground cracks within the segment boundary that are distributed over 15 km². They range in altitude from 2000 to almost

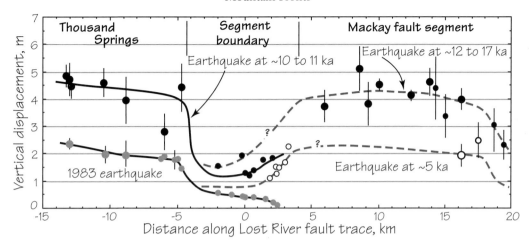

Figure 3.33 Calendric ages and patterns of true vertical displacements for four late Quaternary earthquakes along the Thousand Springs and Mackay fault segments and the intervening Elkhorn Creek segment boundary. Points for the Thousand Springs and Mackay segments are surface ruptures of the range-bounding faults, and points within the segment boundary are the sum of corrected throw for both range-bounding and many internal faults. Black circles are used for faulted landforms that are late glacial in age. Open circles are used for Holocene surfaces faulted at about 5 ka. Gray circles are used for surficial displacements for the 1983 Borah Peak earthquake. Large circles are used where surveys were made of large landforms with an electronic distance meter, small circles for displacements of small stream terraces estimated by eye using Suunto and hand levels. Figure 6.8 of Vincent (1995)

3000 m, and are as much as 2.5 km from the range-bounding fault. One scarp has a vertical displacement of almost 1 m, but most are less than 60 cm; 10 cm scarps and open ground cracks are common. The 1983 ruptures penetrate at least 80% of the way through the fault segment boundary. Extensive ruptures that have faulted and shattered the bedrock of the segment boundary define a much different tectonic style than the range-bounding faulting that characterizes the Thousand Springs and Mackay segments. The apparent slip deficit of about 2 m during post-glacial times in the segment boundary may represent uplift that occurred as distributed tectonic deformation as opposed to discreet and measurable slip on faults.

Which fault-segmentation model best fits the Lost River fault zone? Key aspects include:
1) Earthquake magnitudes appear to be close to the maximum possible, considering the lengths of fault segments bounding the Lost River Range.
2) Displacement patterns are nearly identical for consecutive earthquakes on both the Thousand Springs and Mackay segments.

3) Overlap of surface ruptures from the two segments that is 6 km or less.

The characteristic earthquake model developed by Schwartz and Coppersmith (1984) best describes the recent behavior of the Lost River fault.

3.3.5 Strike-Slip Fault Surface Ruptures

This brief section compares the Idaho study with similar studies of slip variation along two strike-slip faults in the Mojave Desert of California.

By using new approaches to paleoseismology Vincent was able to assess the variability of true vertical displacement along the Thousand Springs segment of the Lost River Fault. Normal-fault slip ranged greatly from 0.5 to 3 m over a distance of 5 km (Fig. 3.31). A histogram of the 38 measurements (Fig. 3.34A) shows slip amounts clustering about peaks at 1.3 and 2.1 m. An investigator finding such results for a fault zone with no known historical earthquakes might erroneously conclude that the two peaks record two prehistorical events of roughly the same magnitude.

About 85 km of the Mojave Desert was ruptured by a right-lateral slip event during the magnitude Mw 7.3 Landers earthquake of 1992 (Sieh et al., 1993). McGill and Rubin (1999) measured many offset stream channels and vehicle tracks along 5.6 km of the central Emerson fault (Rubin and Sieh, 1997). About 60 measurements were made across the main fault zone, where right-lateral slip ranged from 1.5 to 5.3 m. Locally, variations of slip as large as 1.5 m occurred in only 30 m (Fig. 7 of McGill and Rubin, 1999). The unusual density of offset features in their study area confirmed earlier reports of variable

surface slip along strike-slip faults during earthquakes (Toksoz et al., 1977; Sharp, 1982; Thatcher and Lisokowski, 1987). Some distributions of horizontal slip along strike-slip faults are nicely unimodal, one example being the surface ruptures resulting from the Hector Mine earthquake of 1999 (Scientists, 2000, Fig. 3). The strongly bimodal distribution of right-lateral offsets along the Emerson fault (Fig. 3.34B) may be due to:

1) Distributed shear and warping in unconsolidated basin fill that is not recognized when measuring an offset.

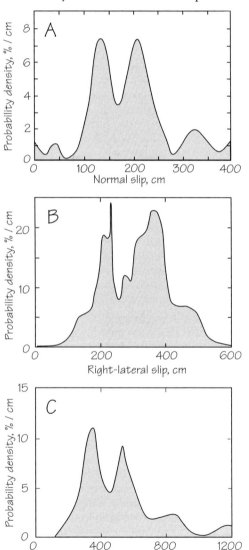

Figure 3.34 Probability density plots constructed as summations of Gaussians representing individual measurements of offset by normal and strike-slip faults. The area under the curve between any two abscissa values describes the relative abundance of geomorphic features offset by that range of slip.

A. Strongly bimodal plot of normal fault offsets along Thousand Springs segment of the Lost River, Idaho fault shows that slip varied greatly during the 1983 Borah Peak earthquake. The means of the two statistically significant component peaks of this distribution model best as 130 and 210 cm. Data from Figure 3.26.

B. Strongly bimodal plot of right-lateral offsets along the central Emerson fault, central Mojave Desert of California, shows that slip varied greatly along the surface rupture of the 1992 Landers earthquake. From Figure 10 of McGill and Rubin (1999).

C. Strongly bimodal plot of prehistorical left-lateral offsets along the Garlock fault, northern Mojave Desert of California, suggests that two prehistorical earthquakes occurred, each capable of displacing stream channels by 2 to 3 m. From McGill and Sieh (1991).

2) Bends and offsets that change the style of faulting and cause local uplift and subsidence along strike-slip faults.

3) Part of the slip occurring outside of the main trace, or on subsidiary faults. McGill and Rubin made 200 measurements across secondary fault traces up to 1.7 km from the main fault. Sums of right-lateral slip on these secondary faults were as much as 1.1 m.

4) Replication tests indicate moderately large uncertainties for some measurements.

In any case, it seems unlikely that fault slip varied this much at depths of more than 1 km.

McGill and Sieh (1991) measured prehistorical left-lateral offsets of stream channels along the Garlock fault (Fig. 3.34C). The strongly bimodal nature of the plot suggests that the most recent earthquake had a left lateral slip of 2 to 3 m, and that some landforms were also offset approximately the same amount by the penultimate earthquake. This assumes that variation in slip amounts for each earthquake is described by a unimodal distribution in a probability density plot. Of course McGill recognized that conclusions of two characteristic prehistorical earthquakes on the Garlock fault were not supported by her later analysis of the 1992 Landers surface rupture. Less than 10% of the Garlock fault displacement measurements were 10 to 18 m, but displacements of this size should be considered as being the result of two or three earthquakes.

Surface rupture offsets of landforms are not a flawless way to estimate slip on a fault plane. Vertical displacement of a stream terrace tread or horizontal displacement of a small stream channel is obvious and is nice because we can include the uncertainties of the field measurement in the estimate. But the spatial variations of apparent slip along a normal and strike-slip faults zone (Fig. 3.34A, B) show that caution is needed. Displacements along thrust fault zones are even more variable (Fig. 1.8). Suppose you have measurements for two offset landforms for a fault zone that you are studying. The resulting quandary is that you do not know if a Figure 3.31 type of distribution is present, and where your two data points would fall. Also, surface-slip measurements may not be representative of the magnitude or range of fault slip at depth.

The geomorphic procedure to decipher slip distributions for prehistorical earthquakes that are not unimodal (Fig. 3.34C) is to assess the relative ages of the faulted alluvial surfaces. By using a soils

chronosequence for Holocene and late Pleistocene faulted surfaces (Bull, 1991, 1996b, 2000), one does not need expensive and time-consuming surface-exposure dating methods. It is best to have at least one alluvial geomorphic surface formed between each earthquake. Situations of insufficient geomorphic surfaces are analogous to stratigraphic sites where earthquakes occur during times of nondeposition. I conclude that both geomorphic and stratigraphic studies may be needed to assess the completeness of an earthquake record.

Alluvial surfaces along a given fault zone may be roughly synchronous because adjacent drainage basins tend to have similar aggradation-event response times to climate-change perturbations, and because faulting may create a base-level fall that causes synchronous stream-channel downcutting and preservation of terrace treads.

Soils are climate-controlled low temperature geochemical systems, so one can expect major differences in both strength and character of soils developed under glacial and interglacial climates (Bull, 1991). Soils on a 2 ka surface may have only incipient development but should contrast nicely with nearby weakly developed 4 ka soils (Bull, 1996b). Without soils input, the paleoseismologist assumes that approximate multiples of the most recent slip event (such as 3, 6.5, and 9 m) represent repeating characteristic earthquakes (Wallace, 1968; and many subsequent workers). The paleoseismologist with soils information groups her or his slip measurements according to their relative ages in a chronosequence of faulted alluvial surfaces. Then one can proceed to determine if the slip events are characteristic or not.

3.4 Summary

Chapter 3 discussions focused on mountain fronts because that is where paleoseismologists should concentrate their assessments of earthquake hazard potential. Geomorphic evolution of mountain fronts is part of the much broader subject of hillslope evolution whose erosion is strongly affected by cumulative base-level fall.

Lofty, rugged mountain fronts result from prolonged uplift along either thrust or normal faults, and the fresh or degraded triangular facets at the range fronts are surprisingly similar. Straths form reaches upstream from either style of active range-bounding fault when climatic controls favor lateral beveling after tectonically induced downcutting returns the longitudinal profile to the base level of erosion between mountain-front uplift events. Renewed valley-floor incision creates strath terraces, and heights of straths above the active channel indicate the amounts of cumulative tectonic base-level change between times of strath formation.

Streams that cross a belt of active thrust faults flow through alternating embayment and gorge reaches. Streamflow quickly erases recent surface ruptures in valley floor alluvium, but sub-alluvial fault scarps remain until removed by the longer-term process of valley floor downcutting. Seismic refraction surveys can assess alluvial thicknesses beneath stream channels to determine which thrust faults are active. Stream gradient (*SL*) indices are markedly different for trunk stream channels with different rates of uplift along their range-bounding faults.

Both thrust and normal styles of faulting create mountain front fault scarps. Holocene scarps may not be as obvious for thrust faults because overthrusting of alluvial materials leads to immediate collapse and slumping. Multiple thrust fault splays create complex scarps whose apparent offset and shapes are tricky to analyze.

Erosion of a new thrust fault scarp can locally bury and terminate soil-profile formation on a stream-terrace tread. The elapsed time since rupture of the buried terrace tread is the stratigraphic age minus the time needed to form the post-depositional terrace tread soil profile. Using only ages of stratigraphic layers, as determined by radiocarbon or other methods, will overestimate the age of a surface rupture by an amount equal to the time needed to form the soil profile. A multiple-rupture event fault scarp at the Loma Alta site on the southern front of the San Gabriel Mountains provided unequivocal evidence for increased levels of earthquake hazard for the adjacent Los Angeles metropolis.

Scarps of normal faults undergo a more systematic retreat of the surface-rupture generated free face (Chapter 5), and small grabens characterize minor faulting of the hanging-wall block. Apparent throw amounts are best estimated by projecting planar alluvial surfaces across each fault zone.

Piedmont deposition is different for these two styles of faulting. Thrust faults are prone to migrate, thus creating a sequence of active mountain fronts during the late Quaternary. The newest structural block is the piedmont foreland, or "foreberg". Paleoseismology of piedmont forelands is challenging to the tectonic geomorphologist, but these low subtle scarps are where the action is. It may take ~100 ky of gradual transfer of slip from an internal front out to the new range-bounding fault, which eventually becomes a prominent tectonic landform.

Thicknesses of basin-fill alluvial fans are much different for active thrust and normal faults, and are a function of net base-level fall. Thrust faulting raises the mountains above the piedmont but this base-level change is reduced by the amount of concurrent stream-channel downcutting. Little piedmont deposition will occur if rapid stream-channel downcutting occurs after uplift along a range-bounding thrust fault. Thrust-fault migration further reduces the potential for piedmont deposition, and the newest fault-bounded block generally is tilted upstream. In contrast, much of the basin subsidence associated with normal fault displacements is available to quickly accumulate impressive thicknesses of basin fill. This may exceed half of the vertical tectonic displacement at the mountain front where tectonically induced stream-channel downcutting absorbs part of the base-level change. The proportion of base-level change that induces basin-fill accumulation is greatest where high rock mass strength slows the rate of channel downcutting at the mountain front.

The premise of the segmentation model of fault behavior is that each section of a range-bounding fault tends to rupture repeatedly in similar-size (characteristic) events that approximate the maximum possible magnitude earthquake. The characteristic earthquake model (Schwartz and Coppersmith, 1984) best describes the behavior of the Lost River fault in Idaho. Kirk Vincent's study is the first field verification of this conceptual model. His corrections to amounts of apparent throw generally are minor to moderate, but are large where fan slopes are steep. His accurate dataset of variation of true throw on the range-bounding faults also explains apparent anomalous behavior in the boundary between two segments of the Lost River fault. The segment boundary is characterized by broadly distributed faulting which contrasts with simple range-front faults of both adjacent fault segments.

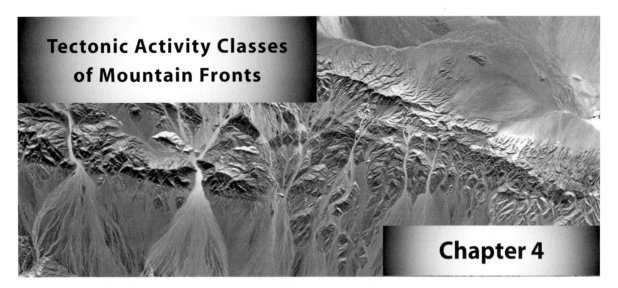

Tectonic Activity Classes of Mountain Fronts

Chapter 4

The broad boundary between the Pacific and North American plates has domains of mountain fronts with different styles and rates of tectonic deformation. Mapping the geomorphic tectonic-activity characteristics of mountain fronts in a region is valuable for understanding regional tectonics or assessing earthquake hazards.

Chapter 4 describes how vertical tectonic perturbations result in distinctive mountain-front characteristics. Different rates of base-level fall influence landforms as small as valley floors. Diagnostic tectonic landforms are described and used to define five classes of relative uplift. Regional maps depicting relative tectonic activity of associated groups of mountain fronts are a useful reconnaissance tool for tectonic geomorphologists and paleoseismologists.

4.1 Tectonic Setting of the North America–Pacific Plate Boundary

The onshore part of the San Andreas transform boundary between the North America and Pacific plates in the southwestern United States is a 200–800 km wide transition zone extending from the Pacific Ocean into the Basin and Range Province. Two features have controlled many secondary tectonic structures of the transition zone. The San Andreas fault presently is the primary plate-boundary fault zone. This right-lateral continental transform fault slices through batholithic complexes to create the Peninsular and Transverse Ranges of southern California,

passes through the Coast Ranges, and turns out to sea at Cape Mendocino in northern California to join the Mendocino fracture zone (Fig. 4.1).

The Sierra Nevada microplate is equally important. The 650 km long Sierra Nevada was created by batholithic intrusions in the Mesozoic. The mountain range is huge but the microplate is immense because it also includes the adjacent Central Valley of California (Fig. 4.1). This tectonic block has undergone minimal internal deformation but the eastern side of the Sierra Nevada was raised recently in dramatic fashion. An impressive escarpment rises 1,000 m in the north and 2,000 m in the south. The microplate was much larger just 5 My ago. Its batholithic rocks extend east into the Basin and Range Province whose extensional style of faulting continues to encroach into the eastern side of the microplate. A major tectonic event – detachment of the Sierra Nevada batholithic root – referred to as *delamination* affected other tectonic elements in much of the broad transform boundary.

We should note apparent coincidences of timing of many important Neogene tectonic events in the boundary between the Pacific and North American plates. The Cretaceous batholith had a thick residual root, but an Airy-type crustal root is no longer present to support the Sierra Nevada. Xeno-

Tectonically active and inactive mountain fronts of the northeastern Mojave Desert, California. From U.S. Air Force U2 photo supplied courtesy of Malcolm Clark, U.S. Geological Survey.

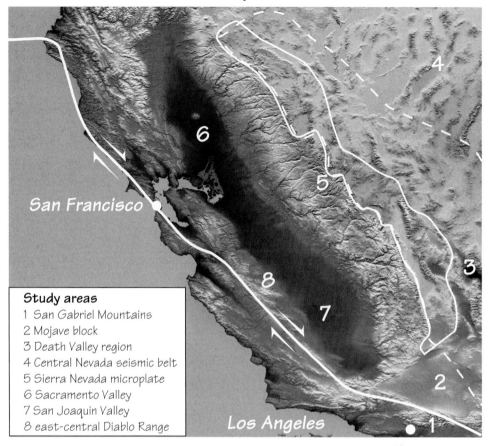

Study areas
1 San Gabriel Mountains
2 Mojave block
3 Death Valley region
4 Central Nevada seismic belt
5 Sierra Nevada microplate
6 Sacramento Valley
7 San Joaquin Valley
8 east-central Diablo Range

Figure 4.1 Tectonic setting of southwestern United States. The Sierra Nevada microplate was wider at 5 Ma because it extended west almost to the San Andreas fault, and extended east into the Basin and Range Province. Area within the solid line outlines the area of accelerated extensional faulting since 5 Ma (Jones et al., 2004, Figure 1). This area and the area inside the dashed line approximate the province of the Walker Lane–Eastern California shear zone, that has accommodated ~10% of the plate-boundary dextral shear since 5 Ma.

lith composition changes (Ducea and Saleeby, 1996, 1998) indicate that a former dense root beneath the Sierra Nevada crest was convectively removed before 3 Ma.

Mantle lithosphere is now abnormally thin beneath the Sierra Nevada and Panamint Ranges (Jones et al., 1994; Wernicke et al., 1996; Ruppert et al., 1998; Saleeby et al., 2003; Saleeby, and Foster, 2004; Zandt, et al.2004). The batholithic remnants seem to be largely supported by a buoyant upwelling of hot asthenosphere. Zandt (2003) refers to this flow as "mantle wind" that shifted this part of the detached Sierra Nevada batholithic root – a mantle drip – to the southwest.

The detached batholithic root sank rapidly to the base of the athenosphere and it is the cold downwelling trail of the drip that is revealed by seismic tomography (Zandt, 2003). It appears that southwest directed mantle flow shifted the post-4 Ma Sierra Nevada mantle drip so it is now beneath the eastern San Joaquin Valley and the foothills of the Sierra Nevada (Zandt, 2003).

Xenolith studies suggest recent removal of garnet bearing rocks triggered a brief (3.5 ± 0.25 Ma) pulse of potassium-rich basaltic vulcanism in a 200 km diameter circular area in the central Sierra Nevada (Manley et al., 2000; Farmer et al., 2002; Jones, et al.2004). This event indicates the most likely timing and location of the main delamination event. Zandt (2003) points out that the highest peaks of the Sierra Nevada occur in the area where the delamination event began; the maximum strength of the geophysical perturbation coincides with maximum uplift. I like their interpretation that recent uplift of the Sierra Nevada also could be this young, and that both volcanism and uplift have a common delamination triggering mechanism.

The interval between 5 Ma and 3 Ma fits a model of regional synchronous changes in many plate-boundary tectonic processes, and of landscapes that record changes in tectonic base-level controls. Recognition of the lack of a crustal root beneath the now lofty Sierra Nevada has led to much creative thinking about tectonics of the region.

The brilliant synthesis by Jones et al. (2004) recognizes the late Pliocene foundering of the Sierra Nevada crustal root as a most important perturbation that changed regional plate tectonics. They reason that this crustal delamination event increased the total gravitational potential energy of the lithosphere (Jones et al., 1996), thus increasing both extensional strain rates in the western Basin and Range Province and the altitude of the eastern Sierra Nevada. They also conclude "an increase in extensional displacement rates must be accommodated by a decrease in rates of extension or an increase in rates of shortening somewhere in the vicinity of the Sierra Nevada" (p. 1411). The timing of recent thrust faulting that created the Coast Ranges bordering the Central Valley coincides nicely. And, "Lithospheric removal may also be responsible for shifting of the distribution of transform slip from the San Andreas Fault system to the Eastern California shear zone" (p. 1408).

The tectonics and landscape dynamics of the southwestern corner of the Basin and Range province were much different after introduction of large amounts of right-lateral slip (Dokka and Travis, 1900a, b; Burchfiel et al., 1995; Burchfiel and Stewart, 1996; Lee et al., 2001). Important fault zones now may run through the valleys as well as along rugged mountain fronts. The more prominent strike-slip faults are shown in Figure 4.2.

Cenozoic tectonism of western North America is related to the Mendocino plate boundary triple junction, which migrated northward creating the San Andreas – a continental right-lateral transform fault system (Atwater, 1970; Atwater and Stock, 1998) that presently includes the Maacama and Bartlett Springs faults. They conclude that no discernible change in rates of motion of the Pacific–North American plate boundary has occurred since 8 Ma. This is a key assumption of the Jones et al. model where horizontal velocities across the Sierra before and after the delamination event match the boundary condition of Pacific–North American plate motion. Local acceleration of extensional encroachment in a belt east of the Sierra Nevada may match increased rates of Coast Ranges shortening to the west. The Sierra Nevada microplate has become narrower as a result of both processes.

Several tectonic events have a timing that is coincident with or shortly following delamination. Crustal extension accelerated into the eastern margin of the Sierra Nevada, thus making the Basin and Range Province ever broader. See Figure 4.1. Part of the San Andreas fault style of dextral shear split off to create the seemingly diffuse Walker Lane–Eastern California shear zone in eastern California and western Nevada.

Compression near the San Andreas fault created a fold-and-thrust belt that is encroaching northeastward into the microplate. The result of this increased rate of crustal shortening has been a regional reversal of sediment-transport direction, with thick Diablo Range alluvial-fan deposits being laid down over basin fill derived from the Sierra Nevada. The present eastern edge of the Coast Ranges was not formed by a synchronous pulse of uplift, as was the eastern front of the Sierra Nevada. A southeast to northwest migration of thrust-faulted mountain fronts profoundly influenced landscape evolution and has continued to the present.

4.2 Appraisal of Regional Mountain Front Tectonic Activity

4.2.1 Geomorphic Tools For Describing Relative Uplift Rates

The above brief summary of Neogene plate-boundary tectonics suggests an intriguing variety of mountain fronts to be studied by tectonic geomorphologists. These include Basin and Range Province normal faulting, Mojave Desert transtensional faulting,

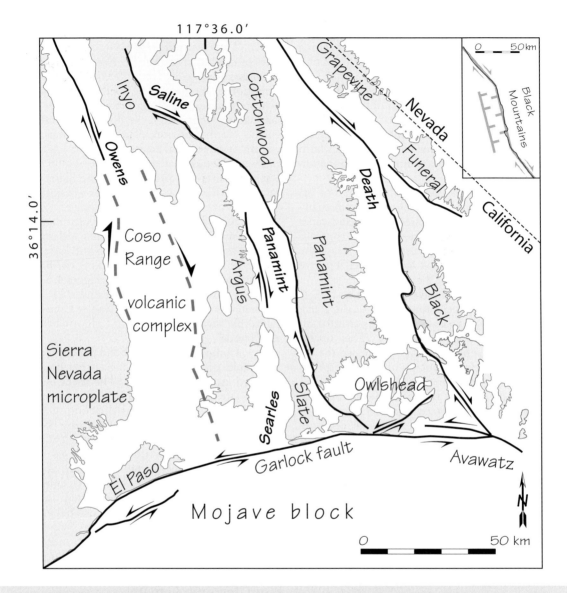

Figure 4.2 Important Neogene strike-slip faulting in the southwest corner of the Basin and Range Province. Mountains are gray areas and intervening valleys are named. The Garlock and associated sinestral faults provide necessary block rotation to impart north–south continuity of the Walker Lane–Eastern California shear zone (Dickinson, 1996). The Death Valley fault system has a releasing right step that results in rapidly subsiding pull-apart basin (upper-right inset map). Owens Valley fault is here included with the broad shear couple proposed by Monastero et al., (2005) who consider the resulting Coso Range volcanic complex to be a nascent metamorphic core complex. The sharp releasing bend in Saline Valley is responsible for the impressive mountain front shown in Figure 2.20A. Panamint Valley has active strike-slip faults on both sides of the valley.

Transverse Ranges thrust faulting, and folding in the margins of the Coast Ranges.

Long escarpments consist of a sequence of mountain fronts. Adjacent fronts with contrasting topography, structure, or rock type may represent fault-zone segments with different styles, rate, and magnitudes of displacement. I use landform assemblages to define and map classes of relative uplift for the mountain fronts in the broad region east of the San Andreas fault. Mountain-front tectonic activity ranges from vigorous to dormant. Dormant landscapes occur next to rapidly rising mountain fronts.

Much has changed since Bull and McFadden introduced this approach to tectonic geomorphology in 1977. Their simple introduction to landscape responses to mountain-front faulting took advantage of obvious differences between rugged mountain fronts that could be designated as belonging in two active and one inactive class of tectonic activity. We now recognize the presence and importance of the Walker Lane–Eastern California shear zone. Strike-slip faulting may create fault scarps that are only a few meters high that commonly are buried by Holocene alluvial-fan deposits. Such highly active "mountain fronts" pose high risk from a paleoseismic perspective and are central in the assessments presented here.

A single working model cannot encompass all styles and sizes of tectonic deformation without becoming unwieldy. The approach here, as previously, will be to emphasize large tectonic displacements. The relevant time span is the Pleistocene because magnitudes of Holocene tectonic deformation do little to change the character of a mountain front.

I emphasize landforms that vary greatly with different rates of tectonic base-level fall. This approach defines broad classes of surface-uplift rates and de-emphasizes the influences of complicating factors. Both erosional and depositional landforms of fluvial systems reflect rates of bedrock uplift as affected by geomorphic processes (Fig. 1.4). Slow erosion preserves tectonic landforms longer in arid than in humid terrains, which can make mountain ranges of arid regions appear to be rising faster than their humid region counterparts. Abrupt local tectonic base-level fall associated with range-bounding faults results in more dramatic landscape contrasts than uplift associated with broad wavelength folding. So, the faulted landscapes of the Walker Lane–Eastern California shear zone are used to introduce and illustrate how to define tectonic activity classes. These mountain fronts range from lofty to low scarps.

We start with numerical descriptions of key tectonic landforms. Then interactions of local base-level processes are used to define classes of relative tectonic activity that are based on definitive landform assemblages. The output consists of maps depicting areal variations of Quaternary uplift of mountain fronts that can be used by earth scientists and engineers. Paleoseismologists use regional assessment of mountain front tectonic activity in reconnaissance investigations. The emphasis is not always focused on structurally complex and fascinating zones of rapid tectonic deformation. Nuclear-power generating plants should be located in the most tectonically inactive sites.

Obvious contrasts in the morphology of landforms in different tectonic settings may be described by simple ratios of topographic lengths and heights. Ratios provide dimensionless numerical indices for describing the states of tectonically induced downcutting of valley floors, and erosional retreat of hillslopes that are useful for defining classes of tectonic activity. Landscape ratios vary in their sensitivity to describe the influences of late Quaternary tectonic deformation. Those that emphasize changes in valley-floor width are good for describing tectonic uplift of the past 10 to 100 ky in arid and semiarid regions and 1 to 30 ky in extremely humid mountains. Parameters that emphasize changes in relief are good for describing drainage-basin evolution over time spans of more than 500 ky in the semiarid American West. Pedimentation operates on an even longer time scale. A kilometer of retreat of a mountain front generally requires more than 1,000 ky. Two ratios have been consistently useful for identification of tectonically active and inactive mountain fronts. These are the sinuosity of the mountain–piedmont junction and the valley floor width–valley height ratio.

Frankel and Pazzaglia (2005, 2006) prefer to examine overall landscape characteristics using digital elevation models that assess the average depths of valleys expressed as a drainage basin volume–drainage basin area ratio. Such metrics and gradients of first-order stream channels (Merritts and Vincent, 1989) do indeed describe significant differences between tectonically active and inactive landscapes. They may be especially attractive for inspection of watersheds deformed by active folding with large wavelengths.

However, the use of simple ratios (Bull and McFadden, 1977) may still be superior for studies of faulted mountain fronts. Tectonic base-level fall emanates from the mountain–piedmont junction so a case can be made to emphasize landform metrics in the immediate vicinity of the tectonic perturbation, including triangular facets, valley floor width–valley height ratios, and mountain–piedmont junction sinuosities. Upstream propagation rates for range-front base-level falls might be best explored through studies of the longitudinal profiles of trunk stream channels that are the connecting link between the mountain-front tectonic perturbation and slowly evolving streams and hills of headwater reaches.

4.2.1.1 Mountain-Front Sinuosity

The straight or gently curving nature of most faults or folds allows evaluation of the degree of erosional modification of a structural landform. Rapid uplift along a range-bounding fault maintains the linear nature of the front. Erosion dominates landscape evolution after cessation of uplift and creates a sinuous mountain–piedmont junction, especially where lithologic resistance to erosion is weak. Intermediate scenarios involve the interplay of ongoing uplift and continuing fluvial degradation, which varies greatly with climatic setting.

Streamflow becomes the dominant process shaping the mountain-front landscape in tectonically quiescent settings. Streams downcut quickly to their base level of erosion by removing small amounts of rock, and then slowly widen their valley floors by removing large amounts of detritus derived from hillslopes. Maximal concentrations of stream power at canyon mouths result in erosional embayments that extend up the larger valleys. The result is a highly sinuous mountain–piedmont junction. Relatively slow uplift may be continuing but fluvial erosion is dominant over uplift. Sections of a mountain front between the principal watersheds are eroded at a much slower rate, which minimizes development of embayments.

The sinuosity index used here (Fig. 4.3) is measured in the same way as for meandering rivers. *Sinuosity* of the mountain–piedmont junction, J, is the ratio of the planimetric length of the topographic junction between the mountains and the adjacent piedmont, L_j, to the length of the range-bounding geologic structure, L_s.

$$J = \frac{L_j}{L_s} \qquad (4.1)$$

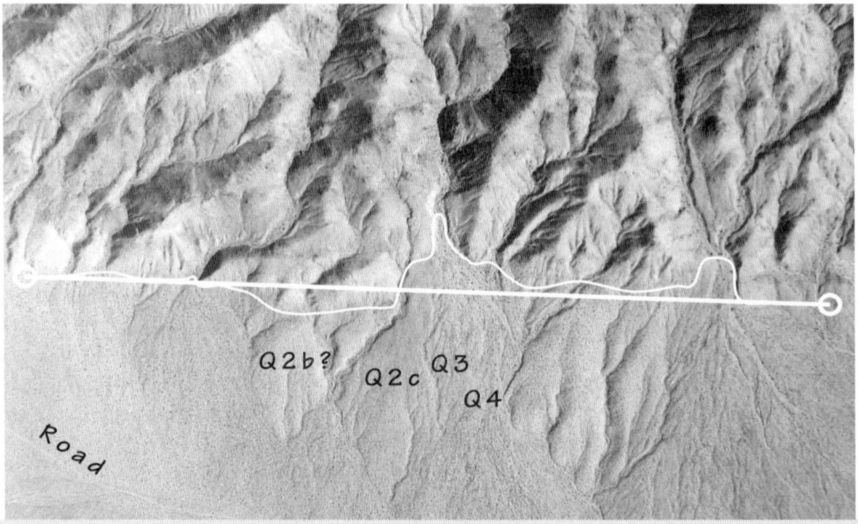

Figure 4.3 Sinuosity of mountain–piedmont junctions.
A. Tectonically active mountain front associated with an oblique right-lateral fault in the central Mojave Desert of California. The length of the thick, straight white line L_s, is the length of the range-bounding fault. The thin, sinuous white line, L_j, is along the mountain–piedmont junction. Its sinuosity records embayments at mouths of watersheds, and minor departures from fault-zone linearity. Single-lane dirt track for scale. Tectonically translocated fan aggradation event surfaces have been offset in a right-lateral sense from their source watershed. Ages of surfaces are summarized in Table 1.2.

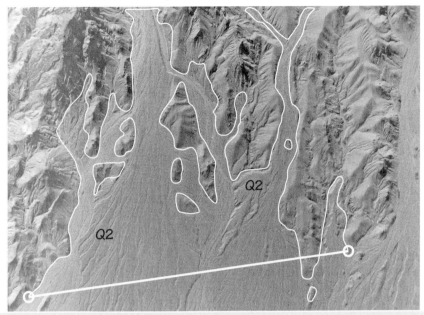

Figure 4.3 Sinuosity of mountain–piedmont junctions.
B. Tectonically inactive mountain front. Advanced stage of erosion leaves few clues as to the actual location of the now dormant range-bounding fault, so endpoints are arbitrary. The thin, sinuous white line, L_j, includes distances around detached parts of the mountain front that are now inselbergs.

J commonly is less than 3, and approaches the minimum value of 1.0 where steep mountains are being raised rapidly along a range-bounding fault or fold. Sinuosity can also be expressed as departures from a value of 1.0, or as departures from mean values for a specific tectonic province.

Consistent procedures should be used when dealing with three potential complications. Wide alluvial valley floors of large watersheds may extend far upstream. To trace the mountain–piedmont junction most of the distance to the watershed divide would produce an unnecessarily large value of J. When tracing such mountain–piedmont junctions, the operator should go only a constant distance up such valleys, such as 0.5 or 1.0 km, or a basin-position coordinate of 0.95, before crossing to the other side of the valley floor and returning to the outer mountain front. Doing this emphasizes the mountain front instead of the sides of anomalously large valleys.

The straight-line distance between two points on the mountain–piedmont junction should be used as the reference distance where the location of the range-bounding fault is not known. Parts of former spur ridges become isolated from the range as

valley floors widen and the mountain front retreats. These erosional outliers are termed *inselbergs*. For pedimented landscapes, the mountain–piedmont junction, L_j, is measured as the sum of the perimeters around inselbergs plus the length of the mountain–piedmont junction (Fig. 4.3B).

Images may be better than maps for measuring range-front sinuosities. The piedmont is smooth in the Figure 4.3A image, except for two areas of dissected terrain. Measurements made on a topographic map would incorrectly suggest that these are part of the overall erosional terrain and should be included with the mountains. Such remnants of old alluvial-fan deposits may have considerable relief but are not included in the measurements because they are basin fill, not mountains. Inclusion of such basin fill would raise the sinuosity value and place this front in a class of relatively less tectonic activity.

Alluvial fanheads extend far up the embayments shown in Figure 4.3B. Mountain-front length focuses on the bedrock-alluvium contact, including remnants that are now isolated from the mountains. Once again, smooth remnants of Q2 age dissected alluvial fans are not included as part of the mountain–

piedmont junction. This distinction can be made on this image, but not on topographic maps.

A variety of topographic information sources can be used to map mountain–piedmont junctions. Replication measurements are precise when the mountain–piedmont junction is viewed on images larger than 1:60,000. Landsat is borderline, but SRTM radar images can be ideal (Fig. 4.3C).

Radar images focus on landscape roughness and piedmonts adjacent to a mountain front commonly show as smooth, nearly featureless plains. Not having the clutter of towns and fields makes radar images especially nice for studying mountain fronts.

Topographic maps with a scale of ~1:25,000 and a 5 to 10-m contour interval yield good results, but 1:250,000 scale maps with 30 m contour intervals are inadequate for estimating the location of the mountain–piedmont junction. Meticulous plane-table surveys of old surveys may provide excellent maps for geomorphic evaluations of mountain–piedmont junctions. The Figure 4.3D example uses a 1.5 m contour interval that provides detailed information. Four domains of contour characteristics are obvious, but domain 4 is where the contour interval changes from 5 feet to 25 feet (1.52 to 7.62 m). The smooth depositional surfaces of the alluvial fans (domain 1) contrast nicely with the erosional topography of domain 2. Domain 3 is also underlain by tectonically deformed Pleistocene unconsolidated deposits but has much more local relief. The sinuous nature of

the internal mountain front suggests the presence of a second thrust fault. The range-bounding front may also record thrust faulting but is not nearly as sinuous because it is in part a function of deposition of adjacent piedmont alluvium.

Different uplift rates (for a given climate and lithology) have distinctive ranges of sinuosity. Sinuosities of highly active mountain fronts generally range from 1.0 to 1.5, moderately active fronts range from 1.5 to 3, and inactive fronts from 3 to more than 10. Sinuosities greater than 3 describe highly embayed fronts. The range-bounding fault along which the mountains were raised may be more than 1 km downslope from the mountain–piedmont junction (Section 4.2.2). Edges of folded mountain ranges whose strata strike parallel to the front may retreat with minimal embayment, and have anomalously low sinuosity values of J. Changes in mountain front sinuosity typically require long time spans, because geomorphic processes change hillslopes fairly slowly. Nonetheless the index complements another index that changes at a faster pace because concentration stream power in valley floors can accelerate local landscape evolution.

4.2.1.2 Widths of Valleys

Another sensitive index to recent and ongoing uplift is the valley floor width–valley height ratio, or for brevity the $V_f ratio$ (Fig. 4.4A). If V_{fw} is the

Figure 4.3 Sinuosity of mountain–piedmont junctions. C. Radar image of the southern end of the North Island of New Zealand. Mountain front of the active Wellington fault (1) has a sinuosity of 1.0. Rimutaka Range front–Wairarapa fault (2) has a sinuosity of 1.1 and ruptured in the magnitude Mw 8.2 earthquake of 1855.
This figure is from NASA/JPL/NIMA image PIA02742 from the Shuttle Radar Topography Mission.

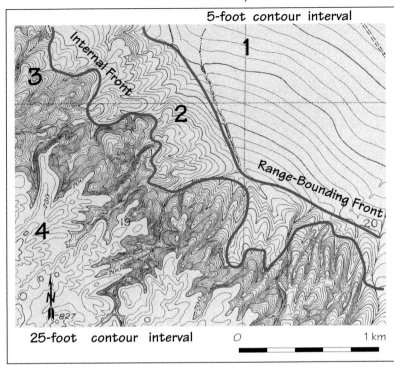

5-foot contour interval

25-foot contour interval   O   1 km

Figure 4.3 Sinuosity of mountain–piedmont junctions.

D. Example of use of detailed topographic mapping for study of a mountain–piedmont junction. From 1926 plane-table mapping of the Levis Quadrangle on the folded eastern margin of the central California Coast Ranges. Contour interval of 5 feet (1.5 m) changes to 25 feet in the southwest corner of the map.

width of a valley floor and A_{ld}, A_{rd}, and A_{sc} are the altitudes of the left and right divides (looking downstream) and the altitude of the stream channel, then

$$V_f = \frac{V_{fw}}{\dfrac{(A_{ld} - A_{sc}) + (A_{rd} - A_{sc})}{2}} \qquad (4.2)$$

Valley-floor widths increase with watershed size, erodibility of rock type, and with decrease of uplift rate. Valley heights decrease with the passage of time after cessation of uplift, but not nearly as fast as valleys widen. The valley floor width–valley height ratio is especially sensitive to late Quaternary tectonic base-level falls because narrowing of a valley floor is accomplished quickly by the downcutting action of streams. Bull and McFadden (1977) found significant differences (at the 0.99 confidence level) in the means of V_f ratios of tectonically active and inactive mountain fronts.

Care is needed in selecting sites to measure valley floor width–valley height ratios. V_f values are more likely to be representative of the relative degree of tectonic base-level fall if determined in similar rock types and at the same basin-position coordinate for a suite of similar size drainage basins along a given mountain front. Rock resistance to erosion may not

change, but annual unit stream power to do the work of erosion increases downstream. Consequently, part of the variation in valley-floor width is a function of drainage-basin size. A limited range of drainage-basin sizes is preferred because stream discharge increases in a nonlinear manner in the downstream direction, particularly for streams of humid regions (Wolman and Gerson, 1978). Measuring valley floor widths can be tricky. Ideally they should be the mean value of

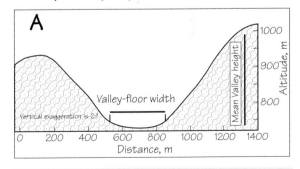

Figure 4.4 Topographic cross sections that illustrate valley-morphology definitions. Starvation Canyon, Panamint Range of California. A. Valley floor width–valley height ratio, V_f.

several measurements made in the field. High quality topographic maps succeed fairly well. Digital sources give erratic results if poor resolution integrates valley floor and footslope relief, thus failing to detect abrupt margins common to many valley floors.

Determine V_f ratios at basin-position coordinates of 0.9, or at a constant distance upstream from the mountain–piedmont junction (for example 1 km). Avoid a basin-position coordinate of 1.0, especially if it coincides with intensely sheared bedrock of a range-bounding fault. This local decrease of rock mass strength is a complication that should be minimized. The mouths of canyons also are the most likely sites for climate-change induced aggradation to greatly widen a valley floor that may be subject to continuing degradation in upstream reaches. One should void mixing data from degrading reaches with reaches whose valley-floor width records episodes of aggradation.

The statistically different widths of valley floors for tectonically active and inactive mountain ranges record much different stream-channel processes. The much larger V_f ratios of tectonically inactive watersheds record widening of bedrock valley floors (type 1 dynamic equilibrium of Section 2.4.3). Such watersheds also have pervasive fill stream terraces created by climate-change induced aggradation events (Bull, 1991). The edges of these valley floors are where valley-floor alluvium meets the footslope.

Valley shape landforms can be selected for a particular time span of interest. The V_f ratio can be used to detect rapid uplift during the Holocene in humid regions and uplift since the middle to late Pleistocene in arid regions. Longer response times are required to change the cross-sectional dimensions of a valley. Valley cross-section ratios (Mayer, 1986), and mountain front sinuosity, would be appropriate for semiarid regions studies of late Cenozoic tectonics and topography.

Mayer (1986) uses a numerical index of valley morphology, V_c, that compares the area of a valley cross section, A_v, with the area of a semicircle whose radius is equal to the height of the lowest adjacent drainage divide, A_C (Fig. 4.4B):

$$V_c = \frac{A_v}{A_C} \qquad (4.3)$$

The **Vc index** is intermediate in sensitivity for describing the effects of tectonically induced downcutting compared to the V_f and V ratios. Ratios of valley width, V, to valley height – for brevity, the **V ratio** – may be useful for describing long-term decreases of relief for various lithologies and climates.

$$V = \frac{2V}{(A_{ld} - A_{sc}) + (A_{rd} - A_{sc})} \qquad (4.4)$$

The V ratio can be used to discriminate between highly active and inactive mountain fronts but is not as useful for distinguishing intermediate classes of tectonic activity. V ratios of 4 to 6 are representative of rising mountains. Inactive mountain fronts generally have values of V greater than 7.

V_f ratios are more sensitive to fluvial baselevel controls than V ratios because changes in the widths of valley floors occur much more rapidly than hillslope relief is reduced. Valleys may be incised in less than 10 ky in response to a base-level fall. Valley-floor narrowing typically is more than the concurrent increase of valley cross-section relief. Then valleyfloor width may double in 100 ky, but >1 My may be needed to reduce valley height by half at the same position in a drainage basin. Highly active mountain fronts commonly have V_f ratios between 0.5 and 0.05. Strath terraces are absent in such narrow canyons if downcutting does not pause (times of attainment of the base level of erosion). The miniscule widths of such valley floors, relative to the adjacent watershed ridgecrest heights, result in Vf ratios that clearly define situations where tectonically induced downcutting maintains the mode of stream operation far to the degradational side to the threshold of critical power.

Moderately to slightly active valleys have flights of strath terraces that record pauses in streamchannel downcutting. Inactive valleys may lack

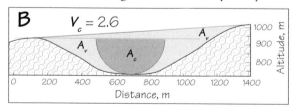

Figure 4.4 Topographic cross sections that illustrate valley-morphology definitions. Starvation Canyon, Panamint Range of California. B. V_c ratio. A_v is the cross-sectional area of the valley. A_C is the area of a semicircle whose radius is equal to the height of the lowest adjacent drainage divide.

strath terraces, but only if they remain at the base level of erosion. The presence of a strath terrace requires not only beveling during a period of attainment of the base level of erosion but also subsequent renewed degradation resulting from climate change or response to long-term isostatic uplift. Isostatic uplift resulting from millions of years of denudation of a tectonically inactive mountain range may eventually favor an episode of renewed stream-channel downcutting that leaves pediments and associated valley floors as strath terraces. Such infrequent episodes of strath-terrace formation probably coincide with climate-change induced changes of bedload transport rate and stream discharge that shift the mode of operation to the degradational side of the threshold of critical power.

4.2.1.3 Triangular Facets

Daunting challenges face paleoseismologists when asked to assess earthquake hazards of vast regions. "Please identify every surface rupture younger than 10 ky in 50,000 km^2 of the Basin and Range Province". "Tell me which mountain ranges are rising faster than 0.1 m/ky". Seemingly not possible but these requests are actually quite straightforward for tectonic geomorphologists. The key to efficiently doing such work is to use simple techniques that quickly discern obvious differences between informative landscape parameters.

Holocene surface-rupture identification is easy. Pleistocene–Holocene climatic changes affected geomorphic processes profoundly in the American southwest, causing piedmont aggradation events (Bull, 1991, Chapter 2). Pedogenic processes and the resulting soils were as different as the global contrast between ice-age and present climates. So the solution is simply "Does the surface rupture disrupt an alluvial fan or stream terrace that has a Holocene soil profile?" Soils geomorphology provides the answer without expensive, lengthy laboratory analyses or modeling. In California the Alquist-Priolo Earthquake Fault Zoning Act (Hart and Bryant, 1997) defines an active fault as one that has "had surface displacement within Holocene time" and no structure for human occupancy is permitted within 50 feet of an active fault.

Heights and stages of dissection of triangular facets (Section 3.2.1, Figure 3.5, Table 3.1) are indicative of relative tectonic activity (Bull and McFadden, 1977). Basal sections of triangular facets may resem-

ble degraded fault planes (Hamblin, 1976; Menges, 1990b; Ellis et al., 1999). Obvious landscape contrasts in the Great Basin of west-central Nevada were used by dePolo and Anderson (2000) to estimate slip rates for hundreds of normal faults. Rapidly rising mountain fronts have 1) fault scarps on the piedmont and at the mountain–piedmont junction and 2) high triangular facets. Tectonically inactive mountain fronts have neither.

Their emphasis was on normal faults because tectonic increase of relief provides the potential energy to erode fault blocks into drainage basins whose characteristics reflect the rates and magnitudes of vertical displacement. Active plate tectonics extensional processes create distinctive fluvial systems.

Pure strike-slip faulting is locally important but does little to change hydraulic gradients of streams. Instead strike-slip faulting disrupts and tears apart the fluvial systems created by base-level fall.

The dePolo and Anderson dataset of 45 mountain fronts provided three classes of estimates of normal-fault slip rates (dePolo, 1998). Type 3 mountain fronts have slip rates of ~0.001 m/ky, which is so slow that these landscapes lack fault scarps and triangular facets. Type 1 mountain fronts have both fault scarps and triangular facets with minimal dissection that rise more than 30 m above the mountain–piedmont junction. All but one Type 1 mountain front has a Holocene surface rupture. Type 2 normal faults

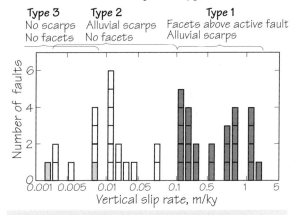

Figure 4.5 Comparison of vertical slip rates for normal faults associated with three types of landscape. Dark boxes are Type 1 faults, white boxes are Type 2 faults, and the two light gray boxes are Type 3 faults. Figure 3 of dePolo and Anderson (2000).

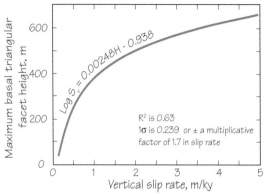

Figure 4.6 General relation between maximum height of basal triangular facets and vertical slip rate. Figure 6 of dePolo and Anderson (2000). S_v is vertical slip rate (m/ky) and H is maximum basal facet height in meters.

have alluvial fault scarps, but lack active basal sets of facets. Relict triangular facets may be present.

Comparison of these simple geomorphic characteristics with known normal fault slip rates was internally consistent (Fig. 4.5). The dePolo–Anderson method highlights Type 1 landscapes with triangular facets indicative of those range-bounding faults most likely to rupture next. They focused on the heights of active triangular facets, which have undergone minimal dissection in the arid Great Basin of west-central Nevada. Triangular facet height for the most rapidly rising Type 1 mountain fronts increases systematically with increasing rate of uplift (Fig. 4.6).

dePolo and Anderson assessed the relative tectonic activity of normal faults for much of the state of Nevada. The most rapid normal fault slip rates occur in the Walker Lane–Eastern California shear zone, perhaps because of recent encroachment of Basin-and-Range extension into the Sierra Nevada microplate, and/or because of local pull-apart basins created by lateral fault displacements.

4.2.2 Diagnostic Landscape Classes of Relative

Tectonic Activity

Local **base-level processes** change streambed altitudes relative to adjacent reaches. Four base-level processes occur at mountain fronts of arid and humid regions. These are the dependent variables of channel

downcutting (cd) in the mountains, aggradation (pa) or degradation (pd) of the piedmont downslope from the escarpment, and the independent variable of uplift (u) of the mountains relative to the adjacent basin. Local base-level processes affect erosion of valley floors and hillslopes in mountains, and deposition and erosion that create piedmont landscapes. Geomorphic processes on alluvial fans and pediments are discussed here in the context of base-level changes and tectonic activity classes (Bull, 1984).

The affects of rapid uplift of mountains relative to an adjacent basin result in a unique landform assemblage in either arid or humid regions. Consider the accumulation of thick alluvial-fan deposits adjacent to mountains where streams are flowing on bedrock (Figs. 4.7A, 4.8A). The stream channel will tend to become entrenched into the fanhead as valley floor degradation continues to lower the stream channel in the mountains. The resulting downstream shifts of the threshold-intersection point are promoted by either channel downcutting in the mountains and/or by fan aggradation on the piedmont. Uplift of the mountains along a range-bounding fault or fold counteracts this tendency to entrench the fanhead. Tectonic elevation promotes sustained channel downcutting in the mountains and piedmont fan deposition, but only when the uplift rate equals or exceeds the sum of the two local base-level processes that tend to cause fanhead trenching.

$$\frac{\Delta u}{\Delta t} < \frac{\Delta cd}{\Delta t} > \frac{\Delta pd}{\Delta t} \qquad (4.5)$$

A tectonically active landscape assemblage may be defined in terms of the three interacting base-level processes of equation 4.5. Unentrenched alluvial fans are present immediately downstream from mountain valleys that have only a veneer of alluvium on narrow bedrock floors.

These are base-level interrelations between geomorphic processes, so equation 4.5 is not generally meant to be used in situations of great thicknesses of basin fill. Basin subsidence below sea level is indeed part of mountain front tectonic deformation, but rarely can be related to present streamflow dynamics. Of course one can change the uplift term from rock uplift ($\Delta u/\Delta t$) to a measure of tectonic deformation, Σtd, in order to accommodate stratigraphic information and a greater spatial vertical scale of mountain-front deformation over longer time spans than usually intended by equation 4.5.

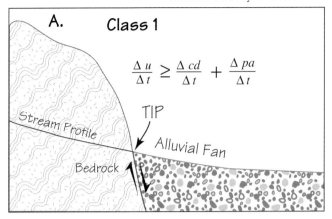

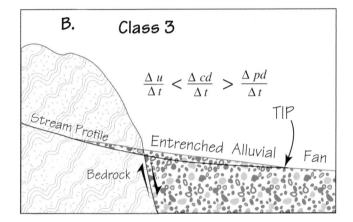

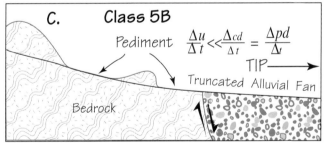

Figure 4.7 Longitudinal profiles of diagrammatic fluvial systems showing the landforms associated with different relative uplift rates. TIP is threshold-intersection point.

A. Alluvial-fan deposition continues next to the mountains where uplift is the dominant base-level process.

B. The apex of fan deposition shifts down fan where stream-channel downcutting is the dominant base-level process.

C. Erosion is the dominant base-level process in the mountains and on the piedmont in a tectonically inactive setting.

Landscape assemblages defined by equation 4.5 are characterized by straight mountain fronts that coincide with range-bounding faults or folds, triangular facets whose younger (basal) portions have undergone minimal dissection, and V-shaped cross-valley profiles with straight or convex footslopes. This suite of characteristic tectonic landforms defines a *class of relative uplift* (class 1) of the mountain front regardless of the prevailing climate and rock types, and with no need to determine the rate of uplift or the time at

which it began. Climate and erodibility of materials greatly affect rates of erosion (Fig. 2.19) and deposition but classes of relative tectonic activity are based on relative rates of uplift, erosion, and deposition.

A low-altitude panorama of the arid western front of the Panamint Range (Fig. 4.9) shows most geomorphic features indicative of rapid base-level fall. A typical abrupt switch between the erosional and depositional subsystems is present. This coincides with the range-bounding Panamint Valley fault

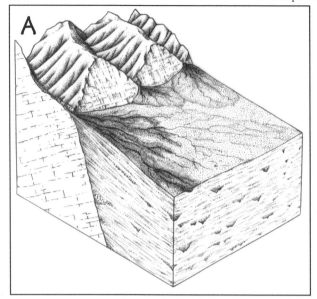

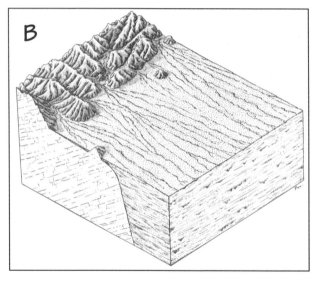

Figure 4.8 Block diagrams contrasting the landscape assemblages associated with active and inactive normal faults. Buried stream channels record episodes of climate-change induced stream-channel downcutting on an aggrading piedmont. Drawings by Bus Royce.

A. Class 1 landscape where rapid tectonic displacement rates are partitioned between creating space for continuing aggradation of alluvial-fan deposits and increasing relief of mountains with narrow valley floors and triangular facets along a straight mountain–piedmont junction.

B. Class 5 landscape where cessation of normal faulting allows fluvial systems to achieve a long-term base level of erosion recorded by a single surface composed of straths in mountain valley floors, beveled pediment surface with inselbergs rising above it, and truncated basin fill. Broad valley floors extend far upstream from a sinuous mountain–piedmont junction lacking triangular facets.

zone, which here is a fairly narrow high-angle oblique normal fault. Rapid deposition of gravel by braided distributary streams has created unentrenched fan surfaces of Holocene age. Base-level rise resulting from this piedmont aggradation has promoted alluvial deposition over the fault zone. Subsequent surface ruptures create piedmont fault scarps. Stream-channel downcutting induced by episodes of tectonic base-level fall preserves several ages of late Quaternary fan surfaces on the edge of the footwall block. Some of these remnants may be preserved as they are raised,

as if on an escalator, into the rising Panamint Range. A Saline Valley example is shown at the left side of Figure 2.20A just above the level of the waterfall.

The fine textured (closely-spaced) drainage density shown in Figure 4.9 suggests the presence of soft rocks. Cichanski (2000) mapped this part of the range flank as a 15° to 35° low-angle normal fault that became inactive when the range-bounding transtensional Panamint Valley fault zone became active. See the Figure 1.7 discussion about how changing from low- to high-angle faulting changed fan deposition.

Figure 4.9 Aerial view of a class 1A mountain front at Bighorn Canyon, west side of the Panamint Range, southeastern California. Landforms indicative of rapid uplift during the late Quaternary include the presence of highly elongate drainage basins, very low valley floor width–valley height ratios, a straight mountain front, stage 1 triangular facets, undissected alluvial fans, and multiple ages of alluvial surfaces separated by fault scarps of different heights. Four faulted fan surfaces range in age from late Pleistocene (1) to late Holocene (4). A remnant of late Pleistocene lake deposits and shoreline ridges is at lower right. Two-lane dirt road for scale at lower left. The width of this view is about 1.9 km.

Parts of the low-angle normal fault are capped with a gently dipping, loose to locally cemented, fanglomerate with a particle-size distribution similar to that of the modern alluvial fans. Highly fractured and altered mylonitic leucogranite orthogneiss below the fault plane is so soft that it is cut by closely spaced rills, much like the Inyo Range example shown in Figure 2.20A. Even these narrow spur ridges have triangular facets (see the left side of Figure 4.9). The relatively undissected triangular facets of the piedmont fault

scarps merely indicate that fan gravels have a much higher infiltration rate than the bedrock, so there is less runoff to erode rills.

Fluvial erosion varies along this uniformly rising escarpment. Stream power is proportional to drainage-basin area. So the drainage basin of Bighorn Canyon in the center of the Figure 4.9 panorama has a deeper valley and a larger alluvial fan than the smaller adjacent fluvial systems. Tectonic controls on canyon cutting are complicated here because Bighorn

Canyon was created in part by base-level falls caused by movements along a sequence of now inactive low-angle normal faults (Cichanski, 2000).

Valley floors upstream from the active range-bounding faults of most class 1 fronts are exceptionally narrow, and the adjacent spur ridges may be high where powerful streamflows have had sufficient time to increase valley relief. Class 1 drainage basins generally are quite narrow relative to their length.

Stream-channel entrenchment into the fanheads of class 1 fans, if present, is temporary and most likely is the result of climate-change perturbations. *Temporary* and *permanent* are easily defined by taking advantage of profound differences between Pleistocene and Holocene climates. Temporary fanhead trenches have formed since the most recent climate-change induced aggradation event. These fan surfaces have Holocene soil profiles. Conversely, the presence of a soil profile on alluvium deposited during the Pleistocene is indicative of permanent stream-channel entrenchment.

Post-glacial aggradation surfaces are younger than 13 ka in the deserts of the American Southwest. These soil profiles are less than 0.5 m thick and commonly lack argillic horizons. Argillic horizons of Late Pleistocene soils are red and the depths and amounts of pedogenic calcium carbonate illuviation are substantially greater than in Holocene soils.

Arroyo cutting and backfilling of basin fill occurs on a still younger time scale of 0.1 to 0.3 ka. Climatic change impacts, locally enhanced by humans, can result in temporary entrenchment (Bull, 1997) of fans composed of cohesive sandy deposits. Equation 4.5 best describes the appropriate set of interre-

lations where such temporary entrenchment has occurred downstream from a mountain landscape with class 1 erosional characteristics. Impacts of long-term and short-term climate changes are present, but the intent here is to classify landforms on the basis of tectonic controls instead of adjustments to changing climates.

Class 1 fronts in weak rocks or unconsolidated alluvium have a different landscape assemblage. Such mountain fronts lack the rock mass strength required for development of narrow, steep canyons in rugged mountains. Triangular facets are common along the front shown in Figure 4.10A, which has a sinuosity that would approach 1.0 (a straight line) were it not for the presence of prominent wide valleys. The larger streams have valley-floor widths that are much wider than their flood-discharge widths because they have been at the base level of erosion for a long time. Small streams, with less unit stream power, still have valley-floor widths similar to their flood-discharge widths. They continue to degrade at the times of infrequent large rainfall events in the Mojave Desert. Deposition of alluvial fans continues immediately downslope from the range-bounding fault. The broad mountain valleys suggest a slightly active (class 3) landscape but the straight mountain–piedmont junction and a piedmont of class 1 alluvial fans implies a much higher level of tectonic activity. The importance of lithologic controls on landscape evolution is acknowledged for the active fault zones shown in Figure 4.10 and such landscapes are assigned to class 1B. Use of class 1B recognizes the importance of rock mass strength on response times of geomorphic processes to tectonic uplift in this region. Class 1B is a convenient way to rank strike-slip faults slicing through the basin fill as belonging in the highest slip-rate class.

Eight different landscape assemblages are used to define the five tectonic activity classes of Table 4.1. Class 1 has been subdivided into A and B parts in order to accommodate the broad valley floors that are typical of noncohesive materials. No difference in uplift rates is implied by this subdivision of class

Figure 4.10 Aerial views of class 1B mountain fronts. A. Cenozoic alluvium and fractured, weathered quartz monzonite have been ruptured by the right lateral strike-slip Calico fault northwest of Hildago Mountain, central Mojave Desert, California. Dirt road in upper-left for scale.

Figure 4.10 Aerial views of class 1B mountain fronts. B. Faulted basin fill along the west side of the Panamint Range, southeastern, California. Large valleys have broad floors indicative of attainment of the base level of erosion, but small streams at the mountain front continue to actively downcut. Note the triangular facets in uplifted alluvium and low fault scarps of late Holocene age.

1. Classes 2, 3, and 4 have permanently entrenched alluvial fans; classification is based on erosional stages of the mountain–piedmont junction that require progressively more time to form. Examples include embayment of valleys to create a more sinuous mountain–piedmont junction, and erosional deterioration of triangular facets. Class 5 consists of landscape assemblages that describe three types of pediments, the piedmont landform indicative of prolonged tectonic inactivity.

The attainment of equilibrium stream channels, as indicated by the presence of straths and strath terraces, varies with tectonic activity class. Class 1A terrains of semiarid regions typically have valley-floor widths that are the same as the widths of peak stream discharges. Strath terraces are not present in such narrow canyons because downcutting is not interrupted by prolonged attainment of the base level of erosion. The miniscule widths of such valley floors, relative to the valley widths, result in Vf ratios that clearly define situations where tectonically induced downcutting maintains a mode of stream operation far to the degradational side to the threshold of critical power. Of course large streams of humid regions have sufficient annual unit stream power to remain at their base levels of erosion nearly all the time, so typically have strath terraces. Class 3 valleys of semiarid regions have strath terraces. Class 5 valleys may lack strath terraces because of insufficient uplift (very slow isostatic) to allow creation of more than one obvious base level of erosion – the beveled bedrock surface of pediments.

The "typical landforms" of Table 4.1 are but a sample of the many landscape responses to uplift. Examples are listed in Table 4.2. Class 1 landscapes are much different than class 5 landscapes with similar total relief, climate, rock type, and drainage-basin area. Class 1 landscapes have more convex ridgecrests, steeper footslopes that in extreme cases are a continuation of the convex ridgecrests, narrower and steeper valleys, less sinuous mountain fronts, predominately young soils on rapidly aggrading piedmonts, and thick accumulations of Quaternary basin fill. Shapes of valley cross sections are useful for identification of tectonic activity classes. Each valley has a shape that is a function of rates of tectonically induced downcutting, with short response times for changes in valley-floor width after a base-level fall, and long response times for ridgecrest relief and convexity. The independent variables of climate and rock type profoundly affect response times to uplift for each particular study area.

The five-part classification of Table 4.1 may be too detailed when there is insufficient time or funds for more than reconnaissance field work in large study regions. A three-part classification may be preferable: highly active class 1, grouping of classes 2, 3, and 4 into a moderately to slightly active second class, and grouping of classes 5A, 5B, and 5C into an inactive third class. Bull and McFadden (1977b) and Bull (1977) used this three-part classification to identify tectonically active mountain fronts and to determine spatial patterns of late Quaternary tectonic uplift in southeastern California.

Class of Relative Tectonic Activity	Relative Uplift Rate	Typical Landforms	
		Piedmont*	Mountain**
Active			
Class 1A - maximal	$\Delta u/\Delta t \geq \Delta cd/\Delta t + \Delta pa/\Delta t$	Unentrenched alluvial fan [0.6-0.9]	V-shaped valley profile in hard rock [1.1-1.4]
Class 1B - maximal	$\Delta u/\Delta t \geq \Delta cd/\Delta t + \Delta pa/\Delta t$	Unentrenched alluvial fan [0.6-0.9]	U-shaped profile in soft rock [1.0-1.2]
Class 2 - rapid	$\Delta u/\Delta t < \Delta cd/\Delta t > \Delta pd/\Delta t$	Entrenched alluvial fan [1.0-1.1]	V-shaped valley [1.1-1.3]
Class 3 - slow	$\Delta u/\Delta t < \Delta cd/\Delta t > \Delta pd/\Delta t$	Entrenched alluvial fan [1.1]	U-shaped valley [1.0-1.1]
Class 4 - minimal	$\Delta u/\Delta t < \Delta cd/\Delta t > \Delta pd/\Delta t$	Entrenched alluvial fan [1.1]	Embayed front [1.0-1.1]
Inactive			
Class 5A	$\Delta u/\Delta t << \Delta cd/\Delta t > \Delta pd/\Delta t$	Dissected pediment [1.1]	Dissected pediment embayment [1.0-1.1]
Class 5B	$\Delta u/\Delta t << \Delta cd/\Delta t = \Delta pd/\Delta t$	Undissected pediment [1.0]	Dissected pediment embayment [1.0]
Class 5C	$\Delta u/\Delta t << \Delta cd/\Delta t < \Delta pd/\Delta t$	Undissected pediment [1.1]	May be like class 1 landscapes

* Unentrenched — entire fanhead deposited recently, or only Holocene fan surfaces are entrenched. Entrenched alluvial fanhead surfaces with Pleistocene soils are entrenched.

** Stream power / Resisting power ratios in [ ] suggest departure from equilibrium value of 1.0.

Table 4.1 Geomorphic classification of Quaternary relative tectonic activity of mountain fronts. Uplift-rate equations are defined in the text. Classes 1A, 3, and 5B are illustrated in Figure 4.7, and applications of the classification to normal, reverse, and strike-slip faults, and folds of arid, thermic, semiarid, and subhumid terrains are shown in Figures 4.21, 4.24, 4.25B, 4.35, and 4.39.

Permanently entrenched heads of alluvial fans (Figs. 4.11, 4.12) indicate that stream-channel downcutting is the dominant local base-level process at the mountain–piedmont junction, relative to uplift and piedmont degradation.

$$\frac{\Delta u}{\Delta t} < \frac{\Delta cd}{\Delta t} > \frac{\Delta pd}{\Delta t} \qquad (4.6)$$

Equation 4.6 pertains to classes 2, 3, and 4 of relative tectonic activity (Table 4.1). The erosional landforms of mountain valleys and mountain–piedmont junction are used to differentiate between these three classes whose piedmont landforms consist of entrenched alluvial fans.

Uplift is relatively less rapid than for class 1 alluvial fans of the same study area. Uplift rates of landscapes described by equation 4.6 may be slow (<0.1 m/ky), where ephemeral streams flow over resistant rocks. Resistant fluvial systems respond slowly to tectonic perturbations, so tend to retain landscape signatures indicative of uplift. Conversely, similar landscape assemblages may be associated with rapid uplift (>2 m/ky) for humid-region streams flowing over soft rocks. These rivers have sufficient annual unit stream power to quickly adapt to tectonic base-level fall.

Stream-channel downcutting shifts the apex of active fan deposition downslope from the mountain–piedmont junction. Deposition can no longer

Class 1A	Badwater fan, Black Mountains, Death Valley National Monument, California; Deer Canyon, Lytle Creek, San Gabriel Mountains, Los Angeles County, California; Panoche Creek, Laguna Seca Creek, Diablo Range, Western Fresno County, California; Wassuk and Stillwater Ranges, of Western Nevada.
Class 1B	Faulted alluvium northwest of Hidalgo Mountain, central Mojave Desert, California; San Andreas fault, Carrizo Plains, California, Death Valley fault zone.
Class 2	West front of Grapevine Mountains and Desolation Canyon area, Death Valley National Monument, California; Eaton Canyon, San Gabriel Mountains, Los Angeles County, California; East front of Toiyabe Range, central Nevada.
Class 3	Hanuapah Canyon, Panamint Range, Death Valley National Monument, California; Little Panoche Creek, Diablo Range, western Fresno County, California.
Class 4	Wilson Canyon, Argus Range, Inyo County, California; 49 Palms Canyon, Joshua Tree National Monument, California.
Class 5A	South front of Whipple Mountains, eastern Mojave Desert, California; West front of Santa Catalina Mountains, Pima County, Arizona.
Class 5B	South front of the Granite Mountains, north front of the Coxcomb Mountains, eastern Mojave Desert, California.
Class 5C	East front Riverside Mountains, eastern Mojave Desert, California; Northeast front of Sheep Mountain, Gila Mountains, western Arizona.

Table 4.2 Mountain fronts whose landscape assemblages illustrate five classes (Table 4.1) of relative tectonic activity in the southwestern United States.

occur on the fanhead, and the strength of soil-profile development indicates how much time has elapsed since the fanhead last received deposits from the source watershed. About 50 to 500 ky may have passed since class 1 conditions prevailed. The strongly developed soil profiles on permanently entrenched fanheads also attest to minimal rates of piedmont degradation. Δpd generally is much smaller than Δu.

Streams of class 2 landscapes downcut sufficiently rapidly to maintain V-shaped cross-valley profiles upstream from the range-bounding fault, whose location generally coincides with the mountain–piedmont junction. Longitudinal stream profiles of such mountain reaches may indicate attainment of type 2 dynamic equilibrium (Section 2.4.3). Deep fanhead trenches, and the dark surficial pavements indicative of Pleistocene age alluvial surfaces (Fig. 4.11) indicate that fanhead trenching is permanent. Minimal

backwearing of the mountain–piedmont junction has occurred, but triangular facets generally are more dissected than class 1 escarpments.

Greater tectonically induced downcutting near the mouths of rising watersheds may result in a drainage basin that is progressively narrower toward the basin mouth. These are referred to as "hourglass" or "wineglass" valleys (Beaty, 1961), and may develop in either class 1 or class 2 fluvial systems. Pleasant Canyon in the Panamint Range is an example.

Class 3 landscapes are indicative of slow or intermittent uplift during the Quaternary. Holocene age fault scarps are rare. Some, like the Hanaupah fault cutting the ~55 ka alluvial fan surface near the toe of the Panamint Range piedmont, are related to adjacent pull-apart basin tectonics, not the now-dormant range-bounding fault. Valley floors are much wider than the floodplain widths, and the ubiquitous presence of U-shaped valleys (Fig. 4.4A) implies pro-

Figure 4.11 Aerial view of the class 2 mountain front along the west side of the Funeral Range, Death Valley National Park, southeastern California. Drainage basins are progressively more narrow downstream. The mountain–piedmont junction coincides with the trace of a normal fault except in fanhead embayments. The fanhead surface may have been created at the time of the Q2b, 125 ka aggradation event (Table 1.2). Two-lane paved highway near the base of the piedmont for scale.

longed attainment of type 1 dynamic equilibrium. Alternating periods of lateral erosion and renewed downcutting result in strath terraces being a typical landform of mountain reaches of class 3 terrains.

The eastern flank of the Panamint Range is an interesting class 3 mountain front. A sinuous moun-tain–piedmont junction (Fig. 4.2) is a short distance upslope from the range-bounding fault zone. Fault scarps cut the oldest Pleistocene alluvial fan. Stream-channel entrenchment has preserved the alluvial sur-faces of three climate-change induced Pleistocene ag-gradation events (Table 1.2). The combined affects

Figure 4.12 Hanaupah Canyon alluvial fan and class 3 mountain front, east side of Panamint Range, Death Valley, southeastern California. Poorly preserved triangular facets mark approximate location of range-bounding normal fault. The fanhead trench is 100 m deep at the fan apex. Approximate ages of fan aggradation event surfaces are: Q4, the active channels; Q3, 7–9ka; Q2c, 55 ka; Q2b, 125 ka; and Q2a, >500 ka.

of Panamint Range tilting continue to lower the base level of erosion for the trunk stream channel. It has been at least 500 to 700 ky since class 1 conditions prevailed, which is a much longer time span than the 125 ky estimate for the class 2 front shown in Figure 4.12. Straths and typical class 3 cross-valley topographic profiles are shown in Figure 4.4.

Class 4 mountain fronts border on being tectonically inactive. Indeed, some may have been inactive since the early Quaternary and their landforms merely reflect long response times to old tectonic perturbations. The numerous embayments of class 4 fronts create a sinuous mountain–piedmont junction that no longer coincides with the range-bounding fault. Wilson Canyon in the granitic Argus Range of southeastern California is an example (Fig. 4.13). Although bordered by steep hillslopes, the broad valley floor of this deep erosional embayment has a U-shaped cross-valley profile. Triangular facets are not obvious and have been notched by deep valleys that extend more than two-thirds of the distance to the apex of the facet (stage 5 of Table 3.1). The wide valley floor has fill terraces, which are clear evidence of temporary reversals of long-term valley-floor degradation caused by climate-change induced aggradation events.

Note the presence of an unentrenched alluvial fan downstream from entrenched valley fill (Fig. 4.13). It should not be considered as evidence for tectonically active (class 1) conditions, particularly in view of the stage of landscape erosion represented by the erosional landforms. Instead this fan is the result of major alterations in geomorphic processes during the change from Pleistocene to Holocene climate. Watersheds underlain by granitic rocks typically yield huge amounts of sediment in response to increases in runoff and decreases in protective plant cover (Bull, 1991, Section 3.3). Response times to this recent climatic perturbation in the Argus Range are 5 to 10 ky. So this fluvial system has yet to switch to a stream-channel entrenchment mode of operation after an aggradation event caused by the Pleistocene–Holocene climatic change.

Class 4 terrains may seem inactive until they are contrasted with class 5 terrains. Compare the granitic landscapes of Figures 4.13 and 4.15A. Class 4 fluvial systems still have well-integrated drainage networks and inselbergs are rare.

Diagnostic landforms of class 5 terrains consist of three types of pediments, each representing a different relative rate of stream-channel and pied-

Figure 4.13 Aerial view of the class 4 mountain front along the east side of the central Argus Range at Wilson Canyon, southeastern California. A highly embayed mountain–piedmont junction, stage 5 triangular facets, and U-shaped mountain valley cross sections all suggest virtual cessation of uplift along the range-bounding fault.

mont degradation. Degradation of the pediment (*pd*) may be less than, equal to, or exceed the rate of downcutting by stream channels (*cd*) flowing across the piedmont. Permanent (Pleistocene) and temporary (Holocene) channel entrenchment of pediments is defined in the same way as for alluvial fans:

$$\frac{\Delta u}{\Delta t} << \frac{\Delta cd}{\Delta t} > \frac{\Delta pd}{\Delta t} \qquad (4.7)$$

$$\frac{\Delta u}{\Delta t} << \frac{\Delta cd}{\Delta t} = \frac{\Delta pd}{\Delta t} \qquad (4.8)$$

$$\frac{\Delta u}{\Delta t} << \frac{\Delta cd}{\Delta t} < \frac{\Delta pd}{\Delta t} \qquad (4.9)$$

Tectonic displacements along mountain-front fault zones ceased long ago in pedimented terrains, or are undetectable relative to the rate of landscape denudation. Broad expanses of pedimented landscapes, such as those in Australia and Arizona, record attainment of the base level of erosion for millions of years (Fig. 4.8B).

Dissected pediments are common. Bedrock surfaces created prior to dissection may be exposed in the banks of permanently incised stream channels.

These pervasive landforms of tectonically inactive regions are the pediment terraces of Royse and Barsch (1971).

Pediments may be regarded as exceptionally broad strath terraces because this landform records attainment of type 1 dynamic equilibrium. As such, pediments can be used as a reference surface to assess isostatic uplift resulting from protracted denudation. Pediment strath terraces converge downstream reflecting decreasing rates of isostatic tilt caused by denudation where streams leave remnants of formerly extensive mountain ranges (Fig. 4.14A).

An erosional base-level fall, such as downcutting of the trunk stream channel to which the pediment drains, causes streams flowing across a pediment to incise. The effects of Quaternary climate change modulate times of incision and burial in much the same way as for strath terraces (see Section 2.6.2).

Aggradation events interrupt beveling of pediments. The piedmont south of the Whipple Mountains of southeastern California (Dickey et al., 1980) has a flight of well exposed dissected pediments beveled across the soft deposits of the 3 to 5.5 Ma Bouse Formation. Episodes of climatic alluvial-fan deposition have mantled successive pediment surfaces with 0.5 to 2 m thick sheets of gravel. Erosional base-level fall of the nearby Colorado River (Bull, 1991, Fig. 2.8) promotes long-term downcutting of tributary piedmont streams. The resulting dissected piedmont landscape is a flight of pediment levels, each representing attainment of the base level of erosion between the times of climate-change induced piedmont aggradation events. The oldest pediments occur as flat ridgecrests between piedmont valleys.

Piedmonts far from the Colorado River (Fig. 4.14B) alternate between modes of incision and aggradation. Aggradation prevails when hillslope sediment reservoirs are being stripped, and incision during climates that favor flash floods from bare rocky hillslopes.

Two types of pediments lack permanently incised stream channels: those described by equations 4.8 (class 5B) and 4.9 (class 5C). Rates of downcutting of trunk streams flowing from the mountains of class 5B landscape assemblages are the same as the long-term rates of pediment downwasting (Figs. 4.15A, B). Bedrock erosion alternates with alluvial mantling of the pediment. Incised channels are temporary, and the mode of operation for these piedmont landforms is the result of latest Quaternary climatic

Figure 4.14 Aerial views of dissected pediments of class 5A mountain fronts.
A. West side of the Tinajas Altas, northern Sonora, Mexico. Entrenched piedmont streams flow below alluvial surfaces with Pleistocene soil profiles. Small triangular facets are present but are deeply dissected. The steep slopes without colluvium are the product of the arid climate and occasional convective storm intense rainfalls. Two-lane highway from lower right to upper left for scale.
Photograph by Peter L. Kresan ©.

Figure 4.14 Aerial views of dissected pediments of class 5A mountain fronts.
B. Mohawk Mountains pediment, southwestern Arizona is subject to aggradation events caused by late Quaternary climate changes.

changes (Mabbutt, 1966). The mountain front does not coincide with the buried inactive normal fault in undissected pediments. This suggests a long-term stable base level, during which mountains are slowly replaced by encroaching pediment.

The highly embayed Granite Mountains (Fig. 4.15A) in the eastern Mojave Desert are a nice example of a class 5B pediment landscape assemblage. The formerly active range-bounding fault is more than 2 km to the left of the view shown. Integrated watershed hillslopes and drainage nets remain only in the higher parts of the range. At lower altitudes streams flow across a piedmont between numerous inselbergs of quartz monzonite. A veneer of alluvium deposited during the Pleistocene–Holocene climatic change aggradation event is being removed. This exposes the weathered surface of the pediment to renewed degradation. Such temporary mantling of granitic rocks by aggradation events creates a moist subsoil environment that accelerates weathering of the granitic rocks

and tends to remove surface irregularities created during times when streamflow removes weathered detritus (Wahrhaftig, 1965; Mabbutt, 1966).

Class 5C undissected pediments (Figure 4.16) are rare. They occur where rates of pediment degradation exceed stream-channel downcutting rates in the mountains. An example occurs along the east side of the Riverside Mountains in the lower Colorado River region of southeastern California. A fairly steep pediment formed in unconsolidated clayey sands of the Bouse Formation is being degraded faster than the valley floors in the adjacent mountains, which are underlain by more resistant schist and gneiss.

Class 5C mountain fronts have characteristics that might suggest the presence of a class 1 ter-

Figure 4.15 Aerial views of the undissected pediments of the class 5B mountain fronts.
A. South side of the Granite Mountains, eastern Mojave Desert. Pedimented bedrock is being exhumed at the left center and center foreground. Dirt road for scale.

Figure 4.15 Aerial views of the undissected pediments of the class 5B mountain fronts.
B. Exhumed pediment of the eastern Mojave Desert, California.

rain. The rugged front of the Sheep Mountain front of the arid Gila Mountains in southwestern Arizona (Fig. 4.16) has a straight mountain–piedmont junction, V-shaped canyons notched into a rugged escarpment with nice triangular facets, and alluvial fans immediately downstream from the canyon mouths. A major fault zone separates the metamorphic rocks of the piedmont from the granitic rocks of the rugged mountains.

Closer inspection reveals some conflicting evidence. Fanhead trenches expose bedrock. So, fan deposition is thin and has been the result of episodes of climatically induced increases in sediment yield instead of sustained tectonic uplift. These are climatic, not tectonic, alluvial fans. Numerous inselbergs of metamorphic rocks protrude through the thin blanket of piedmont alluvium. Channel exposures at the mouth of the largest canyon suggest that the mountain–piedmont junction locally has retreated more than 50 m from the range-bounding fault. The thin stratigraphic section above the range-bounding fault zone contains paleosols that show that fan gravels were deposited by several aggradation events. Active faulting ruptured alluvium deposited prior to 1.2 Ma, but has not ruptured alluvium younger than 0.7 Ma. The broad band of sheared and fractured mafic metamorphic rocks between the mountain

Figure 4.16 Class 5C mountain front northeast of Sheep Mountain in southwestern Arizona. The softer rocks of the pediment have a faster denudation rate than those of the mountains; this erosional base-level fall is responsible for the active appearance of the landforms along a mountain front that has not been faulted for more than 1.2 My.

front and the inselberg at the lower right of Figure 4.16 appears to be downwasting more rapidly than the valley floors in the foliated granitic rocks of these arid mountains. This difference in erosion rates contributes to the erosional base-level fall. The result is a class 5C landscape assemblage with many characteristics of a tectonically active front. The low sinuosity of the tectonically inactive Sheep Mountain class 5C front serves to remind us:

1) Not to rely on a single parameter when collecting geomorphic data about the mountain fronts of a region, and

2) To consider the possibility of lithologic controls on geomorphic processes where landscapes appear anomalous.

The variable of lithology and structure affects rates of landscape evolution by more than an order of magnitude (Fig. 2.19). Consider two fluvial systems with the same uplift rate and climate, one underlain by mudstone and the other by rhyolite. Stream-channel downcutting, hillslope development, and fan deposition will be much more rapid in the mudstone system, which will evolve from class 1 to class 3 faster than the rhyolite system. Fluvial systems underlain by mudstone become pedimented in less than 0.5 My along the eastern margin of the Coast Ranges of California. Fluvial systems underlain by resistant lithologies in other deserts probably require more than 2 My, after cessation of uplift for landscape assemblages to progress through the sequence outlined in Figures 4.7 and 4.8 and for erosional retreat of a mountain front to create a bedrock pediment. For example, Melton (1965) concluded that the rate of retreat of mountain–piedmont junctions in the granitic rocks of the Sacaton Mountains, Arizona has been about 1 km/My. Shafiqullah et al.(1980) and Mayer (1979) used potassium–argon ages to estimate rates of escarpment retreat of part of the Mogollon Rim, Arizona as ranging from 0.5 to 2.0 km/My. Pediments remain the characteristic piedmont landform until base-level changes cause burial or dissection of these beveled rock plains.

Not all landscapes proceed through the orderly sequence suggested by Figure 4.7. Inactive mountain fronts can return to active status when slip is renewed along a range-bounding fault that may have been inactive for several million years. Substantial retreat of the mountain front shown in Figure 4.17 from the range-bounding fault occurred during a lengthy interval of tectonic quiescence. Hillslopes also retreated from the trunk stream channel, resulting in a deep pediment embayment. Then renewed normal faulting associated with development of the Walker Lane–Eastern California shear zone caused rapid deposition of young alluvial fans downstream from the tectonic zone of base-level fall. Both fan surfaces are of Holocene age. The base-level fall also initiated stream-channel entrenchment into the floor of the pediment embayment.

A tectonic geomorphologist with a long time-span perspective would say that this is a class 4 land-scape, maybe even a class 5, because of the extensive pedimentation. Several surface ruptures probably were needed to create this high fault scarp so I prefer a class 1 designation for this mountain front. Holo-cene tectonism makes this a class 1 front from a haz-ards viewpoint. Renewed tectonic ruptures of pedi-mented terrains are common in the Mojave Desert and locally may record the initial stages of re-align-ment of plate boundary fault zones along the Eastern California Shear Zone.

4.2.3 Regional Assessments of Relative Tectonic

Activity

Tectonic activity classes assess regional patterns of Quaternary uplift rates in a variety of tectonic and cli-matic settings. The equations of Table 4.1 purposely define interrelations of base-level processes that can be readily evaluated by field studies, topographic and

Figure 4.17 Aerial view of the effects of renewed faulting on geomorphic processes in a previously inactive mountain front, Owlshead Mountains, north of the Garlock fault. Holocene fault scarp is downslope from embayed mountain front.

digital maps, and aerial photographs and orbital digi-tal images.

Regional assessment of variations in relative rates of uplift of mountain fronts is practical. It iden-tifies potentially hazardous segments of fault zones. It describes areas of minimal tectonic activity in which to locate sensitive engineering structures such as nuclear power plants. Tectonic geomorphology delineates zones of persistent uplift so structural ge-ologists can better understand the kinematic behavior of faults and folds.

Appraisal of rates of erosional and depo-sitional responses to rock uplift is the basis for es-timating uplift rates for specific faulted and folded mountain fronts. Five tectonic activity classes may be thought of as representing five orders of magnitude of uplift rate. Rates of uplift are fast for class 1 moun-tain fronts of semiarid regions and are in the 0.5 to 5 m/ky order-of-magnitude range. Class 2 mountain fronts have uplift rates that appear to be an order of magnitude slower, being about 0.3 to 0.7 m/ky. Class 3 rates may be assigned a rate of approximately 0.05 to 0.3 m/ky. Class 4 mountain fronts are almost inactive. Uplift rates are roughly 0.005 m/ky and in-tervals between surface ruptures can exceed 200 ky (Bull and Pearthree, 1988).

Faster uplift is necessary to maintain the landscape associated with a given tectonic activity class in humid regions where landscapes can evolve quickly. For example, downcutting by New Zealand rivers easily keeps pace with uplift, even where the Southern Alps are being raised at 5 to 8 m/ky. Re-adjustments after each Alpine fault earthquake cre-ate a new stream terrace (Section 6.2.1.4 and Figure 6.18). Class 5 landscapes have been tectonically dor-mant during the Quaternary.

4.2.3.1 Response Time Complications and

Strike-Slip Faulting

The tectonic landscape signatures described above ap-pear straightforward, but should be used in a context of the time spans needed to generate diagnostic land-scape assemblages. Different styles of faulting may create zones of base-level fall that range from narrow to dispersed, thus affecting the location and magni-tude of a tectonic perturbation in a fluvial system.

Two caveats are briefly summarized here as an admonition to use care when applying the tectonic activity classes model to all mountain ranges formed mainly by flowing water. First, landscapes with greatly different relief and area should have a range of response times to tectonic perturbations. Second, strike-slip faulting may not contribute much to base-level fall. Fortunately, secondary faults are quite useful for tectonic geomorphologists.

Application of the diagnostic parameters described in the previous sections is partly a matter of scale and landscape response times to accelerated uplift along range-bounding faults. Class 1 and 2 landscapes alert paleoseismologists to potential hazard. Class 3 mountain fronts may be lofty and impressive. Examples are the 2.4 km high east side of the Sierra Nevada and 3.1 km high east side of the Panamint Range. Late Quaternary range-front faulting does not match their impressive relief. Partitioned strike-slip faulting is an order of magnitude faster than the range-front faulting for the class 3 mountain fronts of both the Sierra Nevada (Le et al., 2007) and Death Valley (Frankel et al., in press for 2007).

Class 4 and 5 landscapes indicate minimal earthquake potential. These assemblages are easily separated because they are at different stages on a long-term landscape evolution scale. Class 4 still has much of the original mountain front characteristics, although in greatly modified form.

Response time needs to be considered, especially when dealing with the slowly evolving landscapes of the arid realm. Here, the tectonic activity classes protocol uses a Quaternary time span. Be ready to change your reasoning when going to an extremely humid rugged mountain range where erosion rates may be two orders of magnitude faster.

By ignoring response times one might assign the wrong tectonic activity class to the landscape shown in Figure 4.17 – class 5 instead of class 1. Geomorphic tools such as valley floor width–valley height ratio and mountain–piedmont junction sinuosity are meant to assess overall Quaternary tectonic deformation, not just Holocene surface ruptures. High quality topographic maps and most imagery would fail to recognize the obvious scarp caused by Holocene surface rupture(s). One needs to recognize the limitations of the maps and images being used for a regional reconnaissance.

Tectonic scale is important too. A format designed for long and lofty mountain fronts cannot be expected to recognize active but small pop-ups and pull-apart basins. Intermediate-scale fronts also may be suspiciously different, but there are limits to what can be done with maps and images.

There really is no alternative to field work to check out presumptions based on office and computer analyses. The complex junction between the right-lateral Death Valley and left-lateral Garlock strike-slip faults (Fig. 4.2) demonstrates this necessity. Adjacent mountain ranges have many characteristics suggestive of being tectonically inactive. Fieldwork by Chris Menges and other U.S. Geological Survey personnel reveals dramatic late Quaternary changes that underscore the importance of the active Walker Lane–Eastern California shear zone.

Mountain ranges, such as the Owlshead and Avawatz, have many late Quaternary surface ruptures whose timing and slip rates are constrained because we know the approximate ages of the faulted alluvial geomorphic surfaces of the region (Table 1.2). Remnants of pedimented fronts in the Owlshead Mountains complex are not really class 5, because fieldwork shows this to be an active transpressional block within adjacent transtensional terrains. Thrust faulting during the late Pleistocene and Holocene on the Death Valley side of the range, normal-fault rejuvenation (Fig. 4.17), and active left-lateral faulting indicate that many mountain fronts are class 1 instead of class 4 or 5.

The bottom line is that one needs to recognize the long response times for active faulting to create diagnostic landforms that are obvious at the scales of the imagery and maps used in a regional reconnaissance. Much time passes before 10 to 100 m of new relief is created by a rejuvenated range-bounding fault. This potential pitfall can be avoided with fieldwork that discerns surface ruptures of only 1 m.

Strike-slip faulting is more complex than normal faulting (Sylvester, 1988). Antithetic and synthetic faults are the norm, as are local domains of extension and contraction. Styles and rates of tectonic deformation for a particular fault may change with time. New faults appear and older structures may become less important in creating tectonic landforms. Complications include:

1) A variety of secondary geologic structures that are associated with a single (primary) strike-slip fault, hence the synonym "wrench fault".

2) The need to understand the consequences of rotation of parallel sets of primary strike-slip faults.

3) Dynamics of strike-slip faulting are affected by rock mass strength contrasts between sheared fault-zone materials and "coherent" blocks between faults. The magnitude of this variation changes as total displacement increases.

4) Vertical and horizontal strains commonly are partitioned amongst several sub-parallel faults.

Such complications are reviewed and assessed in landscapes associated with active strike-slip faults in the Mojave Desert and Death Valley regions.

Tectonic shearing is characteristic of many plate boundaries. Late Cenozoic and historical earth deformation of the broad San Andreas transform boundary results from complex interactions between parallel strands of strike-slip fault systems. Individual strike-slip faults have distinctive structures and landforms, so let us start by reviewing key aspects of wrench-fault tectonics. Wrench faults form where a regional shear couple rotates basement rocks and overlying sediments. Wrench faults can be depicted as situations of simple shear because the orientations of maximum compressional and extensional tectonic stresses are both parallel to the surface.

Each tectonic shear couple results in diverse structures. A shear couple consists of a primary wrench fault and domains of extension and compression that result in normal faults and in thrust faults and folds (Fig. 4.18). Secondary strike-slip faults form too. These may have the same (synthetic) or opposite (antithetic) sense of lateral displacement as the primary strike-slip fault. The style of vertical tectonic deformation of a wrench-fault system depends on the sense of lateral slip on the primary fault and types of departures from a nonlinear trace of a nearly vertical strike-slip fault. Right-lateral slip occurs where the sense of relative motion on the opposite side of the fault is towards the right, and left-lateral slip where the opposite side appears to have moved left.

Each shear couple has a center of rotation. The entire complex rotates, so structures become less favorably oriented with respect to the driving force, which here is the relative movement between the Pacific and North American plates. Both folds and faults have sigmoidal termination bends as a result of progressive shearing about centers of deformation (Schreurs, 1994; Eusden et al., 2005b).

Pervasive shearing and fracturing along the primary fault creates a broad weak zone. While total cumulative displacement along dip-slip faults generally is less than 5 km, strike-slip faults commonly have displacements of 20 to 200 km. Such large amounts of cumulative shear reduce the resistance of earth materials to tectonic shearing and to fluvial erosion. Thus slip may continue on the older faults of a region even though they no longer have an optimal orientation to the stress field.

Wrench faults commonly occur in parallel sets that rotate collectively. A new direction of strike-slip faulting is created when limits of rotation, 20°–45°, are reached and the primary strike-slip fault is no longer oriented favorably with respect to the plate-tectonics shear couple. Theoretically, slip rates should increase on the secondary fault and decrease on the primary fault as primary shearing shifts to a more favorable orientation.

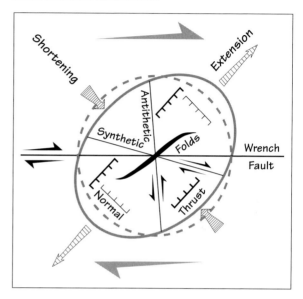

Figure 4.18 Map view of earth deformation associated with a single, straight right-lateral strike-slip fault in homogenous materials. The shear couple shown by the top and bottom half arrows has distorted a circle (dotted gray line) into an ellipse (solid gray line). Secondary structural domains adjacent to the primary fault include right-lateral (synthetic) and left-lateral (antithetic) strike-slip faults, folds and thrust faults perpendicular to the axis of compression shown by the short stubby arrows, and normal faults perpendicular to the axis of extension shown by the long skinny arrows. Axial traces of folds have a sigmoidal pattern (bent in opposite directions) because of rotation caused by the shear couple. A left-lateral strike-slip fault would be a mirror image of this diagram. From Harding (1974) as modified by Sylvester and Smith (1976). AAPG©1974, reprinted by permission of the AAPG.

The initial orientation of faults, relative to driving forces, is a function of the relative strengths of weak fault-zone materials and coherent rock. Blocks bounded by parallel dextral faults rotate counter-clockwise with each surface rupture and paleomagnetic analyses measure the total amounts of rotation. Mojave Desert paleomagnetic work is summarized in Figure 4 of Nur et al., (1993a, b). Total rotations by strike-slip faults can be larger than 45°, but limits to rock mass strength (described by Mohr circle analyses) limit total orientations (block rotations) to modest values of only 20° to 40°, and cannot exceed 45° (Nur et al., 1986). Each increment of rotation decreases effective shear stresses and further slip cannot occur when shear resistance exceeds shear stress.

Continued right-lateral displacements generated by relative movements of the Pacific and North American plates bring the faults of the central Mojave Desert closer to their rotation limit. The primary (oldest) faults may become partially or completely locked. If so, additional regional shearing creates a new, second generation, orientation of primary shearing that splits the primary blocks into smaller wedges (Fig. 4.19). The amount of rotation again is limited to a range of 20 to 45°. Block-rotation generated fault zones may be less important where secondary geologic structures create significant internal tectonic deformation.

The net result is structural and landscape complexity. Primary strike-slip faults commonly have 10 to >100 km of lateral displacement. Slip rates may decrease as second-generation primary faults become more active but their total slip is constrained by widths of fault blocks.

Complexity increases still more when the diverse secondary faults and folds are added to the structural mix. Antithetic faulting occurs as cross faults between the primary faults in a parallel set. They may be more obvious than synthetic faults because of their steep angle to the primary faults (Fig. 4.19). Tectonic geomorphologists need to be alert for landscape characteristics where tectonic deformation is presently occurring in the diverse structural settings created by evolving sets of parallel strike-slip faults. Fault scarps and associated fissures are useful for locating the most recent tectonic deformation.

The degree to which primary faults have become locked needs more study. The way in which the Mw magnitude 7.3 Landers earthquake of 1991 side-stepped along five faults instead of rupturing a single fault zone suggests that rotation of the set of

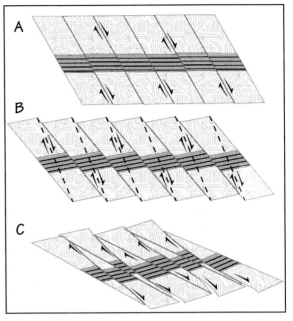

Figure 4.19 Diagrammatic model of set of fault blocks that are rotated counter-clockwise by dextral faulting. From Figure 4 of Nur et al., 1986.
A. Original vertical 30° orientation related to initial angles of shear for a set of blocks created by regional strike-slip faulting.
B About 10° more block rotation brings the dextral fault system to the rotation limit. At the threshold time depicted here the primary faults become locked, and a set of secondary faults starts to rupture the primary blocks. Secondary faulting will continue until they too approach the fault-mechanics threshold, whereupon a third-order set of fault blocks will develop.
C. Second-order faulting creates a complex pattern of strike-slip faults and associated mountain fronts. Sigmoidal characteristics have been added to the cross-fault terminations.

Mojave Desert faults is important, or that the orientation of the stress field has changed. Elsewhere, such as the Marlborough section of the Alpine fault of New Zealand, rupture continues to occur mainly on the primary faults. Although conceptually appealing, the model of Nur et al. (1986) may be difficult to apply to specific field situations.

Using tectonic activity classes to assess the relative displacement rates on strike-slip faults may seem illogical. Many strike-slip faults generate minimal topographic relief and tend to be concealed beneath basin fill. Tectonic geomorphologists use the secondary compressional and tensional structures generated by wrench-fault tectonics, or rely on partitioning of crustal deformation into lateral and vertical domains.

The Blue Cut and Pinto Mountain sinestral fault zones in the eastern Transverse Ranges are active. Both are important elements of the transrotational model of plate-boundary shearing of the Walker Lane–Eastern California shear zone as a single tectonic entity. The Blue Cut fault zone is virtually hidden along most of its length but its tectonic activity during the late Quaternary is revealed by impressive normal-faulted mountain fronts (Fig. 4.20). As elsewhere in the shear zone, faulting of pedimented terrain is creating new mountain ranges. The uplifted pediment shown in Figure 4.20 is 20 to 150 m above actively aggrading alluvial fans. The quartz monzonite of these watersheds is especially sensitive to Pleistocene–Holocene climatic change perturbations, resulting in vigorous aggradation events (Bull, 1991). The ~55 ka Q2c fan surface has a prominent fault scarp where it was not buried by the unfaulted ~7 ka Q3b aggradation event. This front was classified as Class 1 on the basis of a mountain–piedmont junction sinuosity of 1.1 and a mean valley floor width–valley height ratio of only 0.21. I presume that the hidden sinestral-fault component of the partition is also worthy of a Class 1 tectonic activity class.

The hidden sinestral-fault component of the Pinto Mountain fault is much more active than the adjacent mountain-front fault. Trenching revealed five surface ruptures since 14 ka (Cadena et al, 2004).

Thrust and normal faults may also occur as sets where uplift is distributed amongst several faults. Tectonic landforms usually reflect cumulative uplift across a zone of faulting. Tectonically active thrust-faulted mountain ranges commonly have several active fault zones (Section 3.2.3) that define reaches of streams with different uplift rates. This favors analysis of landscapes associated with external (piedmont), range-bounding, and internal faults. Erosional landforms of internal fronts reflect the cumulative influences of downstream base-level falls that are transmit-

ted upstream, as well as the effects of local faulting. Although tectonic perturbations decrease exponentially in the upstream direction (Section 2.2.5), they may keep a stream on the degradational side of the threshold of critical power even adjacent to an active internal fault. Small-scale pockets of deposition (Figs. 3.18, 3.22) may be temporary. Folding results in even broader tectonic perturbations, so presentation of tectonic activity class maps starts out with the simpler situation of base-level changes caused by faulting.

4.2.3.2 Maps of Relative Uplift

Geomorphic analyses of tectonic activity of mountain fronts complement stratigraphic paleoseismic input about timing and magnitudes of surface ruptures provided by trenching of fault zones. Data about the locations, depths, magnitudes, styles, and frequencies

Figure 4.20 Normal faulting of class 1 Eagle Mountains of the eastern Transverse Ranges has raised a pediment. Sinestral Blue Cut fault is parallel partitioned and largely hidden beneath valley fill but is an inferred class 1 strike-slip fault. Ages of alluvial aggradation events are: Q3b ~7 ka, Q3a ~12 ka, and Q2c ~55 ka.

of historical earthquakes provided by seismologists tell us much about historically active fault zones. But we also need a quick, inexpensive way to assess potential earthquake hazards of large numbers of range-bounding faults that are presently locked and may be presumed to be seismically inactive. Scientists, engineers, and planners all need regional assessments of late Quaternary tectonic activity.

This section examines the areal distributions of mountain fronts that range from tectonically active to inactive. Diverse settings include the normal-faulted Basin and Range Province, folded Coast Ranges, thrust-faulted Transverse Ranges, and lateral-faulted Mojave Desert. These maps can be compared with other geomorphic assessments of uplift rates. Published evaluations include the Death Valley region (Smith et al., 1968; Hooke, 1972; Smith, 1975, 1976, 1979; Schweig, 1989; Anderson and Densmore, 1997; Slate, 1999; Machette et al., 2001; Walker et al., 2005), the Great Basin of Nevada (Wallace, 1978, 1987a,b; Fonseca, 1988; Pearthree, 1990; dePolo et al., 1991; dePolo and Anderson, 2000), the Transverse Ranges of California (Rockwell et al., 1988; Keller et al., 1998; Spotila and Anderson, 2004), and the central Coast Ranges of California (Lettis, 1982, 1985).

The state of this art has changed. Studies of several responses of mountain-front landscapes to base-level fall began as a way to class relative tectonic activity of Mojave Desert mountain fronts near proposed nuclear generating power stations. The next step was to study the obviously active Death Valley region, where Bull and McFadden (1977) assessed potentially useful landforms. Their work focused on impressive fronts of lofty mountain ranges, but now we more fully realize that active tectonics is influenced by the pervasive strike-slip faults of the Walker Lane–Eastern California shear zone. Tectonic deformation is partitioned. Strike-slip faults traverse valley floors. Alluvium, instead of rock, is the main material in low piedmont fault scarps and offset stream channels so these strike-slip faults are an important element of tectonic activity class 1B. As for low normal- and thrust-fault scarps, low scarps in Holocene alluvium created by offsets along strike-slip faults catch the attention of ever-vigilant paleoseismologists.

The map of tectonic activity classes north of the Garlock fault (Figs. 4.2, 4.21) has consistent trends. Strike-slip faults are rated in the highest tectonic activity class. About 110 km of the Owens Valley fault zone was the source of a Mw magnitude 7.6 earthquake in 1872. This fault zone may extend all the way to the Garlock fault (Unruh et al. 2002; Monastero, et al. 2005). The 310 km long Death Valley fault zone has a dextral rate of slip of ~ 4 to 9 m/ky (Reheis et al., 1995; Klinger, 1999, 2001; Machette et al., 2001; Frankel et al., in press for 2007). Examples of pull-apart basins include Death Valley by the Black Mountains, Saline Valley, and Deep Springs Valley.

Mountain front characteristics indicate progressively less relative uplift rates from north to south towards the Garlock fault. Bull and McFadden suggested that this was a regional pattern of decreasing tectonic activity. Using the modern transrotational model (Dickinson, 1996) we now realize that the left-lateral Garlock fault and the domains of left-lateral faulted mountain fronts on both sides of the Garlock fault in the northeastern Mojave Desert and the eastern Transverse Ranges accommodate the overall continuity of the Walker Lane–Eastern California shear zone.

The Argus, Slate, Panamint, Black, and Grapevine-Funeral ranges have less Quaternary tectonic activity on their eastern mountain fronts than on their west sides. Part of the Owlshead Mountains is shown here as being class 4, which fits the Quaternary landscape assemblage. But this is another example of rejuvenated mountain fronts: piedmont scarps created by thrust faulting of the piedmont downslope from a pedimented mountain front (Christopher Menges, personal communication). I suspect that most of the Owlshead small mountain fronts should be given a tectonic activity class of 1 – the highest rating from a hazards standpoint.

The Sierra Nevada front is only class 3. The Owens Valley strike-slip fault is class 1 and carries on through the Coso complex all the way to the Garlock fault. Pull-apart basin subsidence that created Owens Lake basin has bounding faults that have encroached into the Sierra Nevada. The result is a local base-level fall and a class 2 mountain front. Then the Sierra Nevada mountain front tectonic activity decreases progressively to the south: 2 to 3, 3 to 4, and 4 to 5. Each of these changes in range-front landscapes is accompanied by a jog in the mountain–piedmont junction (Fig. 4.21). Piedmont alluvial-fan deposition opposite Mt. Whitney consists of climate-change induced pulses of aggradation whose ages appear to match the times of glacial-moraine formation (Le et al., 2007). Instead of being thick tectonic alluvial fans, these are thin climatic fans. Key evidence for

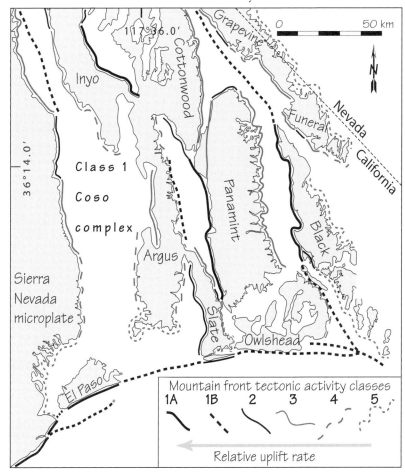

Figure 4.21 Tectonic activity classes of mountain fronts in the Death Valley – Sierra Nevada area.

climatic instead of tectonic fans is the way in which shallow fanhead trenches have cut through thin surficial alluvium to expose buried surfaces of old fans. [10]Be dating of faulted fans by Le et al. indicates a slow rate of uplift of about 0.2 to 0.3 m/ky since 125 ka. Like the eastern Panamint Range, the rugged landscape is mainly inherited from earlier times of rapid uplift. Big mountains take forever to wear down in this climatic setting.

Relative uplift rates of mountain fronts in the extensional terrain of the Basin and Range Province in west-central Nevada were appraised using five tectonic activity classes. Normal faulting has created mountain ranges of volcanic, sedimentary, and plutonic rocks. The climate is moderately seasonal, arid to semiarid, thermic to mesic. So these mountains preserve diagnostic tectonic landforms for a long time, even where rates of tectonic displacement are

modest. This study area is part of the central Nevada seismic belt, a 300 km long zone of surface ruptures that occurred in 1903, 1915, 1932, 1934, and 1954.

These mountain ranges are fairly small (Fig. 4.22), so all drainage basins larger than 3 km^2 were used in the V_f analysis. The Table 4.3 numerical values do not overlap where the three moderately active classes (2, 3, and 4) were lumped together, much like the Bull and McFadden (1977) three-class approach. Overlapping of data is typical when five classes are used, but is minimal here. Even so, it makes sense to use several tectonic landforms to assess relative tectonic activity. Anomalous situations may not be recognized if you use only one landscape statistic.

Only some of the range-bounding faults have been active during the Quaternary. One characteristic of the Basin and Range Province is diverse normal-fault slip rates. Some fronts have been raised during

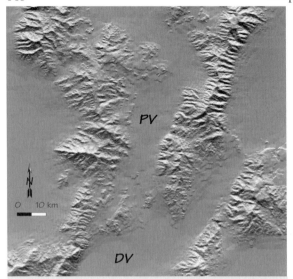

Figure 4.22 Digital image of the Pleasant Valley (PV)–Dixie Valley (DV) region in west-central Nevada.

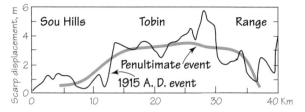

Figure 4.23 Fault scarp vertical displacements that decrease south towards the Sou Hills from the west flank of the Tobin Range for the 1915 A.D. (Wallace, 1984) and early Holocene surface ruptures. From Figure 11 of Fonseca (1988).

historical earthquakes, for example the 1915 magnitude 7.1 earthquake that ruptured several fronts on the west side of the Tobin Range over a time span of several hours (Wallace, 1977). Other mountain fronts, such as those bounding the East Range, have been inactive for more than 1 My. Uplift rates are moderate: class 1 fronts are being raised only about 0.3–0.5 m/ky (Wallace, 1978; Pearthree, 1990). Uplift should be considered relative. Range crests rise ~0.3 m and the intervening basins subside ~0.7 m during magnitude Mw 7 earthquakes (Stein and Barrientos 1985; Stein and Bucknam, 1986). The locations of Holocene faulting in Dixie–Pleasant Valley shift from the east side of the Stillwater Range to the west side of the Tobin Range. The cross-over is in the Sou Hills: a barrier that arrests surface ruptures propagating from either north or south (Fonseca, 1988).

Multiple zones of base-level fall complicate assignment of tectonic activity classes. Streams along the west flank of the Tobin Range cross the active range-bounding fault zone and then flow south, crossing a zone of dispersed faulting in the Sou Hills before descending into Dixie Valley. Minor relative uplift of the Sou Hills, and subsidence of the north end of Dixie Valley, occur with each faulting episode. The net effect is a base-level fall that induces stream-channel entrenchment, which migrates to distant upstream reaches. For this reason, the piedmont west of

the Tobin Range consists mainly of class 2 fans (entrenched), rather than the class 1 fans (unentrenched) that one would expect from the presence of other obvious class 1 tectonic landforms.

Fonseca (1988) describes how the Sou Hills act as a transverse barrier that arrests propagation of fault-ruptures. Total vertical tectonic displacement for the two most recent Tobin Range surface ruptures dies out towards the south where faulting was halted by the Sou Hills transverse barrier (Figs. 4.23, 4.24). The 1915 surface rupture had a normal-fault displacement that decreased from a maximum of 6 m along the Tobin Range to 0.3 to 1.0 m where it terminated as many small scarps in the Sou Hills. The penultimate surface-rupture event had the same magnitude and trend. It is older than the 6.8 ka Mazama volcanic ash (Wallace, 1984).

Was this a characteristic behavior? Fonseca used mountain front tectonic activity classes to see if landforms with distinctive tectonic signatures define a persistent tectonic style during the Quaternary. Total mountain-range relief of both the Tobin and Stillwater Ranges decreases towards the Sou Hills. Geomorphic indicators of Quaternary uplift rates, such as valley floor width–valley height ratio and mountain–piedmont junction sinuosity, decrease towards the Sou Hills from both the north and the south. Climatic and lithologic controls on landscape evolution vary little in this area, so the changes of tectonic activity classes towards the Sou Hills (Fig. 4.24) demonstrate systematic decreases of magnitudes and rates of tectonic base-level fall towards the Sou Hills. Both geomorphic analyses, using short- and long-term landscape characteristics, define a consistent tectonic style.

The National Earthquake Hazards Reduction Program of the U.S. Geological Survey is a ma-

Mountain front of Figure 4.25	Sinuosity of mountain–piedmont junction, J	Mean valley floor width–valley height ratio, V_f	Range of V_f ratios	Triangular facet dissection stages of Table 2.1
	**Tectonically Active Mountain Fronts (Class 1)**			
16	1.08 [-0.92]	0.15 [-1.86]	0.08 to 0.35	1 through 4
18	1.07 [-0.93]	0.06 [-1.95]	0.05 to 0.09	1 through 5
	Tectonically Active Mountain Fronts (Class 2)			
1	1.3 [-0.7]	0.16 [-1.85]	0.08 to 0.51	1 through 5
6	1.2 [-0.8]	0.13 [-1.88]	0.11 to 0.16	1 through 5
7	1.2 [-0.8]	0.13 [-1.88]	0.06 to 0.53	2 through 5
8	1.2 [-0.8]	0.17 [-1.84]	0.07 to 0.25	2 through 5
15	1.09 [-0.9]	0.31 [-1.7]	0.23 to 0.42	2 through 6
17	1.19 [-0.1]	0.51 [-1.5]	0.15 to 0.87	2 through 5
19	1.19 [-0.1]	0.23 [-1.78]		1 through 5
	Moderately Active Mountain Fronts (Class 3)			
2	1.9 [-0.1]	1.7 [-0.31]	0.43 to 3.33	1 through 5
5	1.6 [-0.4]	1.2 [-0.81]	0.23 to 2.59	2 through 4
	Slightly Active Mountain Fronts (Class 4)			
4	2.2 [+0.2]	1.3 [-0.71]	0.36 to 3.53	2 through 5
11	3.2 [+1.2]	7.0 [+4.99]	0.91 to 39.4	4 through 5
12	2.3 [+0.3]	1.2 [-0.81]	0.18 to 3.75	3 and 5
13	2.2 [+0.2]	1.0 [-1.01]	0.15 to 2.25	3 and 4
14	2.0 [0]	1.9 [-0.11]	0.71 to 2.86	3 through 6
20	1.8 [-0.2]	2.94 [0.93]	0.71 to 11.4	4 through 6
	Inactive Mountain Fronts (Class 5)			
3	3.5 [+1.5]	2.5 [+0.49]	1.38 to 3.83	7
9	2.1 [+0.1]	2.0 [-0.01]	1.25 to 2.50	6
10	2.6 [+0.6]	7.8 [+5.79]	1.33 to 16.3	4 through 6

Table 4.3 Data used to assess five tectonic activity classes of mountain fronts in west-central Nevada (Fig. 4.24). Other ways to view the data are shown inside brackets [departure from mean value for entire study area]. Data from Julia Fonseca and John Partridge.

jor source of funds for paleoseismology projects. One was an evaluation of mountain front tectonic activity in the central Mojave Desert (Bull, 1977b), where right-lateral faulting predominates. Two earthquakes occurred after this study, the Mw magnitude 7.3 Landers event of 1992 and the Mw 7.1 Hector Mine event of 1999. Unmodified 1977 results are shown here to better evaluate this geomorphic approach for estimating earthquake hazards.

The results of these studies are combined in the regional assessment of Quaternary tectonic activity shown in Figures 4.25A, B. Tectonically active

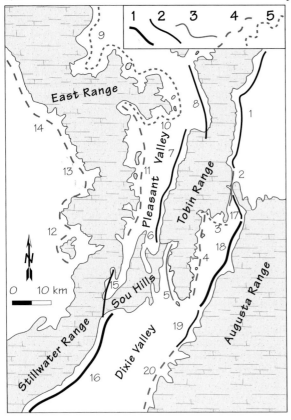

Figure 4.24 Tectonic activity classes (upper right corner) of normal faulted mountain fronts. Bedrock is mainly volcanic and sedimentary rocks with some intrusives. See Table 4.3 for numerical data for the numbered fronts. Data from analyses by John Partridge and Julia Fonseca.

crustal processes in the Walker Lane–Eastern California shear zone.

Some parallel right-lateral faults of the central Mojave Desert are highly active and others inactive (Fig. 4.25B). Uplift rates decrease towards the north: first to class 2 fronts and then to inactive fronts as the fault zones approach the Garlock fault. Jennings (1994) has classed some fronts as active that a tectonic geomorphologist would consider virtually inactive because no surface ruptures appear to have occurred during the past 1 My. The active Garlock fault runs through a region of inactive mountain fronts, with the relative activity increasing to both the north and south (Figs. 4.21, 4.25B). A few short class 1 left-lateral fronts in the northeastern Mojave Desert support the presence of a local tectonic domain proposed by Garfunkel (1974). Ongoing field studies show that this area of active transrotation (Dickinson and Wernicke, 1997) turns out to be much larger and includes both the Avawatz Mountains south of the Garlock fault and the Owlshead Mountains north of the fault (Chris Menges, 2006, personal communication). Active fault zones bound the impressively high Transverse Ranges. These ranges become progressively lower and less active towards the east. They

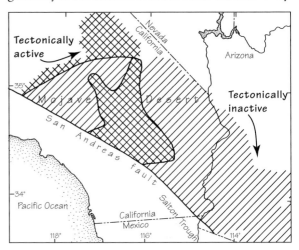

Figure 4.25 Tectonic activity classes of mountain fronts in the arid Mojave Desert and semiarid eastern Transverse Ranges of southeastern California.
A. Regions of tectonically active and inactive mountain fronts in the Mojave Desert of southeastern California.

and inactive regions are bounded by the Garlock and San Andreas faults. The coastal region to the southwest has many active strike-slip faults and the thrust faults associated with the "big bend" of the main trace of the San Andreas fault transform boundary. The large tectonically inactive region of the eastern Mojave Desert includes southwestern Arizona. It extends right up to the edge of the Salton Trough, a rhomochasm connected to the Gulf of California spreading center to the south. The western Mojave Desert is a strange tectonically inactive landscape. The way in which Quaternary surficial faulting appears to die out to the North in the central Mojave Desert describes how unlikely surface ruptures may be, but may not be representative of complex deeper

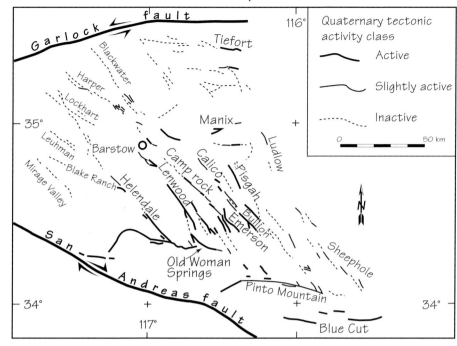

Figure 4.25 Tectonic activity classes of mountain fronts in the arid Mojave Desert and semiarid eastern Transverse Ranges of southeastern California. B. Central and western Mojave Desert. Displacements along oblique strike-slip and normal faults have created mountain ranges of resistant crystalline basement rocks. Barstow is a large town.

still have locally active mountain fronts (Figs. 4.20, 4.26) and are essential for transrotational tectonic connections within the Walker Lane–Eastern California shear zone.

Rather surprisingly, mountain fronts immediately adjacent to the plate boundary Salton Sea trough are as inactive as the mountain fronts of the eastern Mojave Desert and Arizona. Neither area has experienced active range-front faulting during the past million years.

The Landers and Hector Mine surface ruptures sidestepped across several fault zones (Treiman et al., 2002). Spacing of tectonic activity classes of other mountain fronts suggests a similar behavior. An example is shown in Figure 4.26. These mountains are at the eastern edge of the tectonically active central Mojave Desert. The inselbergs and sinuous front of the Calumet Mountains signal long-term tectonic inactivity. However, a series of class 2 fronts describes a left-stepping pattern to the northwest that is suggestive of occasional movements dispersed along a right-lateral shear zone.

How well do the tectonic activity classes of Bull (1977b) compare with the subsequent surface ruptures of 1992 and 1999? The 1977 analysis had most of the Landers surface rupture as discrete class 1 and class 2 mountain fronts, which included obvious fault scarps on desert plains. Each of the five fault zones of the 1992 event had parts that were classed as "most active" in 1977, with intervening sections of lesser earthquake potential. The Landers event involved only mountain fronts expected to be involved in the next surface rupture.

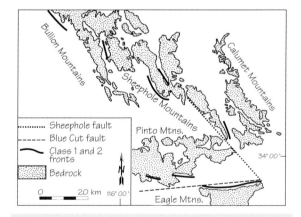

Figure 4.26 Tectonic activity classes suggest a left-stepping pattern of active mountain fronts indicative of right-lateral displacement along several faults of the postulated Sheephole shear zone in an otherwise inactive terrain.

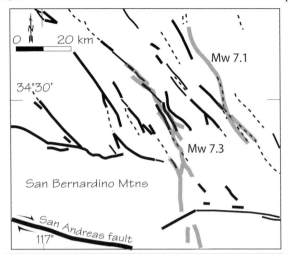

Figure 4.27 Comparisons of 1977 estimates of the most active fault zones of the central Mojave Desert based on mountain-front tectonic activity classes of Figure 4.27B. The complex surface ruptures of the Mw magnitude 7.3 Landers earthquake of 28 June 1992 and the Mw magnitude 7.1 Hector Mine earthquake of 16 October 1999 are shown by gray lines.

The southern part of the surface rupture crosses young alluvial plains (Fig. 4.27) and the 1977 method, like the area north of the Garlock Fault, did not address strike-slip faulting. The same deficiency applies to the analysis of the northern part of the Hector Mine surface rupture, which crosses the plains of the Lavic Lake basin.

The 1999 Mw magnitude 7.1 Hector Mine earthquake produced a maximum slip of 5.2 m on a ~48-km long surface rupture involving the Lavic Lake fault and a portion of the southern Bullion fault. Bull (1977a) did not regard the Hector Mine area as being particularly active, being a mix of active and inactive fronts. The 1999 surface rupture coincided with some mountain fronts, such as along the Lavic Lake fault, that have obvious characteristics of minimal amounts of late Quaternary uplift. The Hector Mine event may be more unusual than the Landers event. This event might have a longer earthquake recurrence interval or the locus of strike-slip faulting might have shifted eastward.

Trench stratigraphy analyses support the geomorphic conclusion that the Lavic Lake fault passes through mountains with a low level of tectonic activ-

ity. The work of Lindvall et al. (2000) is presented here because it underscores the advantages of using soil stratigraphy in paleoseismology (Rockwell, 2000). The "Bullion Fan" trench (Fig. 4.28) revealed a late Holocene surface rupture. But the "Drainage Divide" trench on the Lavic Lake Fault provided a much different and quite valuable fault history. The 1999 rupture cuts steeply sloping alluvial aprons with entrenched stream channels. Massive gravel deposits are weakly stratified, so soil-profile horizons were mapped.

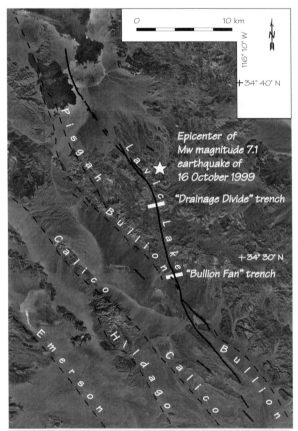

Figure 4.28 Map of the surface ruptures caused by the 1999 Mw magnitude 7.1 Hector Mine earthquake (solid heavy black lines) and the locations of nearby active fault zones (dashed lines) of the central Mojave Desert of southern California. Base map and pattern of faulting furnished courtesy of Tom Rockwell.

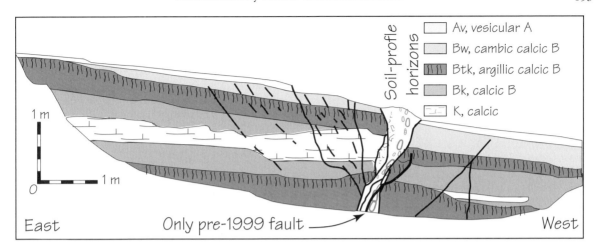

Figure 4.29 Soil stratigraphy for two late Pleistocene aggradation events exposed in the Drainage Divide trench, Lavic lake fault zone.
A. Sequence of faulted soil-profile horizons on both sides of a fault zone fissure filling. All the black lines are fractures and faults created by the Hector Mine earthquake of 1999, except for a single heavy black line for the penultimate event. Figure 2 of Lindvall et al. (2000).

Only soil-profile horizons, not sedimentary strata, were mapped in the wall of the 2-m deep trench. Surficial gravel with pronounced black varnish reveals a lack of Holocene deposition. As throughout the Mojave Desert, pulses of climate-change induced aggradation buried former surfaces with new increments of detritus stripped from adjacent hillslopes. The pulsatory to continuous nature of deposition depends partly on location within a fluvial system (Fig. 2.14). The trench-site location favors incremental accumulation of alluvium during aggradation events. Alluvial thicknesses and the strength and style of soil-profile development at the "Drainage Divide" trench site are nicely detailed in Figure 4.29A. These appear to be typical 125 and 55 ka aggradation events of the Mojave Desert (Bull, 1991; Slate, 1999; Machette et al., 2001).

Evidence for a pulse of Holocene aggradation is not present in the trench, but this is not unusual. Alluvial geomorphic surfaces with black rock varnish are common throughout the Mojave Desert. Each represents a situation where the magnitude of stream-channel entrenchment was deep enough to contain the thickness of aggrading alluvium, thus preventing streams from spreading out and depositing another sheet of detritus stripped from the hillslope sediment reservoir of the watershed.

The two soil profiles of Figure 4.29A do not represent time lines even though pedogenesis began after cessation of the ~125 and ~ 55 ka aggradation events. Each of the two soil profiles of Figure 4.29A records an interval of nondeposition. The Q2b interval is 125 to 55 ka, and would be 125 to 0 ka were it not for the deposition of Q2c, which has been offset by two surface rupture events.

Displacements are several meters right-lateral with a small vertical component. The right side of the faulted section looks like it has been lowered, but this is only apparent, being mainly the result of lateral displacements of stratigraphic units of nonuniform thickness.

The 1999 event produced abundant fractures and faults, many of which extend to the surface. These are fresh and obvious now, but will be much less apparent after 20 to 40 ka of overprinting by pedogenic processes.

The only pre-1999 fault cuts and offsets the argillic soil-profile horizon of the older (Q2b) soil (Fig. 4.29B), which started forming after the 125 ka aggradation event. The penultimate event also offsets the Bk horizon of the younger (Q2c) soil and terminates in the Q2c argillic horizon (Fig. 4.29A). It appears that no fault displacements occurred during the 125–55 ka soil-forming interval. The penultimate

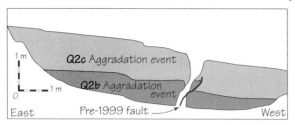

Figure 4.29   Soil stratigraphy for two late
Pleistocene aggradation events exposed in the
Drainage Divide trench, Lavic lake fault zone.

B. Alluvium of the 125 ka (Q2b) and 55 ka
(Q2c) aggradation events based on Figure
4.29A. A single pre-1999 fault displaces the
fissure filling and adjacent deposits but does
not extend to the surface.

surface rupture event may have occurred shortly after
deposition of Q2c alluvium, but before much soil-
profile development had occurred. Holocene surface
ruptures are not present.

Recognizing what might seem as innocuous
"mountain fronts" is essential for testing several models proposed for new orientations of plate-boundary
faulting. Erosion obliterates small scarps in hilly ter-

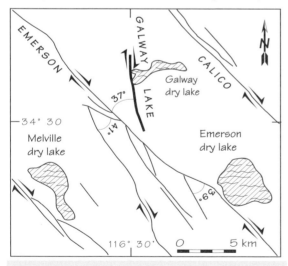

Figure 4.30   Relation of the surface rupture of the
1975 Galway Lake earthquake on a secondary fault
to the orientations of the primary Emerson and
Calico fault zones of the central Mojave Desert.
From Figure 1 of Hill and Beeby, 1977.

rain much faster than on gently sloping piedmonts.
Finding and evaluating Holocene surface ruptures
that cross mountain ranges requires field studies above
a reconnaissance level. More use should be made of
LiDAR (Section 5.6.1) in order to discern diminutive
piedmont fault scarps.

Rotation of the tectonic block between the
Emerson and Calico faults has been sufficiently large
that ruptures now occur on shorter secondary faults
(Fig. 4.30). A Mw magnitude 5.1 earthquake occurred on the secondary Galway Lake fault in 1975
A.D. The angle between the primary and secondary
faults is ~ 37° – quite similar to the 39° and 41° angles noted for other secondary faults in Figure 4.30.

This kinematic style is significant. If the primary direction of faulting is locked because limits of
rotation have been reached (Nur et al., 1986), then
long surface ruptures such as that of the Mw magnitude 7.3 Landers 1991 earthquake can be expected to
sidestep between many faults instead of being along
a single fault zone. Secondary faults might have become more important.

Long-term maximum shearing occurs in the
center of a belt of parallel dextral faults, such as the
Walker Lane–Eastern California shear zone. The
Hector Mine event occurred on the eastern edge of
the zone of active strike-slip faulting in the central
Mojave Desert and only two fault zones were involved. Perhaps the primary faults on the fringes of
the central Mojave Desert set of fault blocks have yet
to become fully locked.

Changes in kinematic styles of faulting may
complicate the use of mountain front topographic
characteristics to classify relative tectonic activities.
Transfer of slip from the Emerson to the Galway
Lake and other secondary faults diminishes the rate
of uplift along the principal Emerson fault, but not
enough to lower its designation from tectonic activity
class 1 to class 2. This is still an impressive, rugged
mountain front. Once again, the response times of
sinuosity of the mountain–piedmont junction, and
the valley floor width–valley height ratio, are sufficiently long that >100 ky is needed to discern decreased uplift rates in landscape characteristics in this
arid realm.

What are the mountain-front characteristics
for the Galway Lake mountain front? Almost none!
There is no mountain front for this surface rupture
zone. It is so young that its location cannot be discerned on topographic maps. The 7-km long rupture barely broke the surface. Hill and Beeby (1977)

note at least one previous Holocene surface rupture and a very short remnant of an older 1-m high fault scarp. Perhaps the primary faults have only recently become sufficiently locked to transfer slip to secondary zones of surface rupture such as the Galway Lake fault zone.

Patterns of faulting may change with time. Only 35 m/ky of the present 49 m/ky of relative motion between the Pacific and North American plates (DeMets et al., 1990, 1994) presently occurs along the San Andreas fault (Sieh and Jahns, 1984; Weldon and Sieh, 1985). Roughly half of the remaining 14 m/ky of right-lateral shift between the Pacific and North American plates occurs along faults in eastern California and Nevada (Savage et al., 1990). Geodetic measurements suggest that cumulative right-lateral strain for the faults of the central Mojave Desert is occurring at about 7 m/ky (Sauber et al., 1986). Tectonic rotation of dextral structural blocks with time places them in less favorable positions, relative to the regional principal compressive stress. Paleoseismologists should note the implications of regional structural re-alignment with the passage of time. Nothing remains static in the world of predicting earthquake hazards.

Several structural models can be evaluated by using the tectonic geomorphology data of Bull (1977), Morton et al. (1980); Matti et al., (1985), and Hart et al., (1989). The northwest-trending faults should lose their ability to accommodate crustal deformation as they become progressively more unfavorably oriented according to the model of Nur et al. (1993a, b). They use six historical earthquakes to define the initial stages of development of a new 120-km long fault, and conclude that the Landers event slip is partitioned between the old trend that has been rotated to N40°W and the new N15°W trend. Sowers et al., (1994, Fig. 5) present evidence that further supports the North–South directed shear model. The Figure 4.25B geomorphic evidence does not recognize this speculation. Perhaps more time is needed for new faults to create hills and scarps with a different orientation in a new version of a tectonically active landscape. The structural realignment, if real, has just started to form.

A second model concerns a larger version of the North–South directed shear concept – the Eastern California Shear Zone, extending from the central Mojave Desert into the west-central Nevada area of Figure 4.24. Miller et al. (1993, p. 871) postulate that the major "seismic gap" in the historical

earthquake record north of the town of Barstow is a likely candidate for a future major earthquake. Figure 4.25B does not support this hypothesis. Slightly to moderately active fronts near the town of Barstow grade into an area of mountain fronts that lack evidence of persistent faulting. Most of the terrain closer to the Garlock fault has been virtually inactive during the Quaternary, when compared to either the central Mojave Desert or the Death Valley region.

Garfunkel (1974) recognized that the northwest striking dextral faults of the Mojave tectonic block do not extend beyond the Garlock fault. So his model emphasizes conjugate domains of right-lateral and left-lateral faults. Right-lateral is dominant but he recognized the importance of left-lateral domains in the long-term process of rotation and deformation of the Mojave block. These include the eastern extension of the Transverse Ranges, a left-lateral block in the Northeast part the Mojave Block, and the Garlock fault. Both left-lateral domains have active mountain fronts (Fig. 4.25B). The lack of persistent faulting north of Barstow remains a problem for this model too. A more complete kinematic model is needed, one that accounts for the large expanses of seemingly inactive terrain in the western and northern portions of the Mojave block.

The dextral faults of the central Mojave Desert intersect the thrust faults of the eastern Transverse Ranges. Spotila and Anderson (2004) have studied the kinematics of these complex fault systems. Their model has dextral slip on the Helendale fault accompanied by movements on the range-bounding thrust fault along the western part of the San Bernardino Mountains. This implies a concurrent decrease in faulting east of the junction with the Helendale fault. The result is a quasi-stable triple junction that separates a western domain dominated by thrusting and an eastern domain characterized by a larger component of strike-slip displacement. They speculate that this domain junction has migrated to the west as the Mojave block has been translated southeastward along the San Andreas fault. The tectonic activity classes of Figure 4.25B fit their model, perhaps with the Lenwood fault being the junction. Alternatively the triple junction could be in the process of migrating from the Lenwood to the Helendale fault, or perhaps the triple junction involves an area instead of a single fault intersection.

The presence or absence of coseismic rockfalls is especially obvious in the Mojave Desert because flat ground allows blocks from several events

Figure 4.31 Comparison of rockfalls from seismically active and inactive hillslopes of plutonic rocks in the central Mojave Desert, California.

A. Abundant large boulders of biotite diorite remain on the hillsides of a class 5 mountain front. Only a few cobbles have been washed onto the adjacent piedmont.

B. Earthquakes have dislodged many large boulders of porphyry breccia from the hillside. Boulder at right is 12 m high and has a volume of ~140 m^3.
This is a class 1 mountain front bounded by the right-lateral Lenwood fault.

to accumulate and there are no trees to hide them. The contrast in rockfall block abundance between tectonically active and inactive mountain fronts can be striking (Fig. 4.31). Rockfalls are such a great way for tectonic geomorphologists to study earthquakes that Chapter 6 is devoted to this subject.

The western Mojave Desert is a bit of an enigma. A triangular chunk of Mesozoic plutonic rocks is caught between the very active San Andreas and Garlock faults (Fig. 3.2). The area lacks the rugged mountain fronts of the central Mojave Desert. Instead a pedimented plain with shallow playa basins are the norm. It seems as inactive as the landscape between the Colorado River and Salton Trough of southern California. Closer inspection (Figs. 4.32, 4.33) suggests that the pediment domes may not be entirely erosional in origin. Subtle, long wavelength buckling folding may play a part in exposing much of the inherited pediments of this area. This block may be quite different from both the eastern and central

Mojave Desert, but still seems to be an area of minimal surface-rupture risk.

The thrust faulted south side of the San Gabriel Mountains of southern California (Figs. 3.1–3.3) is part of a large transpressional bend in the San Andreas fault system. The setting is different than either the Mojave Desert or Nevada, where most mountain ranges are small and the climate is arid to semiarid. Thrust faulting has created a rugged mountain range of sheared and altered plutonic and metamorphic rocks. The long, lofty San Gabriel Mountains promote orographic lifting of storms from the nearby Pacific Ocean. Precipitation increases rapidly with increase in altitude. The climate ranges from strongly seasonal semiarid and moderately seasonal thermic (Table 2.1) on the southern piedmont to strongly seasonal subhumid and moderately seasonal mesic in the mountains below about 1400 m. The higher parts of the range are humid and mesic to frigid. A pronounced rainshadow is present on the northern

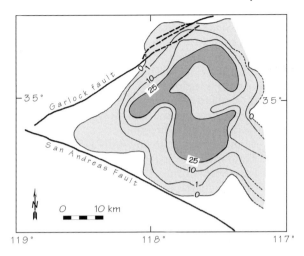

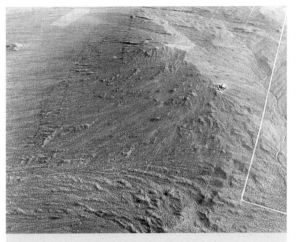

Figure 4.32 Pediments of the tectonically inactive western Mojave Desert.
A. Percentage of exposed pediment in this arid landscape. Dark gray shows area of most common exposure.

Figure 4.32 Pediments of the tectonically inactive western Mojave Desert.
B. Low-sun-angle aerial view of bedrock knobs rising slightly above the surface of a pediment dome of a former mountain range. Two-lane dirt road for scale.

lee side of the mountains, which declines to the arid western Mojave Desert. Long dry summers are followed by mild wet winters. Less than 200 mm of precipitation may fall at a mountain weather station one winter and 1000 mm may fall the next. Years with frequent winter rainstorms are remembered for their devastating floods and debris flows. The resulting landscape of this rapidly rising mountain range has steep, bouldery braided stream channels resembling those of the Southern Alps of New Zealand.

The San Gabriel Mountains fronts were analyzed using five tectonic activity classes. Data for mountain–front sinuosity and triangular-facet stages were adequate, but values for the valley floor width–valley height ratio (Fig. 4.33) had considerable overlap between tectonic activity classes (Karen Demsey, 1986 ,written communication). This scatter is the result of: spatial variations in rock resistance that affects rates of valley-floor widening, and a large range of drainage-basin areas upstream from measurement sites. Small watersheds with low annual unit stream power have downcut less in response to late Quaternary uplift. Consequently they have less relief between valley floors and adjacent ridgecrests than do adjacent large watersheds. The tendency for narrow valley floors of small watersheds is partially offset by lower values of valley floor–ridgecrest relief.

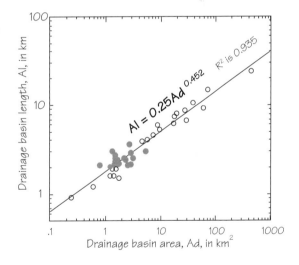

Figure 4.33 Relation of drainage-basin area to drainage-basin length for watersheds draining the south side of the San Gabriel Mountains, California, east of Big Tujunga Canyon. Filled points indicate the narrow range of drainage-basin sizes used to assess valley floor width–valley height ratios.

Variation in V_f ratios caused by both factors was reduced in two ways. First, valley-floor widths were measured at basin-position coordinates of 0.9. This reduced the number of sites where valley-floor width was anomalously wide because of soft, crushed rocks in the range-bounding thrust-fault zone. Secondly, we used only the narrow range of watershed size noted in Figure 4.33. The constants in the power function, A_l, drainage-basin length (straight-line length, not the length along the sinuous valley floor), and A_d, drainage-basin area

$$A_l = 0.25 \, A_d^{0.452} \qquad (4.10)$$

reflect the tectonic environment. Long, narrow drainage basins are indicative of the first stages of watershed development on an actively rising escarp-

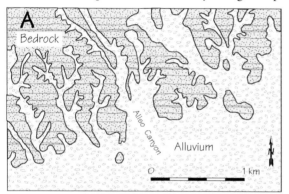

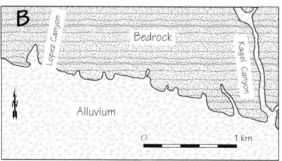

Figure 4.34 Contrasts in mountain-front sinuosity for active and inactive tectonic settings. The bedrock in both cases is soft Pliocene marine sedimentary rocks. Bedrock–alluvium contacts are from the geologic map by Barrows et al. (1973). See Figure 4.35 for locations of front numbers.
A. Inactive front 2B.
B. Active front 7C.

ment. Initial stages of watershed development have lower coefficients and exponents than more circular watersheds indicative of an older or of a less tectonically inactive mountain range, such as John Hack's study areas in the Appalachian Mountains. The exponent of 0.45 is less than 0.5, which would describe geometric similarity (Vacher, 1999) where all length dimensions increase in the same way as they change size in the downstream direction.

Sinuosity varies greatly with age and rate of uplift of mountain fronts. Thrust faults have migrated basinward during the Quaternary, and the outermost fault is the most active at fronts 5, 7, 9, 16, 18, 19, 20, 23, and 28 (See Figure 4.35). Range-bounding fault zone 7C ruptured in the 1971 Mw magnitude 6.8 San Fernando earthquake (Grantz et al. 1971). The youngest fronts are the external (piedmont) faults of the San Gabriel Mountains. They are straight with sinuosities close to 1.0. Range-bounding faults have sinuosities ranging from 1.2 to >10. Sinuosities could be evaluated for only a few internal faults because these former mountain–piedmont junction transition zones have been converted into mountainous terrain. The contrast in sinuosity between class 1 and class 4 fronts is dramatically illustrated in Figure 4.34 where climate and rock type are the same, but uplift rates are much different.

Comparison of relative uplift rates should be considered using a late Quaternary temporal framework because the tectonic signatures in this landscape result from more than 50 ky of uplift. Uplift is rapid for most of the 60 fronts of Figure 4.35. Estimated uplift rates for class 1 fronts range from about 0.5 to >2.0 m/ky and Holocene fault scarps are common. Combined uplift along several overlapping thrust faults suggests that parts of the range are rising faster than 2 m/ky (McFadden et al., 1982; Morton and Matti, 1987). The most active faults are north of the San Fernando Valley and west of where the San Jacinto and San Andreas faults pass through the Transverse Ranges.

The fronts between Big Tujunga and San Antonio Canyons (west-central part of range) appear to be less active. These are class 2 and class 3 fronts and only one Holocene surface rupture appears to have occurred (Fig. 3.23). Front 8 not only has degraded triangular facets but also is the only front with anomalously low stream-gradient indices (Keller and Rockwell, 1984).

The range-bounding fault is not always the most active (Fig. 4.36). Front 2B is highly sinuous

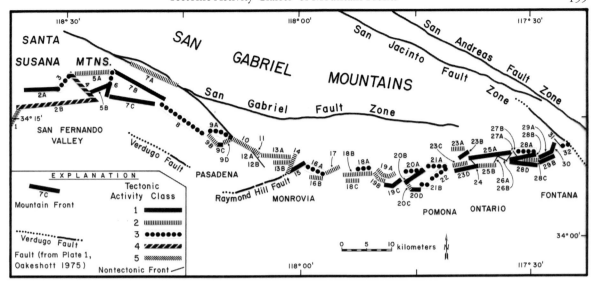

Figure 4.35 Tectonic activity classes of mountain fronts along the south side of the San Gabriel Mountains of southern California. The fronts are numbered consecutively from west to east.

(Fig. 4.34A) and the valley floor width–valley height ratios for Limekiln Wash are large. Note the abrupt increase of V_f ratios just downstream from the active Santa Susana fault zone. The footwall of the Santa Susana fault was raised during the 1994 Northridge earthquake (Yeats, 2001). Dip-slip rates are about 6 ± 4 mm/yr on the Santa Susana fault (Huftile and Yeats, 1996). Small values of V_f ratios in the watershed headwater may be in part a function of stream size as noted in the discussion of Figure 4.33. The piedmont foreland alluvium is old. These gravels were deposited as thin alluvial fans before uplift of front 2B caused dissection of the block between fronts 2A and 2B.

Uplift rates from ongoing geodetic surveys may differ from the rates suggested by tectonic geomorphology studies, which focus on identification of persistent long-term uplift. The mere presence of

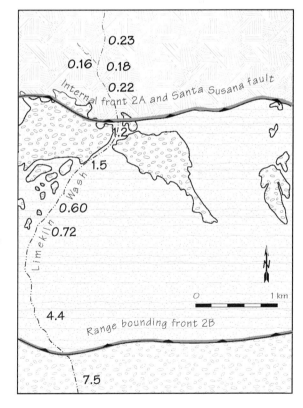

Figure 4.36 Active internal and inactive range-bounding faults of the Santa Susana Mountains. Mountain-front locations are on Figure 4.35. Valley floor width–valley height ratios are in bold numbers along Limekiln Wash. Flat-topped ridges of the piedmont foreland are capped by remnants of alluvial-fan deposits.

low fault scarps in Holocene piedmont alluvium records recent shifts in the locations of active faulting. We should also remember that active thrust faulting is largely hidden beneath the deposits of the Los Angeles basin (Hauksson et al., 1988; Davis et al., 1989; Scientists, 1994; Mori et al., 1995; Hudnut et al., 1996).

Many active faults are relatively narrow linear zones of base-level fall. Complications resulting from secondary antithetic and synthetic faulting may create broad fault zones. Appraisal of landscape response is even more difficult in broad areas of minor uplift. Such upwarps are more akin to areas of regional uplift, and their diffuse border locations of base-level fall may be difficult to discern.

Landscape responses to folding are more difficult to evaluate than responses involving single faults.

Folds, from a tectonic geomorphology perspective, are broad bands of tectonic base-level change between the fold axes (Fig. 4.37A). Magnitudes of cumulative tectonic base-level fall increase with distance away from the axis of an anticline (Figs. 4.37A, D). Rate of increase of base-level fall increases downstream, attains a maximum at the inflection point of the fold curve (location TIP of Figure 4.37A), and then decreases to the axis of the adjacent syncline.

Rock type and climate determine where erosion and deposition occur in landscapes being deformed by folding. Drainage basins underlain by resistant rocks with low sediment yields in a humid

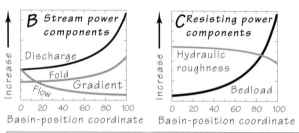

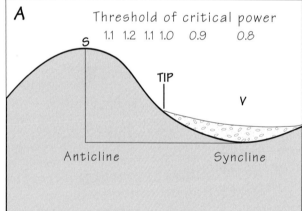

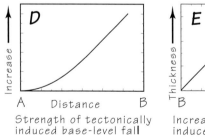

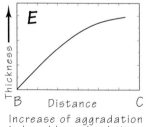

Figure 4.37 Tectonic perturbations in the folded Kettleman Hills landscape.

A. Diagrammatic section of a folded fluvial system. Deposition prevails downstream and erosion predominates upstream from the threshold-intersection point at TIP, which in this case coincides with the inflection point of the fold curve.

B. Downstream trends in factors affecting stream power between S and TIP on Figure 4.37A. Flow gradient becomes more gentle with increase of discharge and steeper as anticlinal folding increases valley-floor slope. Actual stream gradient is a function of these and other interacting variables.

C. Downstream trends in factors affecting resisting power (see Section 2.3, Figure 2.13) between S and TIP on Figure 4.37A. Hydraulic roughness presumably decreases as flow depth increases and size of streambed alluvium decreases. Channel and hillslope incision increases bedload transport rate sufficiently to change the mode of fluvial system operation from degradation to aggradation at the point TIP, the threshold-intersection point.

D. Steepening of the valley floors of consequent streams on the flanks of the anticline increases slowly near the flat anticline axis and is a maximum at the fold-inflection point at TIP.

E. Base-level rise resulting from piedmont aggradation is largest at point V. Rate of thickness increase becomes less nearer the axis of the syncline.

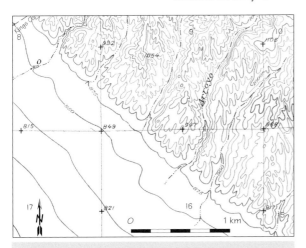

Figure 4.38 Topographic map showing the abrupt, straight mountain–piedmont junction of the North Dome of the Kettleman Hills anticline. Location is on the eastern flank of the Diablo Range, about 2 km north of the town of Avenal and just southeast of the Kings-Fresno county line (upper left). 25 foot (7.6 m) contour interval. Altitudes are in feet on this 1930 survey by the U.S. Geological Survey.

setting have streams that flush sediment into the intervening synclinal valleys and then downstream. Piedmont deposition may be minimal.

Arid settings and soft rocks favor deposition, especially where a piedmont of coalescing alluvial fans creates a base-level rise adjacent to rising folds (Figs. 4.37, 4.38). The proportion of an actively folding landscape exposed to erosion is less than in a humid setting.

Interactions of folding, erosion, and deposition along the western margin of the San Joaquin Valley have allowed deposition by consequent streams to occur well up on the flanks of folds, perhaps to the fold-inflection point (TIP on Fig. 4.37A). However, the transition from mountains to piedmont occurs at the same place in adjacent large and small watersheds. The mountain–piedmont junction (and the reference length, L_j, of equation 4.1) is straight or gently curving in these class 1 landscapes. Base-level rise exerted by piedmont deposition is significant, even where adjacent watersheds have greatly different unit stream power.

The behavior of fluvial systems, such as that diagrammed in Figure 4.37, is best understood using the threshold of critical power (see Section 2.3 and Figure 2.13). Rates of stream-channel downcutting into a rising fold are largely a function of flow discharge and hydraulic slope, both of which increase between locations S and TIP of Figure 4.37A. Offsetting this tendency for streams to downcut are resisting power factors such as hydraulic roughness and bedload transport rate. Bedload input from the hillslopes increases greatly downstream from the headwater divide. Erosion is so minimal in the headwater of the Kettleman Hills that the shape of the growing anticline is visible when looking across the accordant ridgecrests. This is why the assigned stream power/resisting power ratio is only 1.1 in Figure 4.37A.

Introduction of gravelly bedload increases quickly as flow discharge increases downstream to erode deep gorges – the departure from threshold of critical power indicated by a relative value of 1.2 on Figure 4.37A. This is an exponential increase of sediment yield from the mid- to mid-upper part of a non-steady-state landscape. This trend means that eventually the unit stream power of these occasional ephemeral flows is no longer capable of transporting the entire sediment load. The coarsest fraction is deposited downstream from the threshold-intersection point, indicated by the short reach of no departure from a value of 1.0. The mode of operation switches ever more strongly to deposition downstream from the reach that is neither aggrading nor degrading.

Deposition of permeable basin fill is a self-enhancing feedback mechanism. Increased streambed infiltration of water from ephemeral streamflows, compared to bedrock channel floors in the mountains, further decreases stream power. The assigned stream power/resisting power ratio decreases to 0.9, then 0.8. Basin-fill aggradation is a base-level rise that tends to offset tectonic base-level fall or rise caused by subsurface folding. Piedmont deposition maintains a straight mountain front regardless of variable drainage-basin area.

Folding can encroach from steep, rising mountain fronts into the fanhead areas of adjacent piedmonts (Keller et al., 2000). This can change the threshold of critical power values near crests of anticlines, thereby decreasing rates of deposition or initiating degradation. Dispersion of uplift by fold migration and propagation may result in thinner alluvial fans next to the mountain front. This does not indicate a lesser degree of relative tectonic activity. Field studies may be needed to show that such cases deserve a class 1 instead of a class 2 rating because the mountain front is in the process of migrating into the adjacent basin.

The rapidly rising eastern margin of the Coast Ranges of central California is a good place to apply the tectonic-activity-classes approach to folded mountain fronts (Fig. 4.39). This fold-and-thrust belt is near the plate bounding San Andreas fault, about 40–60 km to the west. These monoclines and anticlines of the late Quaternary foothill belt of the Diablo Range are encroaching on the San Joaquin Valley as the result of northeast–southwest compression.

We presume that mountain fronts along the western and southern margins of the San Joaquin Valley have blind thrust faults at depth (Namson, and

Davis, 1988, Wentworth and Zoback, 1989; Namson et al., 1990; Wakabayashi and Smith, 1994; Keller et al., 1998).

Earthquakes occur here. The first of a north-west-to-southeast sequence of earthquakes on the same 100-km long hidden thrust fault (Stein and Yeats, 1989; Stein and Ekstrom, 1992; Lin and Stein, 2006) was the Mw magnitude 5.4 New Idria earthquake of 1982. The Mw magnitude 6.5 Coalinga earthquake in 1983 (Hill, 1984; Rymer and Ellsworth, 1990) emanated from beneath Anticline Ridge, and the Mw magnitude 6.1 earthquake of 1985 from beneath the north end of the Kettleman Hills (Wentworth et al., 1983). The crest of Anticline Ridge rose abruptly about 500 mm during the 1983 earthquake

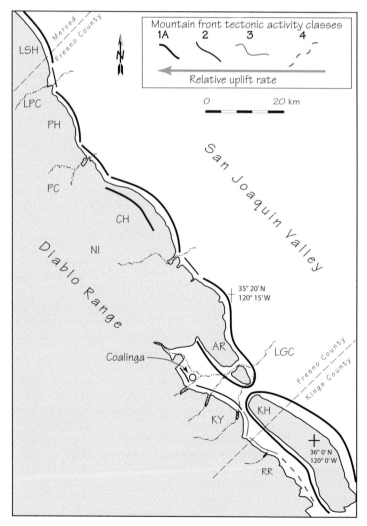

Figure 4.39 Tectonic activity classes of folded mountain fronts along the eastern margin of the Coast Ranges in central California.
AR, Anticline Ridge;
CH, Ciervo Hills;
CN, Coalinga Nose;
KH, Kettleman Hills;
KY, Kreyenhagen Hills;
LGC, Los Gatos Creek;
LSH, Laguna Seca Hills;
NI, New Idria;
PC, Panoche Creek,
PH, Panoche Hills;
RR, Reef Ridge.

(King and Stein, 1983). In hindsight, I now realize that comparison of first-order level-line surveys across Anticline Ridge in March, 1962 and March, 1963 recorded a precursor event to the destructive Coalinga earthquake. Benchmarks rose as much as 24 mm in a pattern that mimicked the topographic profile of Anticline Ridge (Bull, 1975b, Fig. 4). These "hidden earthquakes" (Stein and Yeats, 1989) underscore the continuing active nature of thrust faults in the cores of these folds. This landscape reflects rapid, continuing uplift.

The predominance of Class 1 mountain fronts (Fig. 4.39) describes a highly active tectonic environment, despite the deceptive appearance of smooth, rolling hills underlain by soft marine sedimentary rocks. Large antecedent streams emerge from structural lows between rising folds (Panoche Creek), or are superimposed on rising anticlines (Los Gatos Creek). Even though the larger streams have intermittent flow characteristics, they easily bevel broad strath surfaces in the soft Tertiary marine mudstone and sandstone upstream from the mountain fronts. Alluvial fans downstream from these fronts have increased in thickness by as much as 700 m thick since deposition of the Corcoran Lake Clay member of the Tulare Formation. Sanidine crystals from a tuff in the upper part of the Corcoran have a K/A age of ~625 ka (Janda, 1965). This suggests that mountain-front uplift rates exceed 1 m/ky. Pliocene–Pleistocene stratigraphic formations have been folded on all of the Class 1 range-bounding faults. Stein and Yeats

note "The youngest folded sediments are less than two million years old, which indicates that the fold began to form since that time. If a 75-centimeter growth is typical for events at the Coalinga site, then Anticline Ridge could have been built by roughly 1,000 ancestral earthquakes of magnitudes from M = 6 to M = 7 recurring every 1,000 to 2,000 years."

Class 2 mountain fronts have a less obvious juncture with the alluvial-fan piedmont, and mountain–piedmont junction sinuosity increases greatly. Much broader valleys extend upstream from the mountain front. Class 3 and 4 mountain fronts have another notable characteristic. Differential erosion of the mix of sandstone and mudstone beds results in a mountainous landscape where hard strata form prominent ridgecrests, and soft rocks underlie the broad valley floors of even small tributary streams. Reef Ridge in the southern Kreyenhagen Hills is an example.

Class 1 mountain fronts are being raised so fast that soft mudstone is at the same position in the landscape as the somewhat harder sandstone. Note that the tectonic activity in the Kings County part of Figure 4.39 appears to have shifted eastward with the thrust-fault propagation that resulted in the formation of the Kettleman Hills.

The influence of uplift on landslide distribution is illustrated by the semiarid Arroyo Ciervo fluvial system in the Ciervo Hills (Figs. 4.40, 4.41). This drainage basin is underlain by 60% diatomaceous shale, 8% mudstone and clay, and 32% soft sand-

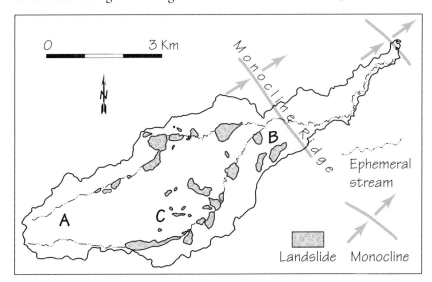

Figure 4.40 Distribution of landslides upstream from tectonically active Monocline Ridge in the Arroyo Ciervo drainage basin, located at CH in Figure 4.39. Oldest landslides are at B and youngest are at C. Area A has yet to experience the pulse of renewed landslide activity.

Figure 4.41 South fork of Arroyo Ciervo where landslides are encroaching into the stream channel from both sides of the valley.

stone and sand. Soft clay-rich rocks occur throughout the basin but landslides are not common near the mouth or near the headwater. Mean annual (wintertime) precipitation increases with altitude from about 110 to 360 mm. An active monocline at the basin mouth is so young that uplift has yet to create high local relief (see Figure 4.3D for 1.5 m contour-interval map of this front), but steep slopes and 300 m high hills are common in the shale and sandstone upstream from tectonically active Monocline Ridge, which is the class 1 internal front on Figure 4.40. Recent accelerated valley-floor downcutting has caused many landslides that occupy 8% of the watershed.

Arroyo Ciervo provides a nice example of how uplift interacts with fluvial processes to influence locations of landslides. Tectonically induced downcutting caused by presumed recent uplift of Monocline Ridge has extended to within 2 km of the headwater divide. Area A (Fig. 4.40) lacks landslides even though the rock type is the same diatomaceous shale as in downstream reaches. Mass movements consist mainly of young to active footslope slumps (Fig. 4.41) that encroach into the two trunk stream channels. In contrast, landslides in that part of the fluvial system immediately upstream from Monocline Ridge (location B) occur higher on midslopes or on footslopes adjacent to tributary streams. The location B landslides are a mixture of ages; some are only a few decades old, but others are so old that the

typical hummocky topography has become quite subdued. Tributary landsliding has just begun as a few small slumps in area C. This landslide succession illustrates the substantial reaction time for increased incidence of landslides as the effects of a tectonic perturbation are transmitted by the stream subsystem to progressively more distant hillslopes.

4.3 Summary

Future work should focus on upgrading the tectonic-activity-class model, which compares relative mountain-front uplift rates. Ideally, we should be making quantitative estimates of total Quaternary tectonic displacement for a mountain front, and of how rates of faulting and folding have changed with time. Such advances require improved understanding of how lithology and structure, and temporal and spatial variations in climate, affect rates of landscape denudation. The roles of these independent variables are fundamental to all geomorphology.

Defining mountain front uplift rates requires descriptions of stream-channel downcutting rates in the mountains and piedmont aggradation rates in order to understand base level controls in those reaches of fluvial systems that cross range-bounding faults. Better dating is essential. Advances in optical and thermoluminescence now are being applied to diverse stratigraphic settings. Terrestrial cosmogenic nuclides now rival standard radiocarbon dating because of their ability to estimate ages of both surfaces and deposits.

Mountain-front denudation is more varied and complex than for low piedmont fault scarps because larger scale makes for more diverse lithologic and climatic controls, and because of uplift may emanate from multiple zones of tectonic base-level fall. The need for better understanding of lithologic and climatic controls of mountain front tectonic activity classes has a counterpart in diffusion-equation modeling of fault scarps.

We now turn our attention to such low, comparatively simple mountain fronts. Fault scarps intrigue earth scientists who use both landscapes and stratigraphy to better understand earthquake hazards. In contrast to the multitude of surface rupture events needed to create an impressive mountain front, fault-scarp studies emphasize the size of the most recent surface rupture and when it occurred.

Chapter 5

Fault Scarps

Tectonic geomorphologists study fault scarps located at mountain–piedmont junctions and on the adjacent piedmonts. Humans use both sites extensively. These simple landforms have much to offer modelers of landscape evolution as they seek to better understand hillslope erosion and the genesis of mountain fronts. A fault scarp may record the latest of many surface ruptures that have created an impressive mountain front, or it may represent its mere beginnings where thrust faulting has propagated out onto a piedmont. Fault scarps are a mainstay of paleoseismology investigations. All the previous chapters have incorporated selected aspects of fault scarps (Figs. 1.9, 2.12, 3.17, 3.23, 3.27–3.29, 4.9–4.11, 4.23, 4.29).

5.1 General Features

I introduce useful terms and concepts and then consider independent variables affecting diffusion-equation modeling of scarps. These include climatic and lithologic controls. An example of fault-scarp processes with an earthquake recurrence interval of >200 ky is presented to balance the emphasis on Holocene fault scarps. The advent of cosmogenic dating creates intriguing possibilities for breakthough studies of fault scarps in alluvium. Cosmogenic-isotope studies of smooth limestone fault planes hold much promise for future studies of bedrock fault scarps. LiDAR (Light Distance and Ranging) radar provides exquisite details of surfaces ruptured by recent faulting.

Uplift along range-bounding faults adds fresh increments to scarps that rupture bedrock but these can be difficult to separate from former increments and analyze in a quantitative manner because of the 1) great diversity of lithologies and structures, and 2) the steep and irregular slopes of mountain-front escarpments are the product of many surface ruptures. Single-event surface ruptures in alluvium that create simple convex–concave curves are much easier to model.

Piedmont fault scarps are the main topic of this chapter because:
1) The surfaces of alluvial fans and treads of stream terraces provide geometrically simple shapes, such as segments of cones and planar surfaces (Section 3.3.4) that are ideal for analyzing tectonic deformation and consequent erosional surface-rupture landforms.
2) Stream terraces and fanhead embayments cross range-bounding faults. Faulting of these planar surfaces records surface-rupture characteristics of one or several prehistorical earthquakes. The ages of adjacent faulted and unfaulted surfaces in a flight of stream terraces bracket times of most recent faulting.
3) The ubiquitous sandy gravels of the piedmonts of the western North America provide an approximately uniform material for studies of fault-scarp denudation and for defining chronosequences of soil profiles of flights of faulted and unfaulted stream terraces.

An exceedingly long time span elapsed between the penultimate and 1887 surface ruptures of the Pitaycachi fault, Mexico. This gently sloping hill has no obvious previous surface rupture, even though it too created a 3 to 4 m high fault scarp.

4) Many piedmonts have a great variety of alluvial-surface ages. Offsets of young and old stream-terrace treads can be compared. Cumulative displacement of older surfaces produces multiple-rupture event fault scarps. Paleoseismologists seek to estimate characteristic earthquake and surface-rupture size and displacement (Section 3.3).

5) Alluvial fans and terrace deposits may contain organic materials for radiocarbon dating, volcanic ash that can be correlated with previously described and dated tephra, fluvial and eolian silt for luminescence dating, and depth-dependent systematic trends of terrestrial cosmogenic nuclides. Surficial boulders can be surface-exposure dated with terrestrial cosmogenic nuclides and weathering rinds.

6) Diffusion-equation modeling of changes in piedmont fault-scarp morphologies is a standard paleoseismology technique. Rates of degradation of uncemented gravelly deposits are sufficiently rapid to allow differentiation of scarps formed by single-rupture events during the late, middle, or early Holocene.

7) Stream terraces and alluvial fans are readily accessible sites for excavation of trenches to study faulted and unfaulted strata that record earthquake recurrence intervals, liquefaction features, colluvial wedges, paleosols, and fault-pond deposits (Sieh, 1978b; Weber and Cotton, 1981, McGill and Sieh, 1993; Pantosti et al., 1993a, b; McCalpin, 1996; Machette et al., 1992, 2005; Machette and Brown, 1995).

Many fault scarps have undergone minimal dissection, which suggests a late Quaternary surface rupture. More than 10,000 km of active faults occur just in the Great Basin part of the Basin and Range Province (Wallace, 1977). Most have low scarps in unconsolidated piedmont alluvium, or in colluvium at the mountain–piedmont junction.

Fault scarps provide a convenient way of describing late Quaternary segmentation of faults (Section 3.3.4). Features such as scarp height, degradation slope morphology, and multiple-rupture-event crests of late Pleistocene and Holocene scarps identify differences between adjacent fault segments (Machette, 1986; Crone and Haller, 1989; Turko and Knuepfer, 1991). These segments can be classed as persistent or nonpersistent where compared to the tectonic activity classes of the adjacent mountain front (Section 4.2.3.2).

By analyzing piedmont fault scarps in 17,000 km^2 of central Nevada, Wallace (1977) estimated that about seven earthquakes of Mw magnitude 7 to 8 have occurred during the past 12 ka. The 1915 earthquake in Dixie Valley created much-studied fault scarps. Historical fault scarps in arid regions tend to have prominent free faces (steep cliffy part) and narrow crests and bases (Fig. 5.1A). The skylined

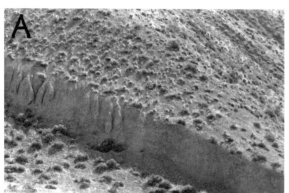

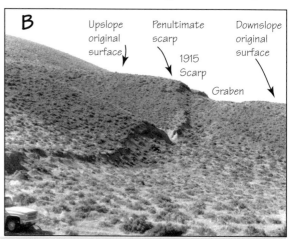

Figure 5.1 Surface ruptures along the range-bounding fault of the Tobin Range, Pleasant Valley, west-central Nevada.
A. This 3-m high scarp was formed by the 1915 earthquake. Left side of view is gullied and scarp-crest sinuosity has increased because of more runoff from hillslope than on the right side of view.
B. Faulted fanhead embayment at Figure 5.1A site. The convex ridgecrest gradually decreases in slope upslope from the graben in the faulted alluvium. This preceding (penultimate) fault scarp now appears as a bevel adjacent to the free face of the 1915 surface rupture. Truck for scale.

scarp (Fig. 5.1B) is the result of two surface ruptures. Wallace found that topographic profiles normal to the fault trace that described scarp steepness and crest sharpness were particularly useful guides to the approximate times of late Quaternary faulting (Figs. 5.2, 5.3).

Substantial progress has been made in understanding the evolution of fault scarps. Diffusion-equation modeling estimates times of formation of scarps with easily described topographic shapes. Pioneering studies by David Nash emphasized simple situations. Single-rupture-event scarps created by normal faulting of nearly planar alluvial surfaces are ideal for modeling. Section 5.2 explores the subject of how scarps become progressively less steep and the crest more rounded with the passage of time.

Complex scarps, such as those created by multiple thrust (Fig. 1.9) or normal faults, are not amenable to modeling. Multiple fault scarps are a composite of overlapping degradation and aggradation. The resulting gently sloping scarp is a composite topographic profile that appears anomalously old.

Fault-scarp terminology emphasizes processes that create distinctive landforms (Figs. 5.2, 5.3). Rupture of a smooth alluvial surface by a normal fault is a tectonic perturbation that abruptly increases local relief and creates a low linear ridge. This new hillslope changes shape with time. The fault scarp initially consists of a prominent *free face* (erosional part of scarp steeper than the angle of repose for loose sandy gravel) and an adjacent *debris slope* (area of colluvial deposition). The debris slope is initially formed mainly by slumps and rockfalls from the free face (Fig. 5.1A). The angle of repose for uncohesive materials ranges from 28° to 37°. The free face retreats upslope and quickly disappears. It may be regarded as a *weathering-limited slope* whose rate of roughly parallel erosional retreat is largely controlled by the rates at which materials are made available for transport by geomorphic processes of rainsplash, slopewash, and creep (Gilbert, 1877). A sharp crestal break in slope (Fig. 5.2) persists as long as the topographic discontinuity of a retreating free face remains. The crest and base define a new hillslope that ranges from 40° to 70° for 10-year old scarps and 3°-10° for 200 ky old scarps. A scarp that no longer has a free face may be regarded as a *transport-limited slope* whose rate of overall slope decline is largely controlled by the ability of erosional processes to remove weakly cohesive soil and alluvium. Fluvial processes broaden the *crest slope* or *scarp crest* (convex slope element between

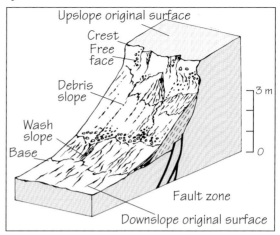

Figure 5.2 Topographic elements of a diagrammatic single-rupture event fault scarp. From Wallace (1977). Each element has a characteristic morphology formed by different processes.

the upslope original surface and free face) and *wash slope* (concave area of fluvial deposition that buries the downslope original surface to form a *new alluvial slope*).

Readers will note a variety of terms used for magnitude of fault-scarp tectonic deformation. Most terms are only approximations of the actual throw as used in a structural geology sense. Scarp height is visibly obvious, but is larger than throw for sloping alluvial surfaces. Projected net separation of the topographic profile across a fault zone (Fig. 3.27) may be close to representing the true displacement but should be regarded as an apparent throw unless corrected for land surface and fault plane dips. Common synonyms for apparent throw include "scarp offset" and "2a", where a is half the apparent throw above and below the scarp-profile midpoint.

Piedmont fault scarps are like range-bounding faults in that they separate reaches of tectonically induced aggradation and degradation. After each surface rupture, the consequences of the tectonic perturbation spread to the adjacent reaches of the fluvial system. Tectonically induced downcutting occurs rapidly in larger *antecedent streams* (streams that were present before faulting), but smaller *consequent streams* (whose channels form as result of the uplift event) also begin to erode headward into newly created scarps (Fig. 5.1A). Sediment derived from

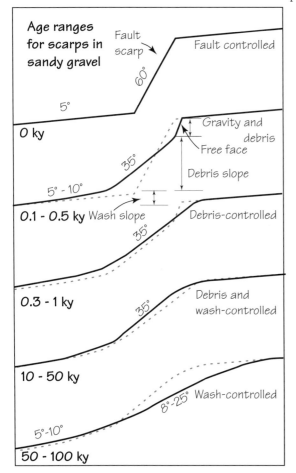

Figure 5.3 Sketches showing stages in the sequence of degradation of a typical normal fault scarp in alluvium. Each dashed gray line represents the solid line of the previous stage. From Wallace (1977).

soil profile is examined. Truncation of a formerly present A soil-profile horizon upslope from the crest reveals where rates of erosion now exceed rates of soil-profile formation. Extent of recent burial of the soil profile downslope from the debris and wash slope defines the limits of the **new alluvial slope** where tectonically induced aggradation now is occurring.

Fault scarp terminology emphasizes processes that create distinctive landforms when piedmont alluvium is ruptured. The terms have a strong genetic connotation (Wood, 1942; Young, 1972; Cooke and Warren, 1973). The same types of processes that shape larger hillslopes control the slope elements and degradational stages of a single-rupture event fault scarp (Figs. 5.2, 5.3). The term "crest slope" implies dominance of rainsplash and sheetflow processes, the "free face" gravity-controlled erosional processes, the "debris slope" a combination of gravity and fluvial depositional processes, and the "wash slope" fluvial processes. These genetic terms are quite appropriate for new hillslopes abruptly created by faulting. However, additional changes in the new landforms of a ruptured original surface also need to be considered.

Crestal convexity and wash slope concavity decrease with time at similar or different rates. The original surface commonly is degraded near the crest, buried near the base, and absent between the fault trace and the crest (Fig. 5.4). The upslope original surface may be subdivided into eroded and uneroded elements. Significant amounts of crestal rounding may occur even when a retreating free face is still present. Although the magnitude of surficial lowering is minor compared to the magnitude of free-face retreat, one or more soil-profile horizons may be removed. A fairly typical example is the Drum Mountain fault scarps in western Utah (Bucknam and Anderson, 1979); their age estimate is 7–10 ka and the original surface seems to have been erosionally steepened by about 1° in the upslope 4 m of the scarp crest. The upslope eroded original surface has been modified by minor erosion.

The presence of a buried or unburied soil profile is helpful in defining the position of the downslope original surface so that estimates of surface displacement can be made. Thin A soil-profile horizons or the boundary between the B and C horizons are useful approximations of planes that parallel the original alluvial geomorphic surface prior to offset by faulting. The presence of an A horizon usually is a good indication of minimal modification of the original surface by fluvial processes. This use of soil-pro-

stream-channel incision is deposited on alluvial fans that radiate downslope from threshold-intersection points on the new alluvial slope. From a geomorphic sense, a new fault scarp is indeed the creation of a new mountain front.

Scarp geomorphic elements may expand in area and change altitude, which changes fault-scarp height (Figs. 5.3, 5.9). **Scarp height** is generally measured as the vertical separation between obvious topographic breaks in slope termed the **base** and **crest**. Geomorphic responses to faulting extend much farther, which is obvious when the pre-surface rupture file horizonation features avoids the pitfalls of 1) estimating displacement of the land surface where there has been erosion of and/or deposition on the original surface and 2) variations in thickness and dip of faulted sedimentary strata, both of which may introduce unknown amounts of error into estimates of vertical fault slip. Describing the soil-profile characteristics in the context of a soils chronosequence constrains the time of the surface-rupture event. Examples of using soils in studies of faulted surfaces and stratigraphy of colluvial wedges include studies by Chadwick et al., 1984; Hecker, 1985, 1993; Demsey, 1987; Forman et al., 1991; Machette et al., 1992, Amit et al., 1995, 1996, 2002; Enzel et al., 1996; McCalpin, 1996, Rubin et al., 1998, Birkeland, 1999; Lindvall et al., 2000; and Phillips et al., 2001.

Erosional terminology should not be used for slope elements downslope from the fault trace because the lower segments of fault scarps typically are depositional areas. The debris slope that is upslope from the fault trace is mainly a slope of transportation or is being denuded, although it may have a temporary veneer of detritus on it. Downslope from the fault trace, the debris slope is aggradational and merges with the depositional piedmont of the new alluvial slope. The new alluvial slope consists of areas of sheetflow deposition and may include small, actively aggrading alluvial fans. The threshold-intersection points represented by the fan apexes tend to

be downslope from the base of the debris slope for antecedent streams, and coincide with the base of the debris slope for consequent streams. Surface rupture, even in the form of 1 m fault scarps, affects local base-level processes as described in Figure 4.7. Part of the downslope original surface is shown in Figure 5.4. This part generally is buried except where the scarp is so young that depositional processes have yet to form an extensive new alluvial slope.

The length and height of historical fault scarps provide general information about earthquake size, but scarp height rarely is the same as throw (vertical fault displacement). Scarp height equals displacement for vertical faults that displace horizontal surfaces. Otherwise scarp height exceeds throw. Scarp height may also be less than vertical displacement where a faulted horizontal surface is raised by aggradation on the downthrown surface.

Several factors cause anomalous scarp heights. The difference between scarp height and displacement increases with decrease in dip of normal faults, increase in slope of an original surface that slopes in the same direction as the fault, downslope migration of the scarp base, and upslope migration of the scarp crest. A common procedure that is illustrated in Figure 3.27 is to use projections of the upslope and downslope parts of the displaced original surface to the estimated position of the fault trace to make an uncorrected estimate of throw. Wallace (1980) devised nomograms for determining the components of slip when scarp height and slopes of the fault plane and original surface are known (Fig. 3.30).

Additional complications to extrapolating surface slopes in order to estimate fault displacements (Section 3.3.4, Fig. 1.9) include
1) re-faulting depositional surfaces that change slope gradually or abruptly,
2) rotation of one or both of the adjacent fault blocks so that pre-faulting surface slopes are changed, and
3) movements along multiple faults. Earth deformation may be concentrated within a few meters of the fault, but commonly it is dispersed within a 10 to 100 m band of shearing and is different in the hanging wall and footwall blocks of normal faults. Fault-zone stratigraphy may define multiple ruptures (Figs. 3.23 and 4.29), give clues regarding frequency of rupture and cumulative displacement, and show how tectonic deformation influences fault-scarp morphology.

Several significant geomorphic, structural, and stratigraphic features of normal faulting are illustrated by the unusually complete exposure shown

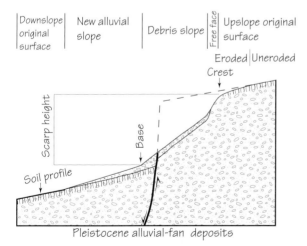

Figure 5.4 Fault scarp terminology and landforms as related to the pre-rupture event soil profile.

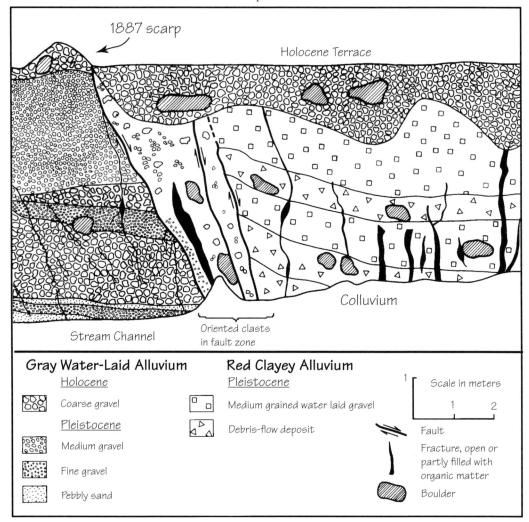

1887 scarp

Holocene Terrace

Stream Channel

Oriented clasts
in fault zone

Colluvium

Gray Water-Laid Alluvium

Holocene

Coarse gravel

Pleistocene

Medium gravel

Fine gravel

Pebbly sand

Red Clayey Alluvium

Pleistocene

Medium grained water laid gravel

Debris-flow deposit

Scale in meters

1   2

Fault

Fracture, open or
partly filled with
organic matter

Boulder

Figure 5.5 Cross section of Pitaycachi normal fault and alluvial deposits exposed in south bank of Arroyo Hondo, northern Sonora, Mexico. Fault planes of the 1887 and mid-Pleistocene surface ruptures are shown. Figure 6 of Bull and Pearthree, 1988.

in Figure 5.5. A 60° to 70° west-dipping fault zone is exposed in a 6 m high streambank. Two markedly different alluvial sequences with different deformational styles have been juxtaposed by faulting and are capped by bouldery Holocene gravel. Gray-brown (7.5YR 7/2; Munsell, 1992), indurated, water-laid alluvium of Plio-Pleistocene age comprises the upthrown block where tectonic deformation is minimal. Drag-folding has not occurred and only a few closed fractures parallel the 1887 rupture plane.

The other two faults and greatly differing alluvial sequences on opposite sides of the fault zone are evidence for multiple fault ruptures and large cumulative displacement. Red-brown (5YR 5/6) alluvium of mid-Pleistocene age exposed in the downthrown block consists of water-laid gravels and debris-flow deposits. The debris-flow matrix contains 20% silt and 22% clay. Two normal faults of the penultimate surface rupture do not penetrate into the capping Holocene alluvium, and drag-folded beds abut the faults.

The downthrown block also has numerous vertical fissures, largely filled with loose sand and fresh-appearing organic debris, that do not extend from the cohesive reddish clayey Pleistocene alluvium into the overlying noncohesive sandy Holocene gravel.

The structures associated with the faulting shown in Figure 5.5 represent three different styles of deformation near a major fault. Minor fractures in the upthrown block that parallel the main 1887 rupture plane were probably generated by tensional stresses during the 1887 surface-rupture event, as the basinward block moved away and down from the relatively stable "upthrown block". Drag-folding associated with Pleistocene faulting in the downthrown block may represent soft-sediment deformation of saturated deposits. The open fissures (shown in black) on the downthrown block represent a more brittle style of deformation during the 1887 event that affected cohesive dry deposits well above the water table. The orientations of these fissures are appropriate for extensional fractures (Davis, 1984, p. 272) if the dip of the fault becomes slightly less at depth. Pervasive fissuring during the 1887 event, and soft-sediment deformation during the prior event, reflect maximal extension of materials close to the fault zone on the downthrown side. They may be major causes of surface subsidence (commonly referred to as back-tilting; Fig. 1.6) near normal faults, which generally are not exposed as well as at Arroyo Hondo.

Many normal faults have a single trace, but the position of the fault trace in the topographic cross section of the scarp is quite variable. The initial position of a fault trace is at the base of the newly created escarpment. Scarp-crest degradation and scarp-base aggradation move the fault trace to a more central location, especially for multiple-rupture event scarps.

Multiple-rupture event fault scarps are a composite of two or more times of surface rupture. Most multiple-rupture event fault scarps have several features indicative of rejuvenated surface rupture. Renewed fault displacement may abruptly steepen the preexisting scarp slope producing a segmented topographic profile. For example, renewed normal faulting might create a scarp with 25° to 35° crest and footslopes above and below a new 70° free face (Fig. 5.1B). Multiple-rupture event fault scarps also may have multiple colluvial wedges and associated paleosols in the debris-slope stratigraphy (Figs. 3.23 and 5.38 discussed later). Scarps formed by frequent surface ruptures may not have sufficient time to create the segmented appearance of multiple-rupture event

fault. Occasionally, scarp segmentation is the combination of tectonic and fluvial processes, for example faulting followed by lateral erosion induced base-level fall that steepens the basal portion of the scarp.

One of the best ways to identify a multiple-rupture event fault scarp is to study stair-step flights of stream terraces that cross the fault zone (Mason et al., 2006, Fig. 2). Terraces may range in age from less than 2 ka to more than 200 ka. The youngest terrace tread may not have been faulted and the fault trace is buried by terrace alluvium. Older terraces commonly have progressively higher scarps with more complex topographic profiles. Uplift history and characteristic earthquake estimates for such a paleoseismology site can be made by estimating the age and displacement and age of each terrace in the flight. A soils chronosequence is an essential ingredient for success, as is having more terraces than surface-rupture events.

The Buttersworth graben has multiple surface ruptures that are revealed by stream terraces, but not by the fault scarp. The graben formed on internal normal faults in a valley fill 275 m above the class 1 mountain–piedmont junction of the Stillwater Range of west-central Nevada (Fig. 4.24 and Table 4.3).

Topographic profiles of the high scarp along the west side of the graben (Figs. 5.6, 5.7) suggest only one 8 to 10 m surface rupture. Profile A has a

Figure 5.6 The 7 m high multiple-rupture event fault scarp at the Buttersworth graben site, Stillwater Range, west-central Nevada. The view is from the east side of the graben. Topographic profiles of the fault scarp at A and B are shown in Figure 5.7. The fanhead trench in which the stream terraces of Figure 5.8 are located is in the background.

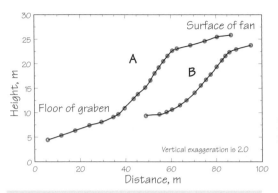

Figure 5.7 Topographic profiles of two-rupture event fault scarp on the west side of the Buttersworth graben, Stillwater Range, west-central Nevada, USA.

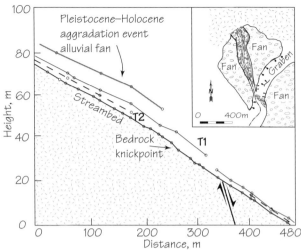

Figure 5.8 Longitudinal profiles of the present and past valley floors at the Buttersworth graben site in the Stillwater Range, west-central Nevada, USA.

break in slope halfway up the scarp that is suggestive of steepening caused by renewed faulting, but has a narrow crest with an abrupt break in slope that is indicative of a young single-rupture event fault scarp. Profile B has no break in slope and has a narrow crest. Rapid slope retreat is favored by moderately fine-grained and uncemented alluvium, which is sandy gravel with few boulders.

Analysis of alluvial surfaces (Fig. 5.8) clarifies the anomalous situation of a high scarp that seems to be the result of only one rupture event. The youngest terrace continues through the fault zone, the intermediate age surface is displaced at the fault, and fan surface ends at the fault. The displaced surfaces indicate that the fault scarp of Figure 5.6 is the result of two surface ruptures, each having a displacement of ~3 m.

The upper surface is an alluvial fan whose soil is a duric natrargid (soils terms after Chadwick et al., 1984, and Birkeland, 1999) with a pH of 9.6 that is indicative of at least an early Holocene, and possibly a latest Pleistocene age. The T1 terrace soil is a natrargid with a pH of 8.3 to 8.5, and has surficial rocks with a thin but brownish-black varnish; these features indicate a mid-Holocene age in the Dixie Valley soils chronosequence of Chadwick et al. (1984). Terrace T2 lacks a soil profile and has abundant driftwood on the tread; it is a recent debris-flow aggradation event. The fan and T1 surfaces diverge from the streambed downstream to the fault. This pattern of tectonically induced downcutting indicates that the fault is active. The T1 terrace is displaced 2.5 m by the fault but only on the west side of the graben; apparently the

ruptures along the east side of the graben occurred earlier. The fan surface ends abruptly at the projection position of the fault, where it is 7 m above the T1 terrace. It seems that an early Holocene surface rupture was followed by a mid-Holocene event, an example of characteristic surface ruptures with vertical displacements of about 2.5 to 3.5 m.

Breaks in slope, or bevels, of scarp crests indicate multiple surface ruptures. Why don't the topographic profiles of Figure 5.7 reveal the multiple surface ruptures? The time span between two surface ruptures has to be long enough to change the scarp profile. Fonseca (1988) uses equation 21 of Hanks et al. (1984) to show that the maximum slope of the older scarp at the Buttersworth site would have degraded to about 33°, only 3° from the 36° angle of repose for these materials. It seems that insufficient change in scarp morphology occurred between the two events to create a recognizable scarp bevel.

Morphologies of fault scarps are also dependent on several important variables: time (Section 5.2), climate and vegetation (Section 5.3), erodibility of surficial materials (Section 5.4).

5.2 Scarp Morphology Changes with Time

The need for paleoseismologists to better understand the frequencies and elapsed times of damaging earthquakes has resulted in a near obsession with the time factor. Standard dating tools, such as luminescence and radiocarbon analyses, are being applied to new geologic settings. Surface-exposure dating methods, using materials ranging from terrestrial cosmogenic nuclides to cobble weathering rinds, are being employed to estimate ages of faulted treads of stream terraces and alluvial fans. They focus on the age of the ruptured surface rather than on the stratigraphic age of the still older underlying deposits. The broad range of precision and accuracy of ways to date earthquakes is discussed in Section 6.1.1.

Modelers of landform change have found the simple landform of single-rupture event fault scarps to be predictable. Systematic changes in scarp morphology provide a way to estimate the elapsed time since an earthquake ruptured a piedmont surface.

One needs to remember that many aspects of fault scarps change with the passage of time, including height, maximum slope, width, and the extent of consequent streams that channel flow to the scarp from the upslope original surface. A surface rupture can also change rates of geomorphic processes, including soil-profile development and infiltration rate; both change the plant community. Late Quaternary changes in precipitation and temperature also influence geomorphic processes and rates of landscape change. Keeping the above cautions in mind, the following discussions evaluate the important constraint of scarp height, and examine the nuances of diffusion-equation modeling.

5.2.1 Changes in Scarp Height

Scarp height – the difference in altitude between the base and crest of a fault scarp – changes even if surface rupture is not renewed. Fault scarps may first increase and then decrease in height, and may have heights that are more or less than the fault displacement. Aggradation and degradation change both the altitudes and the horizontal positions of the scarp base and scarp crest. After the time of surface rupture, t_0, scarp height is increased by retreat of the free face into the upslope original surface (Fig. 5.9). Increase in altitude of the crest at time, t_1, may be large where horizontal shift of free-face retreat is large and the upslope original surface is steep. The position of the scarp crest will continue to rise until scarp retreat eliminates the free face at time t_2. After time t_2, fluvial denudation tends to make the scarp crest broader, less steep, and less convex. The crest retreats more

slowly from the fault trace as the contrast in slope between adjacent parts of the new hillslope decreases. Crest altitude decreases as a function of denudation rate; this concurrent decrease in altitude partly offsets the increase in scarp-crest altitude that accompanies scarp-crest retreat.

An opposite trend is common where aggradation occurs on the downslope original surface. Post faulting altitudes of the scarp base may either increase or decrease and are a function of the relative amounts of aggradation and downslope shift of the scarp base. These two factors tend to offset each other to a greater extent than the opposing base-level processes on the scarp crest. In Figure 5.9 the base has, between times t_0 and t_1, 1) moved downslope from the fault trace as a result of construction of the debris slope, and 2) been elevated above the downslope original surface by aggradation of the wash and debris slopes. The net effect usually is a small increase in altitude of the scarp base.

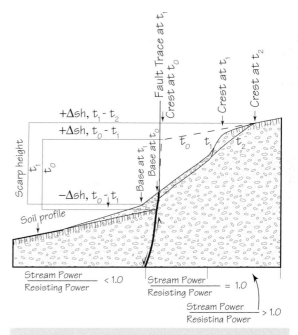

Figure 5.9 Horizontal and vertical changes in the boundaries between slope elements of a fault scarp, and increases (+) and decreases (-) in scarp height, after faulting at time t_0. The morphology of the scarp is shown at time t_1, and the position of the crest upon elimination of the free face is shown at time t_2.

The hypothetical example of Figure 5.9 illustrates progressive increases in altitudes of both scarp crest and scarp base between times t_0 and t_1. The net effect of these nontectonic base-level processes operating upslope and downslope from the fault trace is to increase scarp height to a value that exceeds fault throw. This trend generally is reversed after elimination of the free face.

Geometry of piedmont faulting (Section 3.3.4) and nontectonic changes in scarp height (Fig. 5.9) provide many pitfalls for those tempted to use scarp height as an indicator of earthquake size.

In summary, the base-level processes of vertical fault slip, scarp crest and free-face degradation, and aggradation on the new alluvial slope determine the heights of fault scarps. The downslope change from net erosion to net deposition coincides roughly with the fault trace, which is the position of the tectonic base-level fall. Thus, the location of an abrupt change in mode of fluvial-system operation is tectonically controlled, just as it is for class 1 mountain fronts (Figs. 4.9–4.11). Scarp height approximates uplift only at the time of displacement, and only for vertical faults that rupture horizontal surfaces. Scarp height tends to increase until the free face has been removed by erosional retreat. Then both crest degradation and aggradation of the new alluvial slope may progressively reduce scarp height.

5.2.2 Decreases in Maximum Scarp Slope

Bucknam and Anderson (1979) recognized the importance of scarp height on the steepness of fault scarps of various ages. They compared maximum slope angle and height for scarps formed by piedmont faulting and lacustrine shoreline processes. Slope angles were measured for a range of scarp heights created by the same surface rupture event. Scarp steepness increases with increasing scarp height (Fig. 5.10) but at a rate dependent on scarp age. Free faces were no longer present at their three sites. The climate is moderately seasonal, semiarid, and the lithology is sandy gravel. The times of initial fault-scarp formation are approximately 1, 10, and 100 ka. The three regressions of Figure 5.10 show that maximum scarp slope becomes less with increasing age and decreasing scarp height. For a scarp 3 m high, the maximum slope angles are 27°, 18°, and 10° for the 1, 10, and 100 ka scarps.

Free faces persist longer in higher scarps. Scarps that are 2 to 5 m high expose materials that are more resistant to erosion than do scarps that are less than 1 m high. Even in sandy gravel, surficial soil horizons are less resistant to erosion than the more cohesive or cemented deeper horizons. These factors together with the need to remove larger masses of materials to denude high, as compared to low, scarps may explain much of the variation in maximum scarp

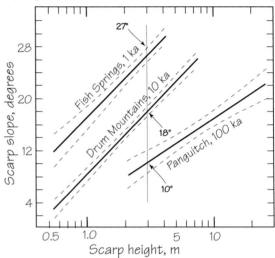

Figure 5.10 Relation of maximum slope to height for Pleistocene and Holocene fault scarps in western Utah. From Bucknam and Anderson (1979). Dashed lines show the regression significance at the 95% level.

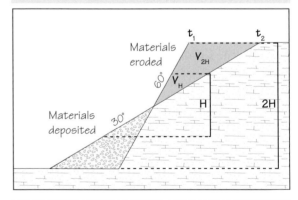

Figure 5.11 Cross sections showing nonlinear fourfold increase of volumes of earth materials eroded from and deposited between times t_1 and t_2 for scarp heights of H and 2H as a linear scarp slope changes from 60° to 30°.

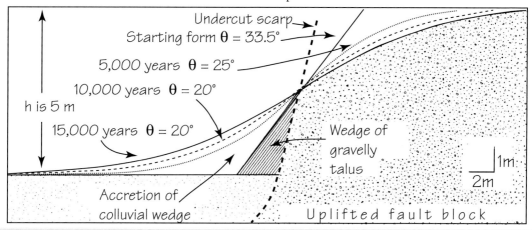

Figure 5.12 Modeled stages in the evolution of a 5 m high scarp in sandy gravel in the 15 ky after the free face has raveled to the starting slope of 33.5°. The scarp profiles that would be predicted by diffusion-equation modeling at 5, 10, and 15 ky use a degradation rate coefficient of 12×10^{-1} m²/ky. Figure 1 of Pierce and Colman (1986).

slopes. A doubling of scarp height greatly increases the volume of materials that need to be eroded for a given change in scarp morphology (Fig. 5.11).

5.2.3 Diffusion-Equation Modeling

Progressive changes in the shape of two fault scarps on different piedmonts would be virtually indistinguishable if the variables of lithology, scarp height and piedmont slope, climate, and vegetation were the same. Only the factor of time would be responsible for differences in steepness and curvature of the topographic profiles of the two scarps. Such reasoning encourages modeling of fault scarp topographic profiles in order to estimate their times of surface rupture. Hold the other variables constant and model the landform evolution.

Diffusion-equation modeling involves matching of surveyed topographic profiles of fault scarps with topographic profiles of different ages as modeled by an equation that treats transport-limited hillslope processes as diffusion. The example shown in Figure 5.12 illustrates the distinctly different predicted topographic profiles, and a decreasing rate of change of scarp morphology, during a 15 ky time span after elimination of the weathering-limited free face.

Accuracy and precision of calibrated modeled fault-scarp ages depend on the degree of variation of the controlling variables within a study region. In a general sense, piedmont fault scarps in the western United States are 1 to 5 m high, and occur on 2° to 20° piedmont slopes of sandy gravel in a semiarid climate. This rough similarity in controlling factors is largely responsible for much of the apparent success of diffusion-equation modeling. However, we should not expect too much from diffusion-equation modeling (Mayer, 1984). It generally cannot distinguish ages of fault scarps that are separated by only 2 ky, but it commonly discriminates between surface ruptures classed as late Pleistocene, and early, middle, and late Holocene. The lack of better precision is attributable mainly to seemingly minor variations of geomorphic variables between sites. For example, sand with 25% gravel has a narrow range of alluvium mass strength, but variation in argillic and calcic soil-profile horizons results in a tenfold variability of alluvium mass strength along with major decreases in diffusivity coefficient.

Such variations are sufficiently important that Sections 5.3 and 5.4 are devoted to climatic and lithologic controls of fault-scarp morphology. Diffusion-equation modeling generally cannot be used where displacement along several faults has created a multiple fault scarp, nor is it usually appropriate for study of multiple-rupture event fault scarps.

Many workers have modeled scarps created by tectonic, lacustrine, and fluvial processes in the northern Basin and Range Province. Scarps formed by lateral erosion of stream channels incised into alluvium provide a great variety of formation times and orientations and thus are well suited to evaluate microclimatic controls (Pierce and Colman, 1986). Scarps formed by highstands or recessional stillstands of lake levels can be assumed to have formed at the same time within a given lake basin (Hanks and Wallace, 1985; Hanks and Andrews, 1989).

Scarps that have resulted from nontectonic processes, such as lateral erosion induced base-level fall, do not have the displaced original surfaces that characterize tectonic fault scarps on piedmonts. Only the upslope original surface constrains the time of the lateral erosion base-level fall to a maximum age. The younger downslope surface is fluvial or lacustrine in origin. Modeling the ages of such nontectonic scarps involves an ***apparent displacement*** of the original surface. This term also is appropriate for fault scarps where the downslope original surface has been buried to create a new alluvial slope.

Characteristics of scarps created during the highstands of late Pleistocene Lake Bonneville in Utah and Lake Lahontan in Nevada are used to test diffusion-equation modeling, and evaluate the importance of lithologic and scarp-height controls. Highstand scarps are notched into a variety of gravelly piedmont deposits. The scarp ages are about 14.5 ka for Lake Bonneville (Scott et al., 1983) and 13 ka for Lake Lahontan (Benson and Thompson, 1987a,b; Adams et al., 1999; Adams and Wesnousky 1999). A graph of maximum scarp slope and apparent dis-

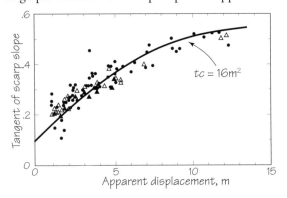

Figure 5.13 Relation of maximum scarp slope and apparent displacement for Lake Bonneville and Lake Lahontan highstand shorelines. The diffusion-equation model curve with a tc value (a constant relative scarp age) of 16 m² implies a single regional dataset. From Hecker (1985) as modified from Hanks and Wallace (1985).

placement (Fig. 5.13) suggests that the two datasets may be considered as a single population. Variations in lithology and microclimate result in some scatter of points. Figure 5.13 also conveys a strong dependence of maximum scarp slope on the magnitude of apparent displacement. Such tests encourage the use of diffusion-equation modeling to estimate ages of Basin and Range Province fault scarps.

Diffusion-equation modeling can be used where transport of surficial materials is directly proportional to the energy gradient. Transport of sediment by flowing water can be described as a diffusion process, but gravity-controlled rockfalls from sandy gravel free faces of fault scarps cannot be considered a diffusion process because they are abrupt mass movements. So, diffusion-equation modeling can be used to describe systematic changes in fault-scarp morphology only after the free face has been eliminated,. This generally requires at least two years (Pierce and Colman, 1986) to several centuries (Wallace, 1977).

Fluvial degradation of scarp crests by overland flow, raindrop splash, and wet or dry soil creep – and the concurrent aggradation of debris and wash slopes – is directly proportional to slope steepness. With the passage of time the convexity of the scarp crest, the concavity of the scarp base, and the steepness of all segments of the topographic profiles decrease (Fig. 5.12). The rate of change in altitude of any point, y, on the topographic profile of a scarp with time, t, can be described by the ***diffusion equation***:

$$\delta y / \delta t = c \, \delta^2 y / \delta x^2 \qquad (5.1)$$

where $\delta^2 y / \delta x^2$ is the slope curvature, and x and y are the horizontal and vertical coordinates of a point, and c (the diffusivity coefficient) represents the rates of fluvial degradation and aggradation, which are assumed to not vary with time or space. The rate at which altitude of a point on the fault scarp changes, $\delta y / \delta t$, is equal to the hillslope curvature at that point, $\delta^2 y / \delta x^2$, times the diffusivity coefficient, c.

The diffusion equation has been applied to a variety of scarps and larger hillslopes (Culling, 1960, 1963, 1965; Hirano, 1968, 1975; Kirkby, 1971; Nash, 1980, 1984, 1986; Colman and Watson, 1983; Hanks et al., 1984; Mayer, 1984; Andrews and Hanks, 1985; Pierce and Colman, 1986; and Nash and Beaujon, 2006). Subsequent applications that use the diffusion equation for modeling slope evolution in larger-scale landscapes and more complex scarps include publications by Begin, 1988, 1992;

Hanks and Andrews, Koons, 1989; 1989; Kirkby, 1992; Martin and Church, 1997; Arrowsmith et al., 1998; Pearthree et al., 1998; Hanks, 2000; Niviere and Marquis, 2000; Camelbeeck et al., 2001, Hilley et al., 2001; Lund et al., 2001; Mattson and Bruhn, 2001: Carretier et al., 2002; and Amoroso et al., 2004. A variety of solutions to the diffusion equation all provide products of age and the diffusivity coefficient — *tc*. Values of *tc* products may be regarded as *relative scarp ages* where the value of *c* is unknown, but is assumed to be constant. *c* can be estimated where *t* is known, and *t* can be made estimated where *c* is known.

The ~15 and ~13 ka ages of the Pleistocene Lake Bonneville and Lake Lahontan highstand shorelines provided an opportunity to estimate values of *c* for sandy gravel. The combined dataset provides an appropriate value of the *tc* product of 16 m² (Fig. 5.13). Substituting the known ages of the lakes in equation 5.1 provides an estimate of ~1.1 m²/ky for the diffusivity coefficient, *c*. Noting a similarity with New Mexico work done by Mike Machette (1982), Hanks et al., 1984 concluded that the ~1.1 m²/ky value might be widely applicable throughout the Basin and Range Province and Rio Grande rift valley.

Values of the diffusivity coefficient may not be as uniform as suggested by lake shoreline studies. Nash's calculations of *c* range from 1.2 m²/ky for sands of wave-cut bluffs of Lake Michigan to 0.44 m²/ky for gravelly alluvium of a dated Utah

fault scarp. It is likely that lithologic controls cause an order of magnitude variation in the diffusivity coefficient of scarp degradation, even when restricting scarp lithology to just alluvium (Section 5.4).

One should be cautious when using an age estimate from diffusion-equation modeling, especially when using an assumed value of diffusivity coefficient, *c*. The instructive Figure 5.14B examples

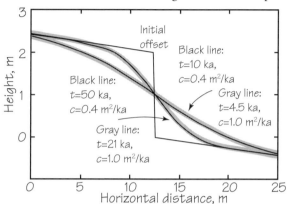

Figure 5.14 Scarp-profile modeling using different diffusivity coefficients.
B. Four modeled scarp profiles for a hypothetical 2 m offset for young and older stages of degradation. Similarity of the black and gray line model results underscores sensitivity of scarp degradation to the assigned value of diffusivity coefficient, *c*. Diffusion-equation modeling does not provide unique solutions for scarp age. Figure 1 from Phillips et al. (2003).

clearly show that tweaking *c* gets much different values for time, *t*. Getting a nice match to surveyed data is not enough. Independent, local determination of the *c* value is advisable; see Fig. 5.18 (Demsey, 1987).

Scarp height also influences age estimates for scarps analyzed by diffusion-equation modeling (Pierce and Colman, 1986). Their diffusivity coefficients, C^*, increase with scarp height and change with the microclimates of different slope orientations.

$$C^* = [1.35\,D_a + 3.03] \times [0.1\ m^2/ky] \qquad (5.2)$$

where D_a is apparent displacement.

The next step is to use appropriate values of *c*, or C^*, for fault scarps in piedmont sandy gravels throughout the same region. The examples used here

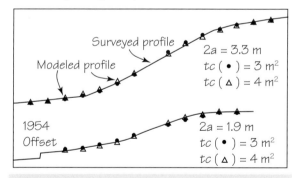

Figure 5.14 Scarp-profile modeling with different diffusivity coefficients.
A. Comparison of surveyed (dashed line) and modeled (points) topographic profiles of fault scarps at the mountain–piedmont junction, Stillwater Range-Dixie Valley, Nevada. Single-rupture event fault scarps using a tc value of 3.0 m² and 4.0 m². From Hecker (1985).

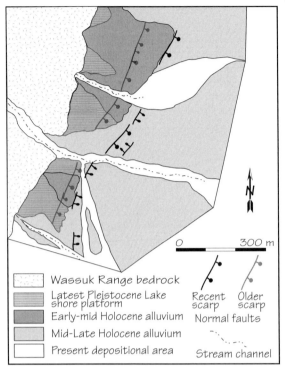

Figure 5.15 Map of piedmont fault scarps 1.5 km north of Copper Canyon that rupture alluvial surfaces postdating the 13 ka Lake Lahontan shoreline, Wassuk Range, west-central Nevada. From Demsey (1987).

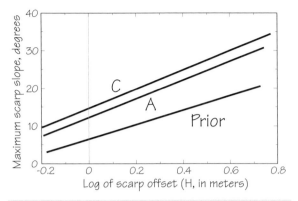

Figure 5.16 Lines for regressions of maximum slope (S in degrees) and scarp offset (H, in meters) for three classes of scarps along the northern Wassuk Range. A is for segment A recent fault scarps. $S = 17.8 (\log H) + 19.1$; n is 17, r^2 is 0.78. C is for segment C recent fault scarps. $S = 17.0 (\log H) + 20.4$; n is 13, r^2 is 0.77. Prior is for segments A, B, C for the penultimate surface rupture event. $S = 16.5 (\log H) + 14.8$; n is 12, r^2 is 0.56. From Figure 7 of Demsey (1987).

are from west-central Nevada. They are based on studies in the central Nevada seismic belt made by Hecker (1985) and Demsey (1987).

Hecker studied piedmont fault scarps along the east side of the Stillwater Range (Fig. 4.26), which is the area of the 1954 Dixie Valley Mw magnitude 6.9 earthquake (Slemmons, 1967). Five of the prehistorical fault scarps can be nicely modeled with a tc product of 3.0 m² or 4.0 m² (Fig. 5.14A). Differences between modeled and surveyed apparent displacements of less than 5% are encouraging.

Diffusion-equation modeling of fault scarps may be used to study fault segmentation. Scarps in the zone of overlap between fault segments may be poorly modeled with a single tc value if they are multiple-rupture event fault scarps. One of Hecker's topographic profiles was modeled crudely with a tc value of 3.0 m². It is nicely modeled with tc values of 5.0 m² and 8.0 m² for the basal and crestal portions

of the profile, respectively, which suggests a second surface rupture in this zone of overlap between two fault segments.

The faulted and unfaulted alluvial surfaces of the piedmont adjacent to the 3,400 m high Wassuk Range in western Nevada have nicely constrained ages. This allowed Demsey to determine the local value of scarp diffusivity coefficient. Adjacent Walker Lake is a remnant of Pleistocene Lake Lahontan. A chronosequence of piedmont soils was described and dated using volcanic ashes and materials for radiocarbon dating from piedmont alluvium, topographic position within the piedmont landscape assemblage, and diffusion-equation modeling of fault scarps. Alluvial surfaces notched by the prominent Lahontan highstand shoreline (Benson and Thompson, 1987a,b Adams and Wesnousky, 1999) are ~13 ka, and alluvium deposited on the shoreline is younger than 13 ka (Fig. 5.15).

The soils-geomorphology work provided internally consistent results. Maximum scarp slope indeed becomes less with time, as described by the Bucknam and Anderson model of Figure 5.10. Parallel regressions A and C of Figure 5.16 indicate the

Soil-profile site	Soil profile development index	Age of alluvial geomorphic surface	Number of surface ruptures
Fan graded to Lahontan shoreline	16.63	13–15 (a)	1
Lahontan shore platform	14.53	11–14 (a)	2
Lahontan shoreline inner edge	11.78	11–13 (a)	1
Fan deposited on shore platform	10.78	7–12 (a, b)	2
Stream terrace at Rose Creek	10.70	>7 (b)	2
High recessional Lahontan shoreline	10.25	11–13 (a)	2
Fan with graben at Copper Canyon	7.70	2–8 (b)	1
Stream terrace at Rose Creek	6.01	2–8 (b)	1
Fan near Squaw Creek	4.06	2–8 (b)	1
Fan near Copper Canyon	2.27	2–8 (b)	1
Fan deposited on lake highstand	2.06	2–8 (b)	1
Stream terrace near Cat Creek	1.68	<3 (b)	none
Fan with graben near Copper Canyon	1.31	2–8 (b)	1
Fan in topographically low setting	1.26	2–8 (b)	1
Surface deposited inside Graben	1.14	<3 (b)	none
Fan with dated tephra layer	0.66	<0.6 (c)	none
Piedmont adjacent to youngest scarps	0.48	<3 (b)	none
Stream terrace with dated tephra	0.32	<0.5 (c)	none
Oldest unfaulted alluvial surface	0.15	<3 (b)	none
Fan in topographically low setting	0.24	<3 (b)	none

Table 5.1 Wassuk Range, Nevada piedmont surfaces and soil profiles ranked by soil profile development index (Harden, 1982). Age estimates are based on relation to the Lake Lahontan highstand, fault-scarp modeling, and radiocarbon dating and tephra correlations. Table 1 of Demsey, 1987.

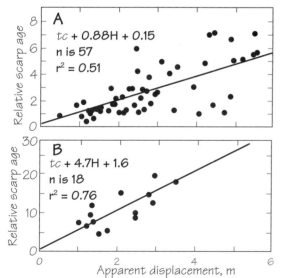

Figure 5.17 Plots of relative scarp age (tc product) and apparent displacement for young fault scarps along the northern Wassuk Range. From Demsey (1987).
A. Both ages of Holocene fault scarps.
B. 13 ka Lake Lahontan shoreline.

range of maximum slope-scarp offset regressions for sets of fault scarps created by the most recent Holocene surface-rupture event in two segments of the Wassuk Range piedmont. The plot for the prior event is beneath the other two and its intercept value describes a greater decrease of maximum scarp slope. Descriptions of the soil profiles on a wide range of alluvial surfaces by Karen Demsey, Oliver Chadwick, and Philip Pearthree provided the detail needed to assess several different diffusion-equation models. Distinct advantages of working in the Wassuk Range study area included having tephra to cross-check radiocarbon age estimates, the latest Pleistocene shoreline of Lake Lahontan, and soil-profile descriptions on surfaces whose ages could also be estimated by diffusion-equation modeling. The information for this tidy package of data is summarized in Table 5.1, where the data has been ranked according to the soil-profile development index of Harden (1982, 1987; Harden et al., 1991). This is done by numerically rating those soil-profile characteristics that change quickly with the passage of time (but at different rates of course). These included percent clay in the B horizon, redness with increase of iron oxyhydroxides, and profile thickness.

The relation between apparent scarp age and scarp height is shown in Figure 5.17 for the Wassuk Range scarps. Variations in local microclimate probably are minimal because the piedmont fault scarps face towards the east. The topographic profiles of piedmont fault scarps with small apparent displacements generally are modeled best with relatively low values of tc products compared to higher scarps of the same age. Numerical age control seems internally consistent as shown in a plot of relative scarp age and apparent scarp height (Fig. 5.17B) for the 13 ka highstand shoreline of Lake Lahontan. The relation between the diffusivity coefficient for the scarps of the Wassuk Range piedmont, $C*w$, and apparent displacement, D_a, is:

$$C*w = (0.36\,D_a + 0.12)\,m^2/ky \qquad (5.3)$$

where $r^2 = 0.76$ and n is 15.

The several approaches to diffusion-equation modeling for Wassuk Range fault and shoreline scarps provide different results (Table 5.2). The modeled ages of the Lake Lahontan shoreline range from 11 to 20 ka, which suggests that the diffusivity coefficient and/or corrections for scarp height may vary between local study areas with the Basin and Range Province.

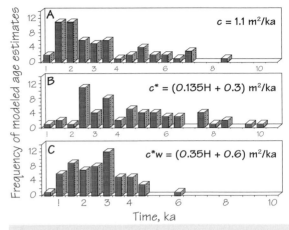

Figure 5.18 Spread of calculated fault-scarp ages based on diffusion-equation modeling of the youngest fault scarps along the northern Wassuk Range, west-central Nevada using different diffusivity coefficients.
A. 1.1 m²/ky of Hanks and others (1984).
B. (0.145 H + 0.3) m²/ky of Pierce and Colman (1986).
C. (0.35 H + 0.06) m²/ky of Demsey (1987).

The $C*w$ diffusivity coefficient seems most appropriate for the Walker Lake piedmont fault scarps because of the small standard deviations of the $C*w$ dataset as compared to the large standard deviations of the datasets when using c and $C*$ diffusivity coefficients (Fig. 5.18). The $C*w$ diffusivity coefficient also agrees best with the independent numerical ages of the most recent surface rupture in the Wassuk Range

Average Estimated Scarp Age, ky				
	Linear diffusion equation			Cubic equation
Scarp set	$c = 1.1$ m²/ky (a)	$C*=0.14(H)+0.3$ m²/ky (b)	$c*w=0.35(H)+.06$ m²/ky (c)	$\lambda = 4$ (d)
North area	2.5 ± 1.8	3.8 ± 2.1	2.3 ± 1.1	1.5 ± 0.6
Segment A	2.4 ± 1.8	3.7 ± 1.9	2.4 ± 1.0	1.4 ± 0.5
Segment B	2.3 ± 1.7		2.1 ± 1.2	1.0 ± 0.1
Segment C	4.5 ± 2.3	2.8 ± 1.3	1.5 ± 0.3	
South area				
Segment D	7.6 ± 1.7	4.6 ± 0.9	2.4 ± 0.7	
Prior event	6.8 ± 3.7	12.3 ± 5.5	6.9 ± 2.3	5.6 ± 3.5
Lahontan shorelines	11.2 ± 5.3	20.1 ± 5.8	13.2 ± 3.6	15.0 ± 6.6

Table 5.2 Comparison of morphologic ages for Wassuk Range scarps using different degradation rate coefficients and diffusion equations. From Demsey, 1987. Sources are a) Hanks and others (1984), b) Pierce and Colman (1986), c) Demsey (1987), d) Andrews and Bucknam (1987).

study area. Karen Demsey's 1987 study indicates a need to critically evaluate and constrain the diffusivity coefficient for each study area. We also need to consider how diffusivity coefficients vary with scarp height, microclimate, and alluvium mass strength.

5.3 Climatic Controls of Fault-Scarp Morphology

Climate may change in either a temporal or spatial sense. Temporal changes range from seasonal fluctuations to the major shifts in air-mass circulation, precipitation, temperature, windiness, and cloud cover associated with shifts from glacial to interglacial climates of the late Quaternary. Spatial changes include microclimate differences on north and south sides of a valley and variations with altitude and latitude. This section summarizes the effects of microclimatic variations on scarp morphology (Tables 5.3 and 5.4) in order to illustrate the overall importance of climatic controls on hillslope processes including fault scarps.

A study of the scarps of late Pleistocene stream terraces of the Big Lost River (Fig. 1.3) in Idaho by

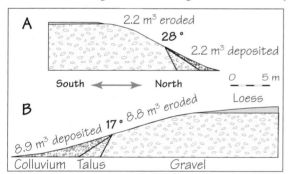

Figure 5.19 Importance of microclimate and scarp height on Idaho scarps with differing orientations.
Contrasts in the morphologies of north- and south-facing 5.8 m and 5.9 m high scarps of the same age. Erosion and deposition are relative to presumed 33.5° starting slope angle. From Figure 3 of Pierce and Colman (1986).
A. 28° maximum slope for north-facing scarp with slower erosion and deposition.
B. 17° maximum slope for south-facing scarp with faster erosion and deposition.

Pierce and Colman (1986) provided essential information about climatic controls on scarp evolution. Their study is fundamental to better understanding of hillslopes in a more general sense. Variables such as lithology, age, scarp height, altitude, and regional climate were held about constant in order to study the effects of microclimate on rates and types of slope processes. Hillslope microclimate at a given latitude is a function of orientation, height, and steepness.

About 100 late Pleistocene scarps of piedmont stream terraces were surveyed. The strongly seasonal climate is arid to semiarid, is frigid (Table 2.1), and has more than 130 freeze-thaw cycles per year. Stream terraces are ideal for studying temporal changes in small hillslopes because each terrace riser is graded to the stable base level of the adjacent lower terrace tread. Stable terrace treads isolate terrace risers from the effects of further stream-channel downcutting.

The Idaho terraces were formed as a result of lateral erosion induced base-level falls by streams incised into glacio-fluvial outwash gravels deposited at about 15 ± 4 ka. Gravelly colluvium deposited at the scarp bases has a silty matrix that is derived mainly from loess that caps the gravels. Rates of scarp degradation are limited only by rates of transport processes. Thus the Idaho stream-terrace scarps are ideal for using diffusion-equation modeling to better understand climatic controls on scarp evolution.

Slope orientation (aspect) has a profound influence on hillslope processes and shapes and is most obvious in asymmetric valleys that have markedly different vegetation on the north- and south-facing slopes (Melton, 1960; Bull, 1964b, Figure 6; Carson and Kirkby, 1972; and Dohrenwend, 1978). Slope asymmetry may be especially prominent north of the

Scarp height, m	South-facing	North-facing	South/North ratio
2	5.4	3.1	1.8
5	16.2	4.5	3.6
10	34	6.8	5.0
15	52	9.2	5.7
20	70	12	6.1

Table 5.3 Effects of scarp heights and microclimate on Idaho terrace scarps. (From Table 3 of Pierce and Colman, 1986.)
Changes in degradation-rate coefficient ($C^* \times 10^{-1}$ m^2/ky) with five heights of north- and south-facing scarps.

January freezing isotherm because of more freeze-thaw cycles on south-facing slopes and longer duration of insulating snow blankets on the north-facing slopes (Russell, 1909).

The striking contrast in scarp morphology for north- and south-facing slopes in the study area of Pierce and Colman is shown in Figure 5.19. Degradation of the scarp crest and aggradation of the scarp toe, after the attainment of the assumed starting angle of 33.5°, are four times larger for the south-facing as compared to the north-facing scarp of the same age. Contrasts in rates of fluvial processes on south and north-facing scarps are a function of scarp heights. For 2 m high scarps degradation rates on south-facing scarps are twice those on north-facing scarps, but for 10 m high scarps the difference is fivefold instead of being twofold. The increase of degradation rate with increasing scarp height is largely a function of sufficient increase in slope area to cross a threshold to more effective erosional processes, such as slope wash and gully erosion. The effects of scarp height and microclimate on slope processes for the Idaho stream terraces are summarized in Tables 5.2A and B. These factors have a considerable influence on the value of the degradation-rate coefficient, C^* (equation 5.2).

The independent variable controlling fluvial contrasts between south- and north-facing scarps is solar radiation. South-facing scarps tend to be much more perpendicular to the sun's rays than north-facing scarps. The percentage of potential solar-beam radiation on slopes of a given orientation and inclination – the solar index – can be interpolated from the work of Frank and Lee (1966). The solar index for 23° slopes at latitude 44° N will be only about 30% for north-facing scarps and about 58% for south-facing scarps.

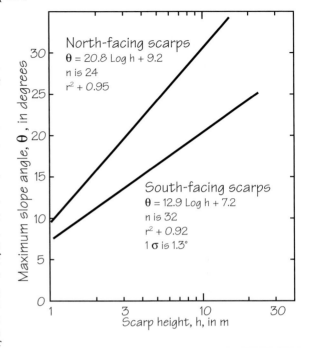

Figure 5.20 Importance of microclimate and scarp height on Idaho scarps with differing orientations. Relation between maximum scarp slopes and scarp heights for north-, and south-facing scarps. For a given height, the south-facing scarps degrade most rapidly and the effects of scarp orientation are more for higher than for lower scarps. From Figure 7 of Pierce and Colman (1986).

Height (size) attributes	Orientation (microclimate) attributes
	Soil wash more effective on south-facing, drier, less-vegetated slopes
Soil wash increases with slope length and height	Freeze-thaw cycles greater on steeper, south-facing slopes
Efficiency of creep may increase with surface gradient	Cohesion resistance to creep less on drier, less vegetated slopes
Lesser infiltration and greater runoff on steeper slopes	Winter snowpack protects north-facing slopes from winter erosion and freeze-thaw and enhances soil moisture and hence vegetation cover
	Steeper slopes drier because they intercept less precipitation per unit of horizontal area

Table 5.4 Effects of scarp heights and microclimate on Idaho terrace scarps (from Pierce and Colman, 1986). Scarp height and microclimate attributes that may affect the C^* degradation-rate coefficient. Attributes listed in order of estimated importance.

Contrasts in solar index for south- and north-facing scarps are reflected in plants and soil profiles. High north-facing scarps have a prairie-grassland type of plant community with only 10% bare ground; the root-bound sod is resistant to erosion. High south-facing scarps have a desert shrub type of plant community with 70% bare ground; the cohesionless silty surficial soils are mantled with a layer of frost-heaved stones. Such profound microclimatic differences on scarp morphology have important implications for fault-scarp modeling. Regional as well as local climatic differences should also have a strong influence on the diffusivity coefficient.

Several Pierce and Colman illustrations summarize the affects of microclimate on the Idaho fault scarps. Maximum scarp slope for a given age increases with scarp height, but increases more rapidly for north- than for south-facing scarps (Fig. 5.20). Slope angles for west-facing scarps occupy an intermediate trend. Volumes of material degraded from upper halves of the scarps (Fig. 5.11) increase with increasing scarp height, but much more rapidly for south-

than for north-facing scarps (Fig. 5.21). The diffusivity coefficient, C^*, increases rapidly with increases in scarp height for south-facing scarps and slowly for north-facing scarps (Fig. 5.22). Pierce and Colman normalized all scarps to west-facing, where the solar index does not change appreciably with variations of scarp slopes. The line with C^* values halfway between those of south- and north-facing scarps defines the relation between C^* and scarp height for due-west facing scarps. Figure 5.22 has trends that can be compared with the solar index. The trends of the residuals of the diffusivity coefficient, C^* result when values of C^* are subtracted from the values along the normalizing line for each scarp height. For south-facing scarps, residual C^* is positive and becomes more positive with increases in scarp height. For north-facing scarps, residual C^* is negative and becomes more negative with increases in scarp height. These trends are similar to solar-index trends for south- and north-facing scarps at 44° N.

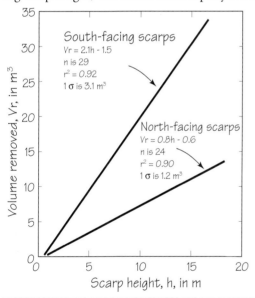

Figure 5.21 Importance of microclimate and scarp height on Idaho scarps with differing orientations.
Contrasts between volume removed from the upper halves of scarps and scarp heights for north- and south-facing scarps. From Figure 11 of Pierce and Colman (1986).

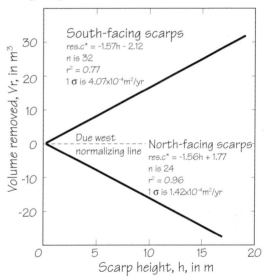

Figure 5.22 Importance of microclimate and scarp height on Idaho scarps with differing orientations.
Trends of the residuals of the diffusivity coefficient that remain after C^* values are subtracted from the due west line that is normalized for the effects of variable scarp height. The negative and positive trends are similar to those for the solar constant. From Figure 13 of Pierce and Colman (1986).

The Pierce and Colman study clearly shows how careful one must be when using diffusion-equation modeling to estimate ages of single-rupture event fault scarps. Scarp morphologies may be a useful tool to estimate ages, but only if one accounts for the effects of climate and scarp height, both of which may have a profound influence on fault-scarp evolution even where alluvium mass strength is constant. Although the influence of variations in solar index is large, scarp height has a still greater influence on the diffusivity coefficient.

Temporal changes in climate may have an equally profound effect on rates of evolution of fault scarps and on hillslope morphology (see chapters 2 through 5 of Bull, 1991). We live at a most interesting time in geologic history from a paleoclimatic viewpoint. Marine records reveal that the major glaciation and sea-level low at 18–21 ka and the subsequent Holocene warming and sea-level rise of 130 m represent by far the most extreme swing in climatic extremes during the past 100 ka. The key point in regard to fault-scarp studies is that rates of hillslope and scarp degradation, the diffusivity coefficient of equation 5.1, may have varied by more than tenfold during the past 20 ka. Mean degradation rates for Holocene fault scarps may be much different than the mean degradation rates for fault scarps that have been subject to the effects of both glacial and interglacial climates.

Age estimates may be in error by a factor of two (Pierce and Colman, 1986, p. 883) if the fundamental controls of scarp evolution are not taken into account for a particular study site. Diffusion-equation modeling, like much other modeling, has the drawback of being difficult to calculate precision and accuracy of the results. It should not be used as a sole way of dating the fault scarps of a study region. Larry Mayer (1984) was the first to emphasize caution. Results should note field measurement errors and residual unexplained variance of model results. Evaluation of possible errors is discussed by Colman, 1987; Mayer, 1987; Nash, 1987; and Avouac, 1993. The remarkable Pierce and Colman study will encourage workers in other regions to carefully appraise the effects of climate. In the next section we shall see that lithologic control is an equally important variable affecting rates of fault-scarp degradation and diffusion-equation modeling age estimates.

Recent studies suggest caution in using the standard diffusion-equation approach. Pelletier et al. (2006) in their study of prehistorical Lake Bonneville

shoreline scarps in Utah concluded that neither scarp orientation nor microclimatic controls had a discernible effect on diffusivity values. Nash and Beaujon (2006) prefer a power-function instead of a linear diffusion model when examining how scarp gradient affects downslope flux of debris. They found an ideal flight of late Quaternary degradation terraces to test the linear diffusion model, which was not the best predictor of scarp morphology. Instead, scarp morphology, and changes related to scarp height, is modeled best when downslope flux of debris is proportional to slope gradient raised to the 3.4 power, results that are similar to the 3.0 power-function conclusion of Andrews and Bucknam, 1987. Modelers should be cautious in using initial results that provide reasonable comparisons, because better fits to the data may be possible. David Nash continues to make pioneering advances in the modeling of scarps, and readers are encouraged to take advantage of his web site[*].

5.4 Lithologic Controls of Fault-Scarp Morphology

Changes of alluvium mass strength that occurs with time or in space are complications for paleoseismologists estimating ages of fault scarps with diffusion-equation models. Sandy piedmont gravel is not a truly uniform lithology. Seemingly minor spatial variations in cohesiveness may be a function of small amounts of clay, and of spatial variations in the abundance and size of boulders. Temporal changes in sandy gravel occur as pedogenic clay and calcium carbonate accumulate in soil-profile horizons. All these variables may affect rates of scarp degradation. An important question is "how wide a range in the variables of climate and lithology can be tolerated and still have a valid comparison of fault-scarp degradation rates between different study areas?"

Variation in resistance of earth materials to erosional processes that change fault scarps is equally as important as climate and scarp height in determining scarp morphology. Two types of field areas are

[*] Nash's computer program SlopeAge II2 for morphologic dating can be accessed at http://homepages.uc.edu/%7Enashdb/SLOPEAGE/slopeage.htm.
Enter the observed profile and assumed initial profile data for the entire scarp profile and the program calculates tc_i, which is the age of the hillslope, t, times the diffusivity, c_i.

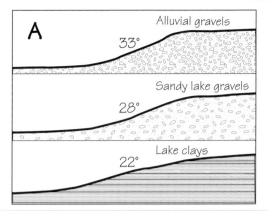

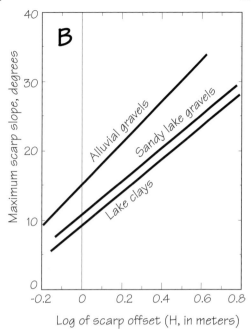

Figure 5.23 Lithologic controls of fault-scarp morphologies, Black Rock fault, northwestern Nevada.
A. Maximum scarp slopes and topographic profiles of 1.4 ka fault scarps in alluvial gravel, lacustrine gravel, and lacustrine silty clay with scarp heights of 3.6, 3.5, and 4.0 m. Figure 6 of Dodge and Grose (1980).
B. Regression lines of maximum scarp slope and scarp height for different deposits. From Figure 4 of Dodge and Grose (1980). Angular alluvial gravels, y = 15.29 + 29.73x, n is 18, r^2 is 0.80. Sandy lake gravel, y = 10.70+ 24.48x, n is 12, r^2 is 0.83. Lake clays, y = 9.36 + 24.11x, n is 11, r^2 is 0.88.

used here for evaluation of lithologic controls; both are of single-rupture event normal fault scarps of late Holocene or historical age. Some surface ruptures cut materials that range in texture from gravel to clay. An example is a Holocene fault scarp in the Black Rock Desert of northwestern Nevada (Fig. 5.23) that was studied by Dodge and Grose (1980). A second approach is to study fault scarps in sandy gravels that have changed their resistance to erosion through the actions of pedogenic and fluvial processes. An example is the Pitaycachi fault in Mexico (Fig. 5.24) where pedogenesis between very infrequent earthquakes makes it excellent for studies of lithologic control of fault-scarp morphology.

5.4.1 Fault Rupture of Different Materials

Dodge and Grose (1980) made regressions of maximum slope and scarp offset for single-rupture event fault scarps in lithologies that range from alluvial

gravel to lake clay, which were ruptured along the Black Rock fault in northwestern Nevada at about 1.4 ka. Scarps in less resistant materials have topographic profiles that appear older than scarps of the same age and height in more resistant basin-fill lithologies (Fig. 5.23A). Slopes are not as steep and scarps have undergone more crestal rounding in lake clays than in angular gravelly alluvium. Regressions of maximum scarp slope angle and scarp height for these three types of deposits (Fig. 5.23B) suggest the magnitude of this lithologic control. The lesser slope for the lake-clay regression, as compared to the alluvial-gravel regression, also suggests that clays degrade more rapidly than alluvial gravels. Maximum slopes for 1–2 m fault scarps are about 23° in alluvial gravel but only 15° in lake clays, and maximum slopes for 3.5 m fault scarps are about 31° in alluvial gravel but only 19° in lake clays.

Watters and Prokop (1990) realized that diffusion-equation modelers prefer initially cohesionless soils, but these are not present at many field sites. They compared fault-scarp characteristics with standard geotechnical engineering properties of shear strength, bulk density, Atterberg limits, and particle-size distribution. Laboratory testing of slightly cohesive materials involved carving large, undisturbed, approximately homogeneous, blocks out of fault scarp free faces in Dixie Valley, west-central Nevada. These

186   Chapter 5

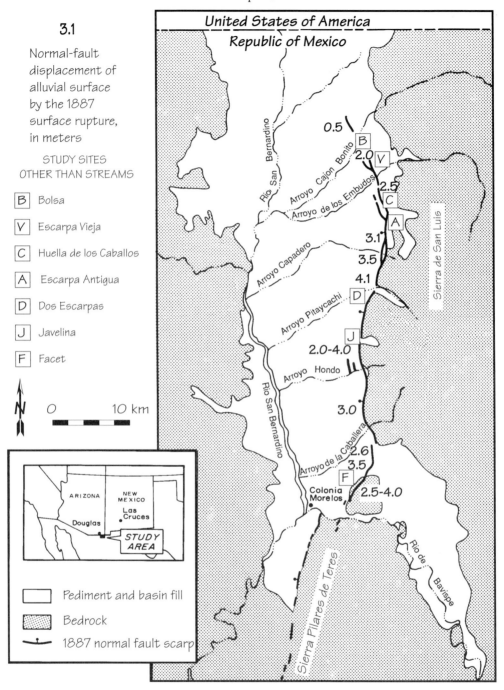

3.1

Normal-fault
displacement of
alluvial surface
by the 1887
surface rupture,
in meters

STUDY SITES
OTHER THAN STREAMS

B Bolsa

V Escarpa Vieja

C Huella de los Caballos

A Escarpa Antigua

D Dos Escarpas

J Javelina

F Facet

O   10 km

Pediment and basin fill

Bedrock

1887 normal fault scarp

Figure 5.24 Map showing informally named study sites (in squares) and heights of 1887 scarps in m (bold lettering). Modified Figure 1 of Bull and Pearthree, 1988.

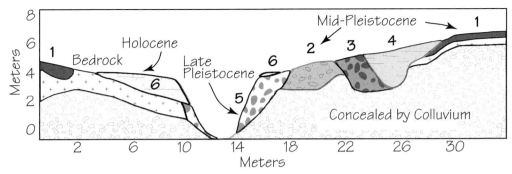

Figure 5.25 Valley-fill stratigraphy exposed in the fault scarp at the Javelina site. Comparison of the maximum hues of soils and fill units listed in Tables 5.4 and 5.5 suggest that fill 6 is a Holocene aggradation event, fill 5 was deposited during the late Pleistocene, and fills 1–4 are early to mid-Pleistocene in age. Figure 13 of Bull and Pearthree, 1988.

engineering soil characteristics influence the ratio of total scarp height to the amount of free face present.

Assessing lithologic influences on fault-scarp processes and morphology cannot be done with standard engineering tests for common alluvium with ubiquitous cobbles and boulders, or for pedogenic overprinting by the formation of calcic and argillic soil-profile horizons. So, next we attempt to devise other ways of appraising lithologic controls on fault-scarp morphology.

5.4.2 Lithologic Controls on an 1887 Fault Scarp

The unusual Pitaycachi fault is interesting to earth scientists and planners. It is unusual because the great Sonoran earthquake of 1887 occurred in a vast region otherwise devoid of historical surface-rupturing earthquakes. The nearest historical surface rupture is a distant 600 km to the west in California. It is interesting because the 75 km long rupture is exceptional for normal faulting events. The 1887 surface rupture occurred in the San Bernardino valley of Sonora, Mexico (Fig. 5.24) which lies on the western fringe of the semiarid Chihuahuan Desert. The central 35 km of the 1887 scarp bounds the Sierra de San Luis, an embayed tectonic activity class 4 mountain front.

Sandy to bouldery piedmont gravel is derived from granitic and volcanic rocks, and limestone. Di-verse alluvial materials and soil-profile characteristics result in variable alluvium mass strength. Piedmont surfaces are gently sloping, generally being 1° to 3°. The pediment immediately upslope from the Pitaycachi fault at the Javelina site is 0.5–1 km wide, and the pediment embayment formed on carbonate rocks farther north is several km wide. Valleys have been incised into the pediment only to be backfilled by climate-change induced aggradation events. Aggradation of valleys cut into a dissected pediment upstream from the Pitaycachi fault has occurred repeatedly at the Javelina site (Fig. 5.25, Table 5.5).

Correlation of the Pleistocene–Holocene stratigraphic contact exposed in the scarp face with that exposed in the fanhead trench downslope from the fault indicates an 1887 apparent throw of 3.1 m. Before 1887 the fault scarp was only 2.5 to 3.5 m high. But unlike most western North America sites, the penultimate surface rupture is very old. Instead of an earthquake recurrence interval of 0.1 to 10 ky, it is on the order of 200–400 ky. Many, extremely infrequent but very large, prehistorical earthquakes have occurred here, dating back into the Tertiary.

Valley-fill stratigraphy underscores the rarity of tectonically induced downcutting events. Five valley fills exposed in the Pitaycachi fault scarp face at the Javelina site represent episodes of about 2 to 4 m of downcutting and subsequent backfilling (Fig. 5.25). The fills have a great variety of lithologies, colors, and sedimentary structures (Table 5.5). These reflect different combinations of climate-controlled changes in weathering and episodes of stripping of the upstream hillslope sediment reservoir, plus diagenetic changes after the aggradation events. The older fills have redder hues, whereas the younger are grayer. The time span required for the numerous climatic changes associated with the six periods of valley degradation and aggradation, and the subsequent diagenetic reddening and cobble weathering, exceeds

Figure 5.22 Aggrada-tion event	Lithology	Dry Color of Weathered Alluvium	Weathering of Grano-diorite Cobbles
6	Sand, silty	Brown (7.5YR 4/3)	Unweathered cobbles to incipient iron oxide stains along fractures in solid cobbles
5	Gravel, clayey	Bright reddish brown (5YR 5/8) to bright brown (2.5YR 5/8)	Iron oxide stains along fractures in solid to punky cobbles
4	Sand, clayey cross-bedded	Reddish brown (2.5YR 4/8)	No cobbles present
3	Gravel, clayey	Reddish brown (5YR 4/8 to bright brown (2.5YR 4/8)	Punky to grussy cobbles
2	Gravel, sandy	Dull orange (7.5YR 7/4)	Punky to grussy cobbles
1	Gravel, clayey	Bright reddish brown (5YR 5/6) to reddish brown (2.5YR 4/6)	Grussy cobble remnants

Table 5.5 Characteristics of Quaternary valley fills at the Javelina site. See Figure 5.22 for the interrelations between these valley fills and their estimated ages.

400 ky. This time span includes only the 1887 and the penultimate surface-rupture events.

The penultimate event was similar in size to the 1887 event, where the height of the much eroded scarp is still preserved on nearly flat flights of ancient stream terraces. Few paleoseismologists would have spotted the trace of an active fault on even gently sloping terrain, where the scarp of the penultimate earthquake is obscure at best. The banner photo on the first page of this chapter and Figure 5.26 are examples. The 1887 apparent displacements at the Facet site range from 2.5 to 4 m. An obvious bevel of the fault-scarp crest is no longer apparent because most of the semiarid hillslope has been slowly eroded to almost a uniform slope . Presumably, a spectacular record of many late Cenozoic events is contained locally in the bedrock fault scarps of the two facets of more resistant rock at both sides of Figure 5.26.

Figure 5.26 Aerial view of the Facet site showing hillslope planation between the penultimate and 1887 surface-rupture events. The 10–35 m high wedges of relatively more resistant lava flows and carbonate cemented colluvium immediately upslope from the 2.5 to 4 m high 1887 scarp record large amounts of cumulative late Cenozoic uplift. The lack of pre-1887 fault scarps between these isolated remnants of former hillslopes indicates an exceptionally long time span between surface-rupture events. Figure 11 of Bull and Pearthree, 1988.

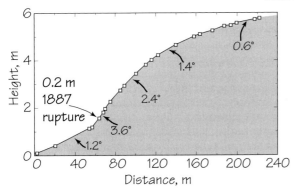

Figure 5.27 Topographic profile of an ancient multiple-rupture event fault scarp. The 1887 surface rupture is the 0.2 m increment in 3.2 m high scarp in clayey gravels at the Bolsa site. From Figure 2 of Bull and Pearthree, 1988.

The northern termination of the Pitaycachi fault scarp has much smaller displacements for both the 1887 and prehistorical events. The sequence of small late Cenozoic surface ruptures is recorded locally by darker soil on the footwall block. Virtually no fault scarp is present (Fig. 8, Bull and Pearthree, 1988). Many small surface ruptures at the Bolsa site have created a broad, curving scarp crest rising above a miniscule 1887 offset (Fig. 5.27).

Pediments were created during the Tertiary when denudation rates exceeded low rates of uplift, and perhaps before the creation of extensional stresses responsible for Plio-Pleistocene normal faulting. This low tectonic intensity landscape-evolution model characterizes parts of Arizona (Pearthree and Calvo, 1987) and New Mexico (Machette, 1986).

The scarp of the Pitaycachi fault rupture is ideal for evaluation of lithologic controls on fault-scarp degradation in gently sloping bouldery alluvium with varying degrees of soil-profile formation. Rupture time is known and throw is fairly uniform. Free

Stage	Diagnostic Morphology	Distribution of $CaCO_3$
Calcic Soils		
I	Thin, discontinuous coatings on pebbles, usually on undersides.	Coatings sparse to common
II	Continuous, thin to thick coatings on tops and bottoms of pebbles.	Coatings common, some carbonate in matrix, but matrix still loose.
III	Massive accumulations between clasts, becomes cemented in advanced form.	Essentially continuous dispersion in matrix
Pedogenic Calcretes		
IV	Thin (<0.2 cm) to moderately thick (1 cm) laminae in upper part of horizon. Thin laminae may drape over fractured surfaces.	Cemented platy to weak tabular structure and indurated laminae. Horizon is 0.5–1 m thick.
V	Thick laminae (>1 cm) and small to large pisolites. Vertical faces and fractures are coated with laminated carbonate (case-hardened) surface).	Indurated dense, strong platy to tabular structure. Horizon is 1–2 m thick.
VI	Multiple generations of laminae, breccia, and pisolites; recemented. Many case-hardened surfaces.	Indurated and dense, thick strong tabular structure. Horizon commonly is >2 m thick.

Table 5.6 Stages in the morphogenetic sequence of soil carbonate accumulation in gravelly alluvium (after Machette, 1985, p. 5).

faces are still present on fault scarps formed in cohesive materials. Our measurements were made 94 to 99 years after the great Sonoran earthquake of 1887. Macroclimate is similar throughout the region. The west-facing aspect of the fault scarps eliminates most local microclimatic variations (Section 5.3).

A key factor is that at least 200 ky elapsed between the prior surface rupture event and the 1887 event. Although multiple-rupture event fault scarps are present locally, a great variety of alluvium is ruptured only by the 1887 event.

Soil-profile characteristics were used to estimate the approximate ages of the alluvial geomorphic surfaces. The Las Cruces, New Mexico soils study (Bachman and Machette, 1977; Gile and Grossman, 1979; Gile et al., 1981), 250 km to the northeast, is relevant because the climate, parent materials, topography, and biota are similar to those of the Pitaycachi area. Rates of input of atmospheric dust are important in the genesis of argillic and calcic soil horizons, and morphologic stages of calcic soil-profile horizons (Table 5.5) are essential for dating and correlation.

The central two-thirds of the 1887 scarp is 2.5 to 4 m high, but south of Rio de Bavispe and north of Arroyo de los Embudos the scarp typically is only 0.5 to 1.5 m high. The 1887 fault trace is especially prominent where graben formation or subsidence due to fissuring of the downthrown block has concentrated local runoff and promoted dense vegetation. Apparent throw can be determined at many localities by measuring the vertical displacement of soil-horizon boundaries, alluvial strata, or terrace treads across the fault. Scarp heights approximate maximum displacements on these gentle piedmont slopes. Representative scarp heights are noted on Figure 5.24, but throw may be less than scarp heights where local subsidence or warping of the hanging-wall block materials has occurred.

Our goal here is to quantify how different earth materials influence types of geomorphic processes, and the amounts and styles of fault-scarp degradation and aggradation. The scope of work done was threefold. First measurements of various earth materials and geomorphic properties were made in a partially successful attempt to quantify differences in erodibility. Topographic profiles of fault scarps were surveyed in order to compare amounts of degradation and aggradation of scarps in different lithologies. Third, new descriptive techniques were developed to better relate fault-scarp morphologies to lithologic controls. These include indices that describe scarp-crest sinuosity and the proportion of free face present a century after surface rupture.

5.4.2.1 Geomorphic Processes

Both tectonic and fluvial processes shape fault scarps. Tectonic processes include rupture of the land surface and alluvium to create an obvious topographic discontinuity. This initial fault scarp may be quickly changed immediately by aftershock seismic-shaking events. Initial fault-scarp morphologies also are dependent on materials strength and cohesiveness that control the initial proportions of free face and debris slope that are present. A general lithologic sequence of increasing tendency to develop and preserve free faces is loose sand, gravelly sand, sandy gravel and clayey sand, clayey gravel, and cemented gravel and bedrock. New scarps in weak materials may consist only of debris slopes or fluvial processes may remove vestiges of free faces in a year. Stronger scarp materials for a given scarp height have a larger initial free-face component that requires >100 years to erode by fluvial processes that also construct the debris and wash slopes.

Lithologic controls of fluvial processes include:
1) Amount of silt and clay in the alluvium. Fine particles tend to plug the interstices between larger particles as part of the process of raindrop splash on bare soil; thereby reducing infiltration and increasing runoff for a given rainfall event.
2) Transmissivity rates for unsaturated flow in porous media.
3) Shear stresses needed, for a given slope, to entrain sand and gravel by flowing water.
4) Abundance and type of pedogenic clay that forms cohesive argillic soil horizons in sandy gravel parent materials. It is subject to shrink-swell processes and may decrease in strength when wetted.
5) Abundance and distribution of pedogenic cementation by oxides of iron, aluminum, and silicon, and by calcium carbonate.
6) Cobbles and boulders increase the alluvium mass strength of the deposits and may accumulate as a surficial lag deposit, or form a desert pavement that armors alluvial surfaces.
7) Soil characteristics that influence the type and density of vegetation whose leaves and stems intercept rain and whose roots increase soil shear strength.
8) Particle-size distribution that favors bioturbation by roots and rodents.

Free faces and debris slopes are the dominant topographic elements of the single-rupture
event fault scarps formed by the 1887 earthquake.
There has been little modification of the upslope
and downslope original surfaces of the Pitaycachi
fault scarp in a century. Incipient rounding of the
scarp crest by sheetflow and rillwash has steepened
a few meters of the upslope original surface from
2° to 3°– 5°. Downslope original surfaces are little
changed. Scarps in cohesive gravelly soils provide a
scattering of cobbles, and some reach the downslope
original surface (Fig. 5.28A, B). Sandy gravel is more

Figure 5.28 Views of 1887 Pitaycachi fault scarps in different materials.
A. Holocene sandy gravel in 3.3 m high scarp near Arroyo Pitaycachi. Weakly cohesive alluvial-fan deposits
still have a free face, below which is a bouldery debris slope.
B. Late Pleistocene sandy gravel with strongly developed argillic and weakly developed calcic soil horizons
at the Dos Escarpas site. Post 1887 free face retreat has made a debris slope that is progressively
coarser-grained in the downslope direction (Fig. 5.27). Hat for scale.
C. The 4 m high scarp in massive carbonate-cemented gravel south of Arroyo Capadero is dominated by a
large free face. Blocks of cemented alluvium lie on the debris slope. Figure 3 of Bull and Pearthree, 1988.

conducive for deposition of a few centimeters as a new alluvial slope that consists of freshly deposited sand derived from the adjacent debris slope and free face. Post-1888 streamflow has notched the fault scarp and deposited alluvial fans (Fig. 2.12).

Free-face slopes reflect lithologic controls, including the soil-profile characteristics summarized in Table 5.6. The Pitaycachi fault dips 50° to 80° at the land surface, and the slope of the initial free face probably was 60° to 90° in clayey and cemented lithologies. Free faces are virtually unmodified in carbonate-cemented limestone gravel (Fig. 5.28C), but are no longer present in clayey gruss derived from quartz monzonite. Free faces are still common in

Soil classification[1]	Site	Age	Bt soil horizon				CaCO$_3$ depth (cm)	CaCO$_3$ stage	Scarp height[4] (m)	Tectonic offset[4] (m)
			Thickness (cm)	Maximum Hue[2]	Chroma[2]	Maximum % clay[3]				
Typic torriorthent	Arroyo Hondo	Holocene	Not old enough for a Bt horizon				None		1.5*	
Ustollic haplargid	Embudos	Late Pleistocene	120	5 YR	6	25	140	I	2*	1.5 to 3.0
Ustollic haplargid	Embudos		320	5 YR	6	25-30	350	I	2*	1.5 to 3.0
Typic haplargid	Huella de Caballos		truncated	5 YR	8	25-30	>90	III–IV	3.1*	
Ustollic haplargid	Arroyo Hondo		93	2.5 YR	8	33	71	II–III		
Ustollic haplargid	Embudos	Mid-Pleistocene	90	2.5 YR	6	60	66	I	4 to 7	4 to 6
Ustollic haplargid	Javelina		>105	2.5 YR	8	45-50			6 to 8	4 to 6
Petrocalcic ustollic haplargid	Arroyo Hondo	Early Pleistocene	>163	2.5 YR	8	45	68	III–IV	15	9 to 13
Petrocalcic ustollic haplargid	Bolsa		200	2.5 YR	8	50	210	IV	3 to 6	
	Escarpa vieja	Ustollic haplargid	116	2.5 YR	8	50	44	I	14 to 16	
Petrocalcic paleargid	Arroyo Hondo	Late Tertiary	54	2.5 YR	6	40	10	V		
	Escarpa antigua		Not described					VI	43	39

* Surface ruptured only by 1887 event
1 Taxonomy is from Soil Survey Staff (1975).
2 Munsell 1992 notation.
3 Gravelly parent materials have less than 5% clay.
4 Scarp height is the vertical measurement from base to crest. Displacement (apparent net vertical throw) is measured from offset stratigraphic units and soil horizons, or more commonly by projection of the slopes of the upper and lower original surfaces to the fault trace.

Table 5.7 Summary of soil-profile characteristics of the Pitaycachi fault terrace chronosequence.

sandy gravel but have retreated 2 to 8 m upslope from the fault trace. Debris slopes typically are 24° to 38° and have been sites of net erosion, not deposition. Amounts of debris-slope retreat can be estimated where roots of mesquite trees are exposed.

5.4.2.2 Scarp Materials

Alluvium mass strength characteristics of fault scarps should be described in both a surface and a subsurface sense. The diverse materials ruptured by displacement along the Pitaycachi fault are summarized in Tables 5.6 and 5.7.

Several characteristics were measured in an attempt to describe alluvium mass strength numerically. Seismic velocities increase with increases in cementation and abundance of unweathered boulders and cobbles of rhyolite. Seismic velocities are a fairly good tool for describing deeper subsurface materials but are not appropriate for describing loose fine-grained materials that are subject to rockfalls, rillwash, and sheetflow. Data obtained with a spring-loaded penetrometer seem reliable for describing weakly cohesive sandy and silty alluvium, but not for other lithologies. Gravels could be tested only by penetrating into pockets of the materials between cobbles. Argillic soil horizons with more than 40% clay were difficult to test. Even when moist, the smallest penetrometer head was unable to penetrate these tough soils at the moderately low design applied stresses of this standard geotechnical engineering tool. The K constant in the Universal Soil Loss Equation (Wischmeir and Smith, 1965, 1978) was quite low whenever gravel was abundant.

David Taylor improvised a test that worked well. He concentrated on measuring resistance to shear of both gravelly and nongravelly surficial materials. By noting the mean depth of penetration during horizontal movements of 6 kg weighted steel tines he made estimates of the shear strength of surficial soils. These shear-strength estimates provided a consistent measure of variations of surficial materials strength that correlated well with the penetrometer data of loamy soils and with scarp morphologic characteristics such as maximum slope and scarp-crest sinuosity (Fig. 5.29). The Taylor rake test succeeded because it measured materials characteristics in a depth increment relevant to geomorphic processes.

This attempt to devise a numerical multi-parameter materials strength index for fault scarps in alluvium was unsuccessful (Table 5.7). Most parameters were inconsistent or were not appropriate for describing lithologic controls of the main fluvial processes responsible for changing scarp morphologies. Quantitative multi-factor indices that describe erodibility of gravelly piedmont alluvium need more testing before use in fault-scarp studies. For now, we relate fault-scarp characteristics to three broad lithologic classes (Table 5.7).

Lithologic resistance to erosion increases greatly in sandy gravel with the passage of time. Sur-

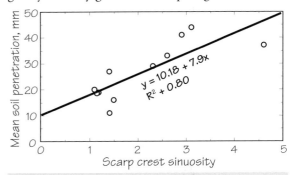

Figure 5.29 Relation of shear-strength index of surficial materials to sinuosity of fault scarp crests in different materials.

Type of alluvium	Particle-size distribution		Shear-strength index, mean penetration, mm	Penetrometer, mean kilonewtons/m²	Mean seismic velocity, m/sec	Mean K factor of the Universal Soil-Loss Equation
	% Silt	% > 2 mm				
Grussy	1 to 7	6 to 15	41	2108	335	0.03
Cobbly	7 to 27	4 to 57	29	4663	354	0.04
Carbonate cemented	8 to 58	30 to 65	20	5390	326	0.05

Table 5.8 Average alluvium mass strength characteristics of alluvial materials ruptured by the Pitaycachi fault in 1887 at 39 fault scarps.

ficial lag gravels accumulate and protect the soil from erosion by rainsplash and sheetflow. Concurrent increases in shear strength are caused by additions of pedogenic clay and carbonate in argillic and calcic horizons.

Erosion of the 1887 scarp resulted in different materials being present on different topographic elements. Figure 5.30 is a sample of particle-size distributions of free-face and debris-slope materials. The free-face source material is finer grained and more poorly sorted than the debris-slope surficial materials. Cobble size increases downslope, and there is an abrupt transition to fine materials at the scarp base that coincides with an incipient wash slope.

5.4.2.3 Scarp Morphology

Topographic profiles of the 1887 fault scarps were surveyed normal to the fault trace to evaluate scarp height, maximum scarp slope, and scarp width. Scarp-crest sinuosity and free-face index also were measured. Each of these topographic characteristics is affected by soil profile and alluvium-mass-strength characteristics. Replication accuracies of a given scarp profile were determined for each survey method and as many as 20 surveys along a single transect were made to compare variations between different surveyors. Mayer (1984) concluded that site selection and careful calibration of survey methods are important.

Profiles for 2 to 3 m high scarps vary greatly as a function of lithology. A century of erosion has reduced scarps in clayey sand to 8° to 20° slopes without free faces. Free faces that appear unchanged typify scarps in carbonate-cemented gravels. They may persist for millennia. Scarps underlain by different materials evolve at different rates so may become more similar, or different, with the passage of time (Peter Knuepfer, personal communication). Scarps in soft, sandy alluvium and clayey gravel are clearly different after a century of erosion, but this distinction will decrease with the passage of time as scarps converge to a common morphology at different lithologically controlled rates. More than 10 to 100 ky may be needed for two lithologic classes of scarps to become indistinguishable. Carbonate-cemented free faces are even more persistent (Table 5.8). They may still be present after 1,000 years.

Morphologic descriptions of young fault scarps should also include parameters that 1) involve both horizontal dimensions and 2) the proportions of topographic elements (Fig. 5.2) present. Here

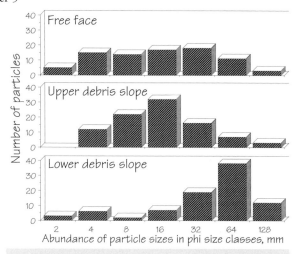

Figure 5.30 Comparisons of particle-size distributions in the free face and the upper and lower parts of the debris slope at the Dos Escarpas site. See Figure 5.25B for a view of the sample site.

we describe scarp-crest sinuosity and the proportion of free face still present in scarp height. Free-face retreat undermines boulders and results in a narrow crest compared to scarps without free faces.

At the Dos Escarpas site, scarp-crest sinuosity was measured by placing a loosely draped tape along the break in slope that defines the scarp crest (Fig. 5.31).

Figure 5.31 View of survey tape laid out along a segment of the crest of the 1887 scarp at the Dos Escarpas site. Instead of being straight, the white tape has a sinuosity of 1.4.

Type of alluvium	Maximum slope, °		Mean slope, °		Scarp width, m		Scarp crest sinuosity	
	Mean	1σ	Mean	1σ	Mean	1σ	Mean	Range
Grussy	28	8	18	4	7.4	1.9	3.8	2.9 – 4.6
Cobbly	44	15	29	8	4.8	1.9	2.4	1.4 – 3.7
Carbonate cemented	80	13	37	10	4.3	1.3	1.3	1.1 – 1.5

Table 5.9 Morphologic characteristics of the 1887 scarps of the Pitaycachi fault as related to the degree of cementation and relative abundances of gravel and clay in the scarp materials.

The *scarp-crest sinuosity index*, J_c, was calculated by dividing the scarp-crest length of the loosely draped tape, L_c, by the straight-line distance between the two survey endpoints, L.

$$J_c = L_c / L \qquad (5.4)$$

All parts of the scarp crest were included, even a large consequent gully. That segment, had a scarp-crest sinuosity of 1.9. Representative sites for measuring scarp-crest sinuosity should not include erosion by antecedent streams. Surveys should be restricted to segments of scarp crests that are being eroded only by local rilling and sheetwash caused by runoff generated on the slope just upslope from the scarp crest. Relations of scarp-crest sinuosity to scarp height and to lithology are summarized in Tables 5.9 and 5.10.

Several topographic profiles were measured across each of seven 20 m segments of scarp crest at the Dos Escarpas site to include ranges of scarp heights. Typical scarp-crest sinuosities at the Dos Escarpas site range from only 1.2 to 1.4 despite the tenfold range in scarp heights of 0.2 to 2.3 m. Thus, scarp-crest sinuosities appear to be influenced primarily by lithology, time, and erosional process. This conclusion was not visually anticipated because the contrast between the crest slope and free face appears to be much more abrupt on the higher and steeper scarps.

Scarp-crest sinuosities vary markedly with lithology. Erosional modification apparently proceeds much more rapidly in nongravelly alluvium than in bouldery fan gravels. The scarp-crest sinuosity index seems to be a useful guide for describing the influence of lithology on scarp morphology, especially for historical fault scarps. Measurements of scarp-crest sinuosities of prehistorical scarps and some historical scarps may be facilitated by use of large scale, low sun angle aerial photographs and laser swath mapping.

Another index describes the proportions of topographic elements present and is useful for studies of historical fault scarps. The *free-face index*, F_f, is defined as:

$$F_f = 100 \, (H_f / H_f + H_d) \qquad (5.5)$$

Range of scarp heights, m	Scarp-crest sinuosity
0.2 – 0.6	1.4
0.9 – 1.3	1.3
1.0 – 1.3	1.2
1.2 – 1.5	1.4
1.5 – 2.3	1.3
1.9 – 2.3	1.3
2.2 – 2.3	1.9 (one gully)

Table 5.10 Comparison of Pitaycachi fault scarp height and sinuosity.

Lithology	Scarp-crest sinuosity
Sheared granodiorite	1.0 – 1.1
Massive cemented gravel	1.0 – 1.1
Sheared and altered volcanic rocks	1.2 – 1.4
Bouldery gravels	1.2 – 1.6
Pleistocene clayey gruss	1.5 – 3.5
Holocene silty sand	1.5 – 3.5

Table 5.11 Comparison of Pitaycachi fault lithology and sinuosity.

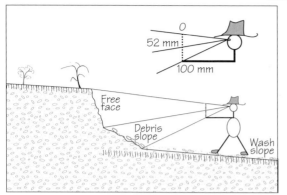

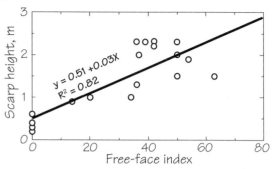

Figure 5.33 Relation of the free-face index to scarp height.

Figure 5.32 Sketch of the procedure used for measuring the free-face index. Invert a 10 cm scale and position it so that lines of sight through the 0 and 100 mm marks coincide with the top of the free face and the bottom of the debris slope. Record the millimeter mark for the line of sight to the junction between the free face and debris slope. The free-face index in this example is 52.

where H_f is the relative height of the free face and H_d is the relative height of the debris slope.

Measurements of the free-face index are easy to make because only the relative proportions of the free-face and debris-slope topographic elements are needed. One simple method is illustrated by Figure

Lithology	Free-face index
Sand	0
Clayey gruss	0–10
Sandy gravel	5–20
Gravel with argillic soil horizon	20–40
Gravel with argillic and calcic soil horizons	35–65
Sheared granodiorite, sheared and hydrothermally altered volcanic rocks	40–70
Massive carbonate-cemented gravel	60–90

Table 5.12 Comparison of Pitaycachi fault lithology and free-face index.

5.32. The index may also be estimated from photographs, preferably from stereo pairs or ground-based LiDAR (Light Distance and Ranging).

The free-face index is controlled mainly by lithology (Table 5.11), elapsed time since surface rupture, magnitude of tectonic displacement (Fig. 5.33), and climate. Local variations in the index are common. One should avoid sites where trees, bushes, and boulders increase the proportion of free face present, or where trails created by grazing stock decrease it. Values of the free-face index increase with increasing scarp height for a given lithology.

5.5 Laser Swath Digital Elevation Models

An important way to collect exceptionally detailed topographic information about fault scarps has been developed by NASA. The preferred way to examine topography now, even in rain forests, is LiDAR (Light Distance and Ranging, also known as Airborne Laser Swath Mapping or ALSM). It uses an airborne narrow laser beam to scan through trees to collect data for splendid digital elevation models of the ground surface. Parts of each laser pulse reflect off foliage at several levels as well as off the ground. Sophisticated algorithms are used to separate out laser returns that describe vegetation height and density characteristics and the ground surface. Tripod mounted ground-based LiDAR quickly obtains three-dimensional topographic images of specific landforms. LiDAR is fast becoming an essential tool in the search for fault scarps. It quickly and efficiently surveys bedrock and alluvium, so is especially useful where late Quaternary climatic perturbations have created bedrock or alluvial (Table 1.2) surfaces of approximately synchronous age.

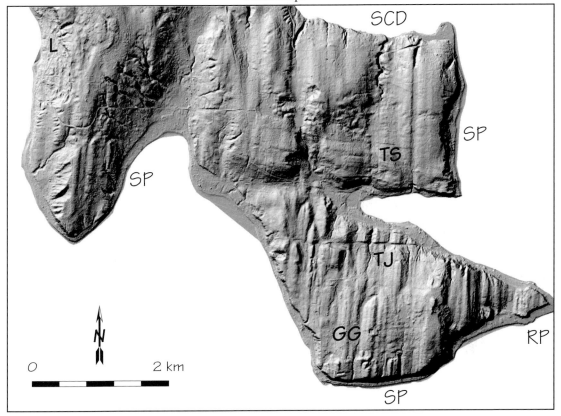

Figure 5.34 A LiDAR survey digital elevation model of the ground surface beneath dense second-growth rainforest on southern part of Bainbridge Island, 5 to 10 km west of Seattle in Puget Sound, Washington, USA. Part of Figure 1 of Harding and Berghoff (2000). This 2X vertical exaggeration image supplied courtesy of Greg Berghoff, Kitsap County Public Utilities District, Washington. GG, glacial grooves and deposits; TS, traces of tilted strata; TJ, scarp of the Toe Jam thrust fault. SP, modern shore platform, and the adjacent raised shore platform and its sea cliff; L, landslide; SCD, possible tectonically induced downcutting of stream channels. RP, Restoration Point study site of Bucknam et al. (1992).

Unmatched resolution LiDAR topography is indeed a bonanza for earth science (Haugerud et al., 2003) as shown by what it revealed beneath the dense second growth forest of Bainbridge Island (Fig. 5.34) in Puget Sound (Location OM on Figure 1.3). Traditional photogrammetry would record a tranquil hilly forest bordering a sinuous shoreline. LiDAR shows much more. The dominant landscape features are prominent north–south grooves (GG) left by latest Pleistocene glacial erosion and deposition. Three types of scarps cut across the grooves (SP, TJ, and L of Figure 5.34). They postdate the 16 ka retreat of the glaciers, but until the advent of LiDAR the fault and landslide scarps were hidden in the forest (see the

landslide study by Glenn et al., 2006). The orientation of tilted strata (TS) can be seen locally.

The LiDAR image reveals diverse clues of recent tectonic activity. Aligned, south facing topographic scarps (TJ) form an east–west lineation suggestive of recent faulting. LiDAR based topographic profiles described a multiple-rupture event fault scarp that is 1 to 5 m high. Subsequent trenching described Holocene thrust-faulting (Nelson et al., 2002). This discovery of an active element of the Seattle fault zone did not fit the assumed tectonically quiescent tectonic setting for the city. The most recent uplift event for Bainbridge Island (Bucknam et al., 1992) occurred at about 1.1 ka and is responsible for the uplifted shore platform and sea cliff that is shown in remarkable detail at SP. The raised shore platform at Restoration Point (RP) is a nearly planar bedrock surface as much as 7 m above the present high tide level. The base-level fall created by the recent uplift event may have contributed to stream-channel downcutting (SCD), which is migrating upstream from the coast. A large fresh landslide (L) may have been triggered by seismic shaking and/or lateral erosion induced base-level fall of the adjacent coast.

An image of two scarps of the oblique dextral Lenwood fault (Fig. 5.35) illustrates the amazing detail of LiDAR digital-elevation models. The larger creosote bushes grow along the broad valley floors of the active washes. They are more stunted on terrace treads. Pleistocene fan surface(s) have been offset by the multiple-rupture event fault scarp at the right side of the view where the incised valley floors are the result of several tectonic base-level falls. Only a late-Holocene surface has been offset by the single-rupture event fault scarp at the left side of the view. Low-sun angle photography would have been needed to study this faint fault scarp prior to the advent of LiDAR, which provides information about pre- and post-depositional stream channels on the fault scarp. This contrast illustrates the resolution of LiDAR for topographic and geomorphic mapping and its potential for measuring fault slip.

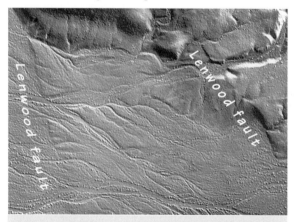

Figure 5.35 LiDAR digital-elevation model of late Quaternary scarps of the Lenwood fault in the central Mojave Desert of southeastern California. Scale provided by the symmetrical 0.5 to 1.5 m creosote bushes (Fig. 4.31). Image provided courtesy of Mike Oskin.

Long-term stream-channel downcutting induced by base-level fall favors preservation of late Quaternary aggradation events as flights of fill terraces. LiDAR digital topographic data generally has a resolution of 50 to 300 cm horizontally and 10 to 20 cm vertically (Shrestha et al., 1999), which allows discrimination of surface characteristics. Surface topography and materials change with time after deposition by ephemeral braided streams. Granitic lithologies weather and disintegrate and volcanic rocks split and accumulate varnish. Soil profiles increase in thickness and strength of development until erosion rates exceed pedogenic rates. Initial post-depositional rainsplash and sheetflow smooth the surface and, together with decreased infiltration rates caused by soil-profile development, gradually result in progressively more channelized runoff. Shallow rills become valleys that broaden and eventually consume even calcic soil horizons formed at depths of >1 m below formerly stable surfaces. LiDAR-based topographic profiles describe such landscape-evolution transitions in amazing detail. Examples presented here include general characteristics of aggradation event surfaces in the Rodman Mountains, and then quantitative assessment of Death Valley surfaces.

Desert alluvial surfaces change with time as a function of particle size and slope. The bar-and-channel characteristics of the active channels (Q4b) have minimal smoothness (Figs. 5.36 and 5.37). Riparian trees, 4 to 6 m high, grow along the active channels. Boulders undergo little change, but sand and fine gravel are smoothed quickly. So even late Holocene surfaces (Q3c) have progressed to the stage of having bar-and-swale characteristics. Desert pavement surfaces may be amazingly smooth after ~60 ky (Q2c), but stream channels have already started to convert fairly planar surfaces into a rounded degradational ridgecrests. Channels have deepened still more by 120 ky and large Q2b areas have noticeably deeper valleys between ridgecrest elements that have yet to undergo much degradation of soil profiles. Much older aggradation event surfaces, conveniently grouped as Q2a, have been so eroded that ridgecrests may be degraded below the depth of calcic soil-profile horizon formation. Q2a hillslopes that are convex from ridgecrests to valley floors (Fig. 5.37) suggest a landscape response to rapid base-level fall caused by stream-channel incisement. LiDAR resolution, as illustrated here, is such that flights of stream terraces and incised alluvial fans can be described to allow precise, efficient mapping of desert piedmonts.

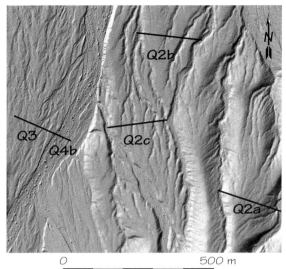

0   500 m

Figure 5.36 LiDAR map of alluvial geomorphic surfaces along Sheep Springs Wash, northern Rodman Mountains in the central Mojave Desert of southeastern California. Alluvial chronology follows the terminology of Table 1.2. Lines show the locations of the topographic profiles of Figure 5.34. Image provided courtesy of Mike Oskin.

The Q2c surface at Sheep Springs Wash has been displaced 900 ± 200 m by the Calico fault. The mean of seven [10]Be dates is 56.4 ± 7 ky for Q2c, so the overall rate of horizontal slip is 1.8 ± 0.3 m/ky (Oskin et al., in press for 2007). This nicely illustrates the combined power of laser swath digital elevation models and dating fault scarps of the Mojave Desert regional alluvial chronology with terrestrial cosmogenic nuclides. Such work will provide new models about the tectonics of the Walker Lane-Eastern California shear zone.

The same alluvial chronology is present farther north in Death Valley (Location DV on Figure 1.3). Frankel and Dolan (2007) used LiDAR to help map the piedmont surfaces along the highly active Death Valley fault zone. They also quantitatively described the roughness contrasts of the aggradation event surfaces summarized qualitatively in Figure 5.37. The sharp LiDAR image has distinct characteristics (Fig. 5.38A) that aid in mapping of the different ages of piedmont surfaces (Fig. 5.38B).

Laser swath mapping indicates ~300 m of late Quaternary right-lateral displacement on the north-

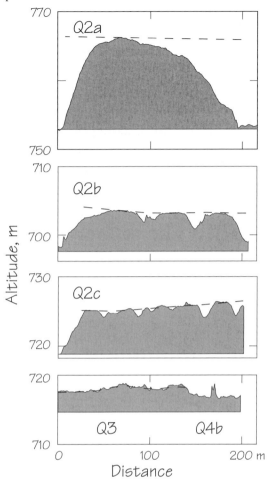

Figure 5.37 LiDAR based topographic profiles of late Quaternary alluvial geomorphic surfaces of Figure 5.33. Dashed lines indicate mean position of depositional surface, where preserved, and amount of stream-channel entrenchment into approximately planar original surfaces. Materials to construct this figure provided courtesy of Mike Oskin.

ern Death Valley fault zone (Fig. 5.38A). Frankel et al. (in press for 2007) conclude that this prominent fault scarp has a slip rate of ~4.5 ± 0.2 m/ky. [10]Be and [36]Cl terrestrial cosmogenic nuclides date the Q2c as ~66 ± 4 ka. This strong aggradation event buried older fan surfaces and the fault scarp. Right-lateral strike-slip fault displacements continued and for the next 60 ky while sediment was flushed further down

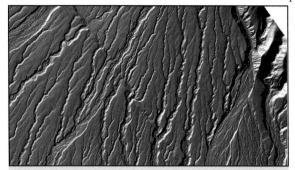

Figure 5.38 Surficial features of a faulted Death Valley piedmont.
A. LiDAR image from Frankel and Dolan (2007) Figure 3A.

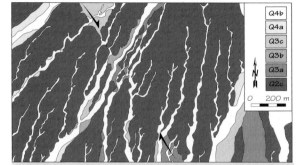

Figure 5.38 Surficial geologic map of a faulted Death Valley piedmont.
B. Late Quaternary alluvial surfaces as modified from map by Klinger (2001). Aggradation-event surfaces are the same as Table 1.2. Arrows point out the trace of the dextral-slip northern Death Valley fault zone. Frankel and Dolan (2007) Figure 3A.

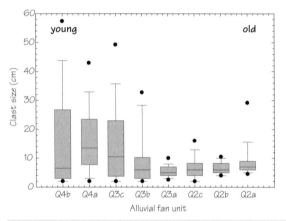

Figure 5.39 Box and whisker plots of clast sizes for each aggradation event surface. Fifty clast sizes were measured at one-meter increments beneath a survey tape. Horizontal bar in the center of the box describes mean clast size. The limits of each box are the 25th and 75th percentiles for clast sizes. Whiskers (bars extending from the boxes) define the 10th and 90th percentiles and the black dots the 5th and 95th percentiles. Frankel and Dolan (2007) Figure 6.

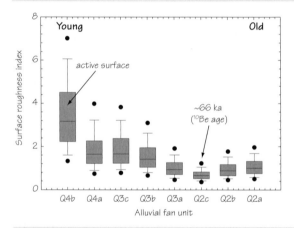

Figure 5.40 Box and whisker plots of surface roughness for each aggradation event surface. The horizontal bar inside each box is the mean. Box limits describe the 25th and 75th percentiles. Whiskers (bars extending from the top and bottom of each box) are the 10th and 90th percentiles, and the black dots the 5th and 95th percentiles. Frankel and Dolan (2007) Figure 8.

the piedmont in incised channels of the ephemeral streams. The Holocene aggradation event, which occurred in three stages (Q3a, b, c), was strong enough to overcome local uplift that had raised the Q2c a few meters. This section of the fault has a scissors style of vertical displacement with the downstream side up at the north margin arrow and down at the south mar-

gin arrow. These base-level changes promoted deposition of Q3c at the top and Q3a at the bottom, and contributed to post-66 ka incision of stream channels into the desert pavements of the smooth Q2c alluvial surfaces.

A high level of image resolution provided Frankel and Dolan with the data for quantitative description of the roughness characteristics of each aggradation event surface. Surface roughness is a function of both the fluvial-system characteristics for a given episode of deposition, and the post-depositional erosional and pedogenic processes. Surficial particle-size distributions vary greatly (Fig. 5.39). Q3a has the narrowest range of the black dots depicting the 5th and 95th percentiles, which makes sense from a climate-change perspective. Q3a,b,c record three stages of the most recent climate-change stripping of accumulated colluvium in the hillslope sediment reservoir of the source drainage basin. About 40 ky of hillslope weathering resulted in fine-grained surficial materials grading downward into slightly weathered joint blocks. The return of intense monsoon thunderstorm rainfalls at about 12 ka (Bull, 1991) would remove the finer surficial materials first (Q3a) and much larger joint-block detritus last (Q3c). Mean particle sizes also increased in the later stages of this aggradation event.

The roughness of these piedmont surfaces also changes with the passage of time. The LiDAR altitude data were aligned in a 1 m grid to create a digital elevation model with 100 cm horizontal resolution and 5 to 10 cm vertical accuracy. Surface roughness was calculated as the standard deviation of slope in a 3 m by 3 m moving window across the 1 m dataset, which discriminated between surfaces at the 3 σ level (the 99% confidence level). Roughness of piedmont alluvial surfaces initially decreases with the passage of time, eventually becoming smooth, planar landforms after ~ 66 ky (Fig. 5.40). A roughness threshold is crossed when stream-channel downcutting becomes more important than surficial splash and creep processes. Stable flat-topped ridgecrests then erode into convex hillslopes as the surface of deposition and its soil profile are removed.

5.6 Dating Fault Scarps with Terrestrial Cosmogenic Nuclides

5.6.1 Alluvium

Terrestrial cosmogenic nuclides create opportunities to expand paleoseismology endeavors at sites ranging from flights of stream terraces to bedrock fault scarps (Zreda and Phillips, 1998). Dating the times of surface ruptures in bedrock or gravelly alluvium

may utilize several terrestrial cosmogenic nuclides. Increasing acceptance of techniques that measure minute amounts of isotopes of He, Be, C, Ne, Al, and Cl created by cosmic-ray penetration into terrestrial surficial materials is the result of improving reproducibility of results. Common quartz-bearing rocks such as quartzite, rhyolite, granite, welded tuff, and sandstone are obvious choices when using ^{14}C, ^{10}Be, and ^{26}Al. Ongoing development of extraction procedures for isolating ^{14}C from quartz and minerals such as olivine (Lifton et al., 2001: Pigati et al., 2005) will have a precision and accuracy of dating that will further enhance the potential and reliability of many applications using terrestrial cosmogenic nuclides. Basic rock types such as basalt contain olivine or pyroxene whose times of exposure to cosmic rays can be evaluated with ^{3}He (Cerling, 1990; Kurz et al., 1990; Cerling et al., 1999; Fenton et al., 2002). Watershed denudation rates can be estimated with ^{21}Ne (Phillips et al., 1998) and ^{10}Be (von Blanckenburg, 2006). These paleoseismology tools will become ever more powerful as new cosmogenic isotopic methods come on line, laboratory measurements become even more precise, and we refine depth-production models for specific rock types, geomorphic settings, altitudes, and latitudes.

Dating geomorphic processes with cosmogenic nuclides requires careful attention to many variables and is the result of collective expertise of astrophysicists, geochemists, and field geologists. Explaining such complexity to students is challenging but the ever-resourceful Ed Evenson of Lehigh University was up to the task. As described by Gosse and Phillips (2001, p. 1485) Ed noted similarities between common sunburning of a human's skin and cosmogenic nuclide dating. The systematic accumulation of both sunburn and cosmogenic nuclides resulting from nonterrestrial radiation can be used to estimate how long skin or rocks have been exposed. The same variables affect both. Consequences of solar radiation and cosmic-ray exposure increase with time, and both vary with altitude and latitude. When removed from exposure tanned skin becomes lighter and cosmogenic radionuclides decay. Rates of nuclide production vary between minerals and sunburn between people. Clothing shields the skin, and atmosphere, snow, and nearby mountains shade landforms from cosmic radiation. For both it is easy to overestimate or underestimate the exposure time resulting from complications such as peeling of human skin or erosion of rocks. The increase in suntan from a second trip to a beach builds on what was inherited from the first trip, and shifting rocks carry their inherited terrestrial cosmogenic nuclides with them.

We can now measure the concentrations of stable and radioactive cosmogenic nuclides in boulders, stream channel straths, hillslopes, and entire drainage basins. Such dating supplies the keys to better understanding rates of deposition, tectonic displacement, frequency and magnitude of recent surface ruptures, and long-term retreat of escarpments.

Much of the discussion about fault scarps in previous sections of this chapter noted the many variables that influence hillslope erosion, including diffusion-type geomorphic processes that systematically degrade fault scarps. It is an advantage to independently constrain the value of the diffusivity coefficient, c in equation 5.1, such as done in the Wassuk Range piedmont study (Demsey, 1987). Cosmogenic isotope geomorphology has come to the fore since then, so let us summarize important major advances in paleoseismology and geomorphology of mountain fronts based on models of erosion that are constrained by distributions of terrestrial cosmogenic nuclides. Putkonen and O'Neil (2006) note that the universal degradation of even gently sloping alluvial surfaces needs to be accounted for in cosmogenic-exposure studies.

The study by Phillips et al. (2003) is an especially nice example. I like it because they use characteristics of surface and buried soil profiles to frame the interpretations of geochemical results and a modeling procedure that incorporates spatially variable rates of degradation of alluvial surfaces.

A Pleistocene stream terrace at Socorro, in the Rio Grande rift valley of southern New Mexico, USA is cut by a ~1.5 ka fault scarp that ruptures a late Holocene terrace. Overall scarp morphology and a trench that exposes the soils-stratigraphic section clearly indicate that this is a multiple-rupture event fault scarp. How many pre-Holocene surface-rupture events are recorded by this fault scarp and stratigraphy? Analysis of Pleistocene events means that the diffusion-equation modeling approach will need refinement of the diffusivity coefficient. If older than the range of radiocarbon dating, we enter the obscure time span not covered well by the radiocarbon dating methods used in most trench-and-date paleoseismology. Phillips et al. used ^{36}Cl analyses to overcome these several obstacles by collecting suites of 6 or 7 samples from three depth profiles on the footwall and hanging wall blocks.

The first was a set of six control samples at 27 m upslope from the fault-scarp crest where the terrace tread is unaffected by faulting and has undergone minimal erosion. About 150 fine gravel clasts were collected from each 10 cm depth increment to average out inherited cosmogenic nuclide signals. Clast lithology was 80% welded rhyodacite to rhyolite containing abundant phenocrysts of quartz, sanidine, plagioclase and biotite. Two splits of the deepest sample gave the same ^{36}Cl concentrations, indicating that the procedure of amalgamating ~150 clasts successfully averaged out the clast-to-clast variability.

The routine cosmogenic modeling results revealed a simple exponential decline with depth of ^{36}Cl for which a time of deposition of 122 ka ± 18 ka was calculated. This late Pleistocene age matches the soil-profile characteristics of a regional chronosequence well, and appears to be an aggradation event that coincides with the major sea-level rise of oxygen isotope stage 5e. This is the time of the regional Q2b aggradation event of Table 1.2. A relatively large amount of pre-deposition ^{36}Cl inheritance, equivalent to 43 ky of exposure to cosmic radiation at the land surface, is similar to that observed by Phillips at other sites in western North America. They suggest that corrections of similar large magnitude may negate cosmogenic dating of Holocene surfaces and deposits.

The locations of the other two depth profiles for suites of samples are shown in Figure 5.41. Horizonation of soil profiles of semiarid regions tends to be sufficiently distinct that photographs show the principal features (Phillips et al., 2003, Fig. 4). Organic matter makes the rapidly-forming A horizon darker than the parent material. The underlying argillic horizon of a Pleistocene-age soil is where clay minerals and reddish iron oxyhydroxides gradually accumulate. Still deeper, the white color of calcium carbonate provides an abrupt color contrast and sometimes cementation whose stages provide age estimates on a regional basis (Table 5.5).

The footwall-block stratigraphy is not complicated. Only a single soil profile is present and is sufficiently old that Stage III carbonate is present beneath the argillic Btk horizon. This pedogenic carbonate horizon slopes downward following a former scarp slope instead of being truncated by it, which suggests a short time span between cessation of deposition of the terrace gravels and the first surface rupture event. Alternatively, this represents erosion of the scarp after the first event.

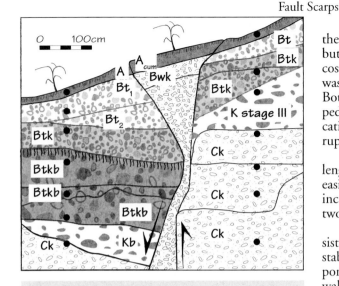

Figure 5.41 Soil log of the Socorro Canyon fault trench. Filled circles are where [36]Cl samples were collected. From Figure 5 of Phillips et al., 2003.

The hanging-wall-block stratigraphy is where the details of repetitive faulting reside. Each downward displacement creates a small local basin of deposition to receive detritus that falls off or is washed off of the adjacent uplifted fault scarp. Soil profiles are not as strongly developed on the footwall-block because they were buried by the next episode of surface-rupture induced deposition.

The hanging-wall stratigraphy has two colluvial wedges. Each colluvial wedge has a basal debris facies formed by free-face collapse. This is overlain by a wash slope facies created by more gradual scarp erosion. The third, 1.5 ka surface rupture event, is so young that it may be represented only by the very weak B soil-profile horizon (Bwk) at the top of the small graben infilling. A small amount of debris eroded off the footwall after a minor late-Holocene rupture filled a tension fissure instead of being deposited as a colluvial wedge.

More than just sandy gravel wash was washed off the scarp and deposited in the colluvial wedges. Some in-place weathering occurred on the stable tread of the stream terrace but most of the pedogenic materials were brought in by dust and rain (Goudie, 1978; Cooke and Doornkamp, 1990; Cooke et al., 1993; Reheis and Kihl, 1995). Changes in amounts of cosmogenic nuclides are mainly a function of time for each depth zone.

The hanging-wall depositional area also has these primary sources, but also has secondary contributions of clay, carbonate, iron oxyhydroxides, and cosmogenic [36]Cl derived from the incoming materials washed off the scarp to create the colluvial wedges. Both colluvial wedges have stage I to early stage II pedogenic carbonate development (Table 5.5), indicating a substantial time span between two surface ruptures that were both Pleistocene in age.

These complications posed additional challenges to the Phillips et al. team. It would have been easier to model the flux decrease (footwall block) and increase (hanging-wall block) for a single, instead of two, late Pleistocene surface rupture.

They succeeded in getting internally consistent results. The [36]Cl concentration beneath the stable surface of the control profile had a smooth, exponential decrease with depth. The degrading footwall profile results had a significant [36]Cl deficit and the aggrading hanging-wall profile an excess of [36]Cl. Re-deposited sediment had already received a dose of high intensity cosmic radiation. These nicely consistent results included a discontinuity in [36]Cl concentration at the contact between the two colluvial wedges shown by the buried soil symbol and nomenclature in Figure 5.41.

All [36]Cl exposure ages and erosion rates in this study were calculated using the depth-distribution equations and production parameters of Phillips et al. (2001). Modeled surface rupture ages simulated only the first two rupture events. They assigned displacements of 1.7 and 1.8 m to the two events, based on similar thicknesses of colluvial wedges. Modeling errors were minimized at 92 + 16 − 13 ka for the first rupture and 28 + 18 − 23 ka for the second. Their slip rate estimate of 0.2 m/ky is only 20% of that calculated from the morphologically based rupture history of this fault scarp (Clark, 1998). The modeled best-fit rupture history has an associated value of geomorphic diffusivity of 0.4 + 0:4 − 0:1 m^2/ky; at 1σ confidence level uncertainties. These clayey welded-tuff gravels have a greater alluvium mass strength than the sandy gravels of the Bonneville shorelines studied by Hanks and Wallace (1985), but are the same as for the Dixie Valley study area of Hecker (1985).

This breakthrough study is important to paleoseismology. Cosmogenic dating of surface rupture events can be done where materials for radiocarbon or luminescence stratigraphic dating techniques are not available, and can be used to date events older than 50 ka. Application of the Phillips et al. method in studies of single-rupture scarps should yield narrow estimated age ranges for the time of a surface rupture.

5.6.2 Bedrock

Faulted limestone is used here to illustrate the potential for estimating the times of prehistorical surface rupture of bedrock fault scarps. Each of the three examples has an independent estimate of the time of the penultimate earthquake. Cosmic rays produce neutrons that strike Ca target elements in the surficial 2 to 3 m of limestone creating a spallation reaction. Neutron flux is a function of altitude and latitude (Lal, 1991). Spallation of ^{40}Ca nuclei by fast neutrons creates ^{36}Cl in limestone. Muons are more likely to penetrate deeper than neutrons. They create a small component of terrestrial cosmogenic nuclides.

The exceptional smoothness of limestone fault planes (Fig. 5.42) enhanced the Giaccio et al. (2003) photographic study. Subsoil limestone surfaces have slickensides created by rupture at depth with superimposed patches of accreted soil carbonate. Limestone surfaces change color and become etched and rilled after exposure to rain and runoff. Distinct color or texture bands ranging in width from a few centimeters to several meters record exposure events, which are presumed to record episodes of tectonic displacement. One needs to be wary that they do not record gradual exhumation or mass-movement abrupt removal of the blanketing materials. Some prominent bedrock fault scarps may merely be the result of prolonged exhumation of soft alluvium adjacent to a siliceous plane of a now-dormant fault zone (Harrington et al., 1999). Tectonically induced burial of fault planes is possible too, especially adjacent to a tectonically active range-bounding fault. Previous chapters of this book have many examples of abrupt transitions from erosion to deposition that coincide with active fault zone base-level falls.

Cosmogenic isotope analyses of prominent bedrock color or texture banding in limestone can date times of fault displacement. This is the structural throw component of normal and thrust faults. The use of terrestrial cosmogenic nuclides in studies of active strike-slip faults may be limited to dating horizontally offset alluvial geomorphic surfaces.

The 1915 displacement of the Pearce normal fault zone exposed an increment of smooth limestone in the Tobin Range of west-central Nevada

Figure 5.42 Contrasts in weathering of limestone fault plane on a splay of the Pearce fault, west side of Tobin Range, west-central Nevada. Lower part of surface was raised above the soil line during the 1915 surface rupture. Older upper part of view is solution pitted and is one-third covered by lichens.

(Fig. 5.42). Pedogenic carbonate accumulated on the subsurface fault plane but now is being removed by acid rain. With time such exposed surfaces develop solution pits, and the texture and color also change (McCarroll and Nesje, 1996; Stewart, 1996). Abrupt changes in such characteristics appear as bands on the exposed fault surface that Wallace (1984) recognized as resulting from multiple faulting events. He compared the relative weathering of the next older band of weathering on the exposed fault plane with weathering of limestone boulders on 13 ka Lake Lahontan shorelines. Wallace concluded that the 2 m high Tobin Range limestone scarp face recorded a single surface rupture event that occurred between 3 ka and 12 ka.

Cosmogenic isotope investigation of the Tobin Range penultimate surface rupture event used in situ cosmogenic ^{14}C and confirmed Wallace's impressions (Handwerger et al., 1999). Limestone associated with the well-dated Pleistocene lake shorelines in Utah was used to measure the rate of in situ production of cosmogenic ^{14}C (18 ± 3 atoms per gram of $CaCO_3$ per year). Specific altitude and latitude correction is essential because cosmic ray input varies between sites. They concluded that the exposed To-

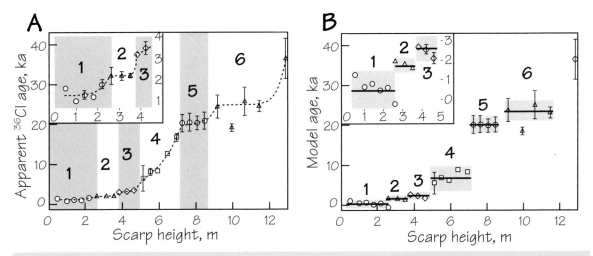

Figure 5.43 Repetitive exponentially decreasing patterns of cosmogenic isotope abundance indicative of surface-rupture induced increases in exposure of the limestone fault plane of the 1959 Hebgen Lake, Montana earthquake. From Zreda and Noller (1998a,b).
A. Six apparent exponential changes in abundance of ^{36}Cl in samples taken from six visibly different bands of the exposed limestone fault scarp. 1σ error bars.
B. Modeled ages of the six groups based on a combination of field and laboratory data. Short black lines are the means for each event and are significant at the 1σ level. At the 2σ level (light gray bands) groups 5 and 6, 2 and 3, and 1 and 2 are statistically indistinguishable.

bin Range fault plane represents just one event with an age of at least 10.6 ± 0.9 ka. This substantial age matches the degradational characteristics of the beveled fault scarp shown in Figure 5.1B.

Patterns of abundance of terrestrial cosmogenic nuclides in a partially buried outcrop vary systematically with depth. A simple exponentially decreasing trend of isotope abundance occurs for a tectonically stable planar rock face where the depth of burial by colluvium remains constant. High-energy cosmic-ray nucleons penetrate only about 2 to 3 m of rock because of progressively more shielding. The chances of termination in a spallation impact increase with cumulative increase of rock mass penetrated. Abrupt increase in exposure depth, and concurrent decrease in subsurface shielding, of the rock face by landsliding, fluvial erosion, or tectonic shift along a fault plane results in a an increase of production of terrestrial cosmogenic nuclides for a given parcel of rock. Multiple abrupt increases, instead of a single exponentially declining trend, may record multiple surface-rupture events (Fig. 5.43). Zreda and Noller (1998a,b) describe six such bands on the limestone

fault plane of the 1959 Mw magnitude 7.5 Hebgen Lake earthquake in Montana.

Cosmogenic ^{36}Cl amounts change abruptly between bands, suggesting abrupt changes in dose of cosmic rays resulting from less shielding material as the hanging-wall block drops away from the footwall block during a surface-rupture event. A nearby diffusion-equation modeling age estimate for the penultimate earthquake of 2.8 ± 1.1 ka (Nash, 1984) supports their cosmogenic age estimate of ~2.6 ka. Assuming only tectonic causes of renewed surface exposure, Zreda and Noller conclude that this fault zone is characterized by two episodes of closely spaced (clustered) earthquakes, at about 24 to 20 and 7 to 0 ka. Such results should encourage parallel paleoseismology endeavors.

Careful field work (Mitchell et al., 2001, p. 4228) is needed to determine how to separate surface rupture induced increases of fault-plane exposure from non-related exposure events such as a landslide-induced increase of the fault-plane exposure caused by seismic shaking resulting from another earthquake source in a study region.

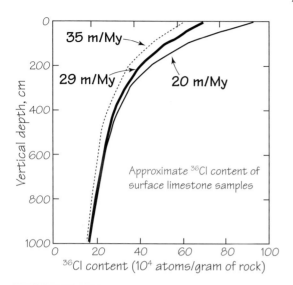

Figure 5.44 Concentrations of cosmogenic ^{36}Cl in the upper meter of a dolomitic limestone fault scarp as a function of three different long-term erosion rates. Samples from an altitude of 300 m and 33° N latitude. Figure 8 of Mitchell et al., 2001.

A detailed numerical model was devised by Mitchell et al. (2001) to assess cosmogenic ^{36}Cl dynamics of faulted limestone in Israel. An accurate model of surface rupture induced perturbations to the accumulation of ^{36}Cl is based on isotope abundance as a function of depth, and acknowledges how depth changes with time as a function of the rate of erosion of exposed limestone (Fig. 5.44). The mean erosion rate for two of their samples is 29 ± 3 m/My, which is faster than for limestone in an Australian outcrop (Bierman and Turner, 1995, Bierman and Caffee, 2002). Having a dip of 51° the Nahef East fault zone will receive less cosmic rays and muons than a horizontal surface. Their model used differences in the arrival mechanics of both. The footwall shielding effect is a function of depth below a horizontal surface. Modeling scenarios were judged at the 95% confidence level. These included tectonic creep, and many to few surface ruptures. Goodness of fit improved when the number of surface rupture events was increased from 1 to 5. The most consistent results from their terrestrial cosmogenic nuclides model yields a rapid, significant displacement on this fault zone at about 6.5–4 ka with smaller displacement events in the 13 to 11 and 2.5 to 0.5 ka intervals (Fig. 5.45). A mean displacement rate for the Nahef East fault, 9 m in 14 ky, is estimated to be 0.7 ± 0.4 m/ky.

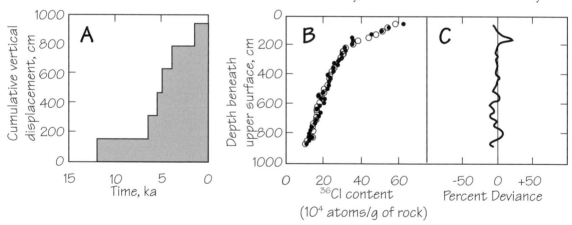

Figure 5.45 Most likely displacement (throw) scenario for the Nahef East normal fault for a postulated six-event series. A. Maximum displacement in the mid-Holocene results in a reasonable best fit for the entire profile. B. Modeled ^{36}Cl amounts (circles) as compared to measured values (dots) in the uppermost meter. C. Residuals calculated as the percent difference between measured and modeled ^{36}Cl values. From Mitchell et al., 2001.

It is always preferable to have independent checks for numerical modeling. The Nahef East fault is part of the Zurim escarpment where broken stalagmites in nearby cave debris yielded a U-Th age estimate of 6.2 ka. The possible range of earthquake Mw magnitudes ranges from 6.8 (1.6 m displacement) to 7.1 (4.7 m displacement).

Both the Montana and Israel models have characteristics suggestive of a major, obvious tectonic displacement plus possible sequences of smaller events. Both seem statistically robust and both have partial support from independent ways of dating the penultimate earthquake. I look forward to more such work, and to definitive statements of the precision and accuracy that can be expected from cosmogenic-isotope studies of bedrock fault scarps.

5.7 Summary

The geomorphic processes and morphologies of piedmont fault scarps are functions of lithology, climate and associated vegetation, scarp height, and the time that has elapsed since surface rupture. Each of these variables may vary by an order of magnitude, which complicates modeling of scarp topographic profiles.

For example, variations of alluvium mass strength along the Pitaycachi fault are in large part responsible for the substantial variations in morphology of fault scarps created in 1887. Scarps formed in loose grussy sand and in clayey sand have been greatly eroded by slopewash and rillwash. A century after the surface rupture, the maximum slopes for 2 to 3 m high scarps typically are only 6° to 20°, scarp widths are >10 m, free-face indices range from 0 to 10, and scarp-crest sinuosities from 2.5 to 3.5. Scarps in gravel erode at a much slower rate. Free faces are still present in loose Holocene sandy gravel, and are prominent in Pleistocene alluvium. It has a surficial protective layer of lag gravel and unweathered cobbles embedded in tough pedogenic clay. Mass-movement processes plus slope and rill wash have resulted in retreat of scarp crests and deposition of debris slopes in such cohesive materials. Maximum slopes typically are 30° to 50°, scarp widths are 2 to 5 m, free-face indices range from 10 to 40, and scarp-crest sinuosities generally are 1.2 to 1.5. Resistant lithologies such as sheared bedrock and massive carbonate-cemented gravels are the least eroded. Gravitational processes such as block collapse and single-rock falls are the main processes of fault-scarp change. Minimal amounts of scarp-crest retreat have occurred and the development of debris and wash slopes is much less than for more erodible lithologies. Maximum scarp slopes approach vertical inclinations, scarp widths are <2 m, free-face indices range from 40 to 90, and scarp-crest sinuosities are only slightly more than the minimum value of 1.0.

Diffusion-equation modeling of fault scarps became fashionable because of a general lack of datable materials in piedmont alluvium. Diffusive hillslope processes cause progressive decreases of scarp-crest convexity and scarp-base concavity in such an orderly manner that computer modeling was the obvious way to estimate the elapsed time since disappearance of the free face of a scarp. Crosschecks against tectonic, fluvial, and lacustrine scarps of known age yielded positive results. But like all hillslopes, fault-scarp erosion is a function of many variables. The net result of many diffusion-equation modeling studies is considerably more insight into factors affecting scarp degradation. These include scarp height and microclimatic controls of geomorphic processes and plant cover for different orientations. Even the sandy gravel piedmont basin fill of a region really does not have a universally applicable alluvium mass strength, so one should expect regional variations in the diffusivity constant. We now realize that such modeling age estimates are merely approximations: early Holocene, mid-Holocene for example. But that is useful where zoning regulations are strict for faults with a demonstrated Holocene surface rupture. It seems that a soils chronosequence for flights of stream terraces and alluvial-fan remnants would yield the same resolution of dating.

The advent of laser swath altimetry provides superb digital elevation models that reveal subtle fault-scarp features. Such LiDAR efficiently helps to identify alluvial geomorphic surfaces of a regional chronosequence created by climate-change induced aggradation events. Recent fault scarps hidden in the dense vegetation of rain forests are easily discerned, underscoring the efficiency of LiDAR as an essential reconnaissance tool.

Better ways to date fault scarps have decreased the emphasis on diffusion-equation modeling. Advances in optical and thermoluminescence procedures now date silt-size materials from diverse depositional settings. Terrestrial cosmogenic nuclides estimate ages of both surfaces and deposits. They may become a standard tool for dating fault scarps, and for estimating rates of offset alluvial geomorphic surfaces by strike-slip faulting. And, best of all, their age range is not limited to the Holocene or to the 40 ka time span of radiocarbon analyses. Cosmogenic dating has expanded from alluvium to bedrock fault scarps, with especially promising results for normal-faulted limestone. This soluble rock is smooth when first exposed, after which surface-texture roughness is a function of exposure time to solution processes.

Breakthroughs in the use of terrestrial cosmogenic nuclides provide superlative opportunities for estimating times of prehistorical surface ruptures and mountain front uplift rates. Multiple Pleistocene shoreline gravels clinging to the rapidly rising Black Mountains of Death Valley in California, and to the Wassuk Range in Nevada, are obvious candidates for cosmogenic dating. Unweathered rhyolite boulders on the old, stable, alluvial geomorphic surfaces along the Pitaycachi fault piedmont await cosmo-

genic study. Isotopic signatures of many prehistorical earthquakes reside in the 10–35 m high bedrock fault scarps shown in Figure 5.26. This extraordinary site awaits the combination of fieldwork, laboratory analyses, and numerical modeling that will reveal the systematic behavior of the Pitaycachi fault. The results could clarify how we view the earthquake hazards in the southern Basin and Range Province.

Let us compare the above advances with other approaches to paleoseismology that have a dating precision and accuracy better than ±10 years. The Pitaycachi fault discussion in this chapter is an end member in regard to length of earthquake recurrence intervals. Get ready to shift gears again! The next chapter focuses on dating and describing frequent earthquakes on a plate-boundary fault zone in one of the world's fastest rising mountain ranges.

Analyses of Prehistorical
Seismic Shaking

Chapter 6

Advances in paleoseismology have greatly improved our perception of earthquake hazards and risks. The previous emphasis has been largely stratigraphic and generally consists of radiocarbon dating of layered deposits that have been ruptured by prehistorical earthquakes.

Studies of geomorphic processes in tectonically active mountain ranges add several new dimensions to paleoseismology. Landscape evolution in tectonic settings with different uplift rates creates landform associations that are so distinct that they can be used as a reconnaissance tool in assessing regional earthquake hazards. Tectonic activity classes of mountain fronts is an example that is useful for zoning purposes. But engineers and planners need more information about the frequency, magnitude, and extent of seismic shaking events. Tectonic geomorphologists can make contributions that complement the fine work of the tectonic stratigraphers.

Diverse studies work best in paleoseismology. This chapter will highlight the influence of earthquakes on hillslope geomorphic processes, emphasizing rockfall types of mass movements. Key collaborative disciplines used as essential cross-checks include tree-ring dating of earthquake damage to individual trees (sometimes to the year), forest disturbance events to assess the intensity of seismic shaking and to estimate times of earthquakes, and of course radiocarbon dating of stratigraphic evidence for earthquakes in trenches (Figs. 3.23, 4.29). It is also useful to date earthquake-induced aggradation events in the valley floors of rivers, and for coastal dunes whose source of sand is a nearby river.

My theme is to underscore the benefits of multidisciplinary paleoseismology. The goals are to address previously unanswered questions about the effects of seismic shaking emanating from the Alpine and Hope faults in New Zealand and the San Andreas fault in California. Chapter headings are geographical, but the topics discussed and examples presented in each section have minimal overlap.

6.1 Paleoseismology Goals

Introductory thoughts include appraisals of precision and accuracy of dating methods. But first, let's set some challenging goals.

No single paleoseismology technique does everything, but it is instructive to make a wish list. My list includes:
1) Ability to date the time of the earthquake itself rather than dating material created before or after the earthquake.
2) Determination of the intervals between several earthquakes for a specific fault.
3) Avoid having to choose between multiple possible dates, such as for radiocarbon age estimates of the past 300 years.
4) A quantitative index of intensity of seismic shaking for either historical or prehistorical earthquakes.
5) Ability to make seismic shaking index maps, and ratio maps for pairs of earthquakes.

Accumulation of large rock-fall blocks in the valley of the South Fork of the Kings River, Sierra Nevada, California.

6) Each application of the earthquake dating method replicates and tests all assumptions, measurement routines, and analytical procedures.

7) Identification of the fault responsible for a specific prehistorical earthquake.

8) Ability to evaluate both precision and accuracy of the earthquake-dating method.

9) Length and direction of propagation of prehistorical surface ruptures.

10) Determination of the amount of slip on a fault plane in order to estimate earthquake moment magnitude, Mw.

11) Ability to date earthquakes over the broad time span of 50 to 50,000 years ago.

The lichenometry surface-exposure dating method is used here. It achieves all but the last two objectives in my paleoseismology wish list, but only for the past 1,000 years.

Precision and accuracy of dating methods describe the uncertainties of the age estimates for prehistorical earthquakes. *Accuracy* is the error between an age estimate and the true age of the event, and is best assessed with historical events or with dates based on tree-ring analyses when dated to the year or season. Knowing the accuracy lets one evaluate assumptions, methods, and equipment used for a specific method of dating earthquakes.

We commonly lack ways of determining accuracy so turn to the unrelated measure of *precision*, which is the statistical spread of replicate procedures or multiple estimates of the age for a particular event. Precision is described as a Gaussian probability distribution of age estimates. Unfortunately, statistical estimation of uncertainties varies from method to method. Luminescence and lichenometry propagate all quantifiable errors at 2σ (95% confidence level), but uncertainties for most radiocarbon dates are limited to just the counting statistics for radioactive decay at 1σ (68% confidence level). Recycling of organic matter (detrital charcoal) into new stratigraphic settings, inheritance of old radiocarbon ($CaCO_3$ in shells), and contamination by young radiocarbon (roots) are difficult to address statistically. Baillie (1995) concludes that radiocarbon age estimates can be misleading unless they are cross-checked by other dating methods, preferably dendrochronology.

Contrasts of precision and accuracy for nine dating methods are shown in Figure 6.1, but only for the recent past. Dendrochronology (A) may date tree death to the season of a year but, as described later in this chapter, the effects of earthquake-induced damage to roots and branches may be delayed several years in surviving trees. The lichenometry example (B) uses the method of Bull and Brandon (1998) where precision is assessed at multiple sites and accuracy is measured by dating historical earthquakes. Varve chronology may be precise to within ±1–2 years but the example used for D has the added complications of a ±40 year radiocarbon-dating uncertainty, plus the uncertainty of how old the wood or charcoal was before it was deposited. Radiocarbon dating (E) that uses the corrected half life of 5,730 years and accounts for variations in the ^{14}C content of the atmosphere with time has good precision and accuracy where the material sampled for dating does not significantly predate or postdate the event being dated. Knuepfer (1988) and McSaveney (1992) developed sophisticated ways of dating rock weathering rinds (F). This surface-exposure dating method dates the event itself but precision is degraded by the need to calibrate rind-thickness growth with radiocarbon dating of landslide-buried trees.

The use of terrestrial cosmogenic nuclides (G) relies on cosmic-ray production of a variety of nuclides accumulating in specific minerals of surface and buried rocks. Accuracy uncertainties are similar for the different methods and center about ways of ascertaining variations in the rates of nuclide production in minerals at different altitudes and latitudes. Another factor to be estimated is how past erosion, exfoliation, and burial during the exposure time span affected incoming cosmic rays for the material sampled. The amino-acid racemization example (H) has nice precision because of good instrumental replication, but poor accuracy because of errors resulting from temperature-history assumptions of variable quality. Strength of soil-profile development (I) has weak precision and accuracy because of spatial variations of parent material and climatic factors. Even after a soils chronosequence has been described, use of parameters such as percent carbonate or a soil-profile-development index based on several properties will have a precision and accuracy uncertainties that exceed 1,000 yr for Holocene soils. Despite such limitations, knowing the approximate ages of faulted and unfaulted alluvial surfaces of soils chronosequence is an essential tool for paleoseismology reconnaissance investigations.

Describing dating methods and applications requires a book. See Noller et al. (2000) for information about the types of materials dated and age ranges for these and many other methods.

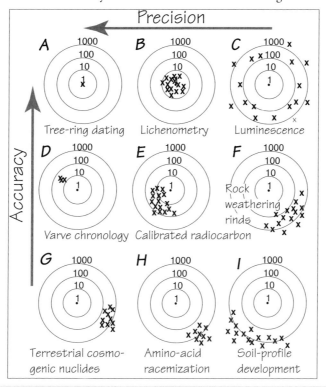

Figure 6.1 Examples of precision and accuracy of dating methods. Logarithmic-scale circles are numbered in years. A. Correlation of annual growth rings between trees yields dates to the year or season. B. Lichenometry is accurate and has tightly clustered precision. C. Luminescence may have good accuracy but poor precision due to a large number of controlling factors for a given sample. D. Varve chronology is precise and accuracy can approach 1 year for short time spans. Accuracy is degraded in older example shown here where radiocarbon dating is needed to estimate the approximate age of glacial-lake deposits. E. Accuracy of moderately precise radiocarbon method is best when calibrated against variations in production of radiocarbon in the atmosphere. F. Rock-weathering-rind thicknesses of a given age vary greatly for most lithologies. G. Accuracy for terrestrial cosmogenic nuclides is lowered by uncertainties in scaling the production rate of a calibrated site to the altitudes and latitudes of other sites. H. Amino-acid racemization has good precision but poor accuracy. I. Strength of soil-profile development has weak precision and accuracy because of spatial variations of parent material and climatic factors. Like weathering rinds, propagation of the uncertainties of methods used to date soils of different ages decreases the precision of dating.

Lichen dating of regional coseismic (earth-quake-generated) landslide events (Bull, 1996a; Bull and Brandon, 1998, Bull, 2003b, 2004, 2005) complements stratigraphic dating of earthquakes. Advantages of the lichenometry approach to paleo-seismology include:

1) Estimated date of the earthquake, not the age of organic material formed before or after.

2) Precision and accuracy of dating better than ±10 years.

3) Ability to make maps depicting patterns of seismic shaking for prehistorical earthquakes that are as good as Mercalli intensity maps for historical earthquakes.

4) Capacity to study seismic shaking caused by distant large earthquakes, including earthquakes that occur offshore, or on blind thrust faults.

Lichenometry deficiencies include inability to measure amounts of coseismic slip, local lack of seismically sensitive topography or suitable slow growing crustose lichens, and absence of lichens older than 300 years on geomorphically active hillslopes. Lichens older than 1,000 years are unusual.

Lichenometry measurement and analytical procedures are described in more detail elsewhere. See Bull and Brandon (1998) for discussions of theory of analytical procedures. See Bull (2003a) for an evaluation of the types of geomorphic processes and landforms that are most suitable for dating with lichenometry. Being a surface-exposure dating method, lichenometry can date the time of creation of a landslide and the times of all subsequent disturbances to the landslide surface (Bull, 2003b). "Further readings" in Bull (2004) lists URLs that have photos of lichen characteristics, and discusses field site-selection and measurement techniques.

6.2 Earthquake-Generated Regional

Rockfall Events

This lichenometry procedure was developed and tested in the Southern Alps of New Zealand. It was then applied to studies of coseismic rockfalls in the Sierra Nevada of California, and to date geomorphic processes in aseismic Sweden (Bull et al., 1995).

This chapter describes efficient ways to better understand prehistorical earthquakes in lofty mountain ranges. Both the South Pacific island country of New Zealand and the North American state of California sit astride transpressive plate boundary fault zones. Thrust and strike-slip faults are common and submarine fault zones are close to shore in New Zealand (Pettinga et al., 2001). Discussion here focuses mainly on dating, seismic shaking, and magnitudes of earthquakes but includes assessment of the precision and accuracy of age estimates, and compares lichenometry and precise radiocarbon age estimates of exceptionally well-documented San Andreas fault California earthquakes.

Description of seismic shaking characteristics of prehistorical earthquakes is a strong point of the Bull–Brandon (1998) approach to paleoseismology. The intensity and extent of a seismic shaking event are assessed by using regional variations in rockfall abundance as an index of landscape response to seismic shaking. Variations in rockfall lichen-size peaks record 1) the decrease in intensity of seismic shaking with increasing distance from an earthquake epicen-

ter, and 2) topographic influences on local response to seismic shaking. Some seismic-shaking index maps suggest the likely direction that an earthquake rupture propagated. Insight about moment magnitudes for prehistorical earthquakes entails comparing sizes of lichen-size peaks and respective seismic shaking index maps for historical and prehistorical earthquakes.

The Chapter 6 menu is diverse both in terms of topics and in geographic settings. The paleoseismology procedure is introduced after a tectonic-setting summary. The main topics are calibration of New Zealand lichen growth rates, precision and accuracy of dates for historical earthquakes, and identification of significant lichen-size peaks produced by large and small seismic-shaking events. Then analyses and locations of recent Alpine fault earthquakes are cross-checked by using several ways to analyse tree-rings. Examining recent earthquakes in the Marlborough region of the South Island, New Zealand, uses lichenometry to introduce new procedures in paleoseismology.

Then I examine how historic and prehistorical earthquakes are recorded by rockfalls in the glaciated granitic landscape of the Sierra Nevada of California. This tectonic geomorphology approach to paleoseismology evaluates how topography affects earthquake-generated mass movements and provides robust crosschecks for stratigraphic paleoseismology. Lichenometry documents both local and distant San Andreas fault earthquakes.

6.2.1 New Zealand Earthquakes

6.2.1.1 Tectonic Setting

The Alpine fault (Fig. 6.2) is part of a transpressional transform boundary between the Australian and Pacific plates (Berryman et al., 1992; Norris and Cooper, 1997, 2001). Approximately three-quarters of the total displacement between the Pacific and Australian plates occurs on the Alpine fault. A single nearly vertical fault zone in the south is characterized by right-lateral displacements. The central section has serial partitioning. Moderately dipping thrust complexes have a more northerly strike, and dextral partitions strike more to the east. Where it is oblique-slip, it strikes ~ 055°. Oblique convergence of the plates raised the Southern Alps at 5 to 8 m/ky years during the past 135 ky with the area of maximum uplift rates coinciding with the highest mountains (Cooper and Bishop, 1979; Bull and

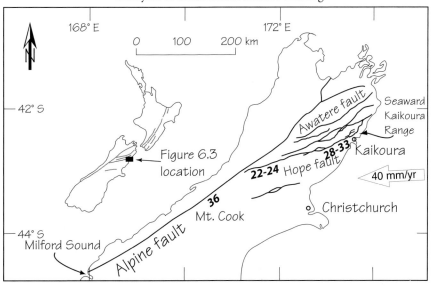

Figure 6.2 Tectonic setting of study area in the South Island of New Zealand. Diagonal zone of faulting is along the transpressional boundary between the Pacific and Australian plates. Faults are shown by broad black lines (Van Dissen and Yeats, 1991). Bold-oblique numbers are mean right-lateral fault displacements in m/ky. Pacific plate motion vector is from De Mets et al. (1990, 1994). Location of Figure 6.3 image is shown by the rectangle on the inset map showing the North and South Islands.

Cooper, 1986, 1988). Zircon fission-track analyses by Tippett and Kamp (1993) suggest a maximum of 9 m/ky (exhumation) over a longer time span of 900 ky. The Alpine fault splays into many faults in the Marlborough region in the northeastern South Island.

The 220-km long Hope fault is the most active splay of the transpressional transform plate boundary in the Marlborough section of the Alpine fault system (Knuepfer, 1992; Pettinga and Wise, 1994; Cowan et al., 1996). The focus here is on the 58-km long Conway segment (Kahutara segment of Van Dissen and Yeats, 1991) extends northeast from the 15-km long Hanmer pull-part basin to a 5 km right step at Mt. Fyffe (Fig. 6.3) near the town of Kaikoura. The range-bounding fault for the Mt. Fyffe block is the Hope fault and the internal Kowhai fault of Van Dissen and Yeats underlies the deeply incised valleys of the Kowhai and Hapuku rivers (10 on Figure 6.3).

Conway segment strain southwest of Mt. Fyffe has a parallel style of fault partitioning between a range-bounding fault that dips 75° to 85° and is predominately strike-slip, and an internal fault with large amounts of vertical displacement. The Kowhai fault is clearly apparent in the topography southwest of Mt. Fyffe. Further southwest, surficial expression of the internal fault zone is progressively less obvious but it may continue as a blind thrust.

The range-bounding fault is as active as the San Andreas fault of California (Sieh and Jahns, 1984). Estimated slip rates are rapid (Bull, 1991, Section 5.1.4); latest Holocene horizontal slip is estimated to be 33 ± 2 m/ky. The Amuri and Seaward Kaikoura Ranges on the hanging-wall block have been rising during the late Pleistocene at about 2.5 to 3.8 ± 0.4 m/ky. Hilly terrain on the footwall block has been rising at 0.5 to 1.3 m/ky.

Estimated magnitudes of Conway segment earthquakes follow conventional procedures. Oblique dextral slip for the most recent and penultimate surface-rupture events are estimated to average 5.5 ± 1 m, based on offset stream channels and stream terrace risers (Bull, 1991; Van Dissen and Yeats, 1991). Rupture length is presumed to be the 58 km between the major right steps in the Hope fault at Hanmer basin and Mt. Fyffe, and seismogenic rupture depth along the fault plane is assumed to be 12 km. The result would be a Mw magnitude 7.3 ± 0.2 earthquake. Earthquakes on other Marlborough fault zones, such as the Clarence and Awatere, have similar magnitudes, but longer return times.

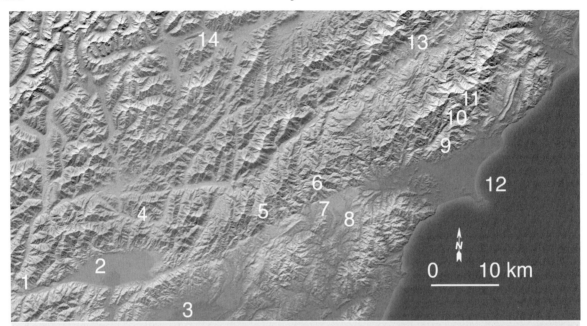

Figure 6.3 The Hope fault stands out on this digital image because of its rapid slip rate. The Seaward Kaikoura Range (9, 10, 11), northwest of Kaikoura Peninsula (12), rises 1,500 to 2,600 m above the Pacific Ocean. Study-site and place name locations: 1, Hope segment of the Hope fault; valley to north is underlain by cross faults shown at the left side of Figure 6.4; 2, Hanmer pull-apart basin; 3, Culverden basin; 4, Stonehenge; 5, Fenceline Crags, Mt. Lyford; 6, Stone Jug, Dog Hills, Amuri Range; 7, Goat Hills; 8, Cattle Gully; 9, Mt. Fyffe; 10, saddle between Kowhai and Hapuku Rivers; 11 , Barretts landslide; 13, Clarence fault; and 14, Awatere fault. Image provided courtesy of Scott Miller and Manaaki Whenua Landcare Research, New Zealand.

The adjacent Hope segment of the Hope fault extends at least 25 km west from the Hanmer pull-apart basin and is the only segment of the Alpine fault system to have ruptured since colonists arrived in substantial numbers in 1840 A.D. This was the North Canterbury Ms magnitude ~7.1 earthquake of 1888 (Cowan, 1991).

Strain partitioning of thrust and strike-slip components of tectonic deformation in the Marlborough fault zone is related to the larger-scale interactions between the overriding Australian and subducting Pacific plates (Nicol and Wise, 1992; Pettinga and Wise, 1994; Eusden et al., 2005a, b). Their work is the basis for a model that explains sustained regional compression at high angles to the primary faults of the Marlborough region, which are dominated by strike-slip displacements. A north-striking hinge zone, with diamond-shaped fault blocks, marks a regional change in fault strike from 080° (translational) to 065° (transpressional) to the northeast of the Hanmer pull-apart basin (Fig. 6.3, location 2). Antithetic cross faults are topographically obvious in the translational domain west of the eastern margin of Hanmer basin because streams have entrenched these fault-controlled valleys (for example, between locations 1 and 5 on Figure 6.3). Their topographic expression is subdued north of the transpressional Conway segment of the Hope fault between locations 5 and 9.

The plate tectonics aspect of their model proposes a crustal-scale "flower structure" in order to explain the along-strike convergence of opposite dipping, subducting plates. This might be considered fault partitioning in a vertical sense. A near-surface regional slab accommodates compression normal to the strike of right-lateral displacements, which are concentrated in the underlying semidetached slab of the plate.

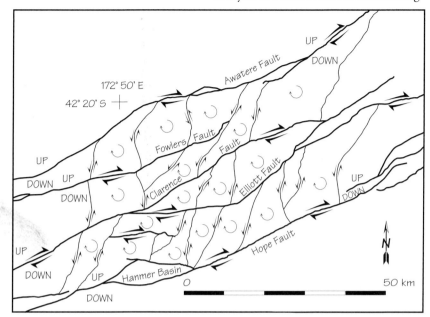

Figure 6.4 Pattern of primary faults (large dextral arrows) and secondary cross faults (small sinestral arrows) of the Marlborough fault zone. Arrowed circles show sense of fault-block rotation about vertical axes (Eusden et al., 2005a, b). Map supplied courtesy of J.D. Eusden.

Faults of the Marlborough fault zone (Fig. 6.4) are linked by antithetic cross faults (defined and discussed in Section 4.2.3.1.2) that transfer strain between the primary right-lateral fault zones – the Wairau, Awatere, Clarence-Elliot, and Hope faults. North-striking cross faults as long as 40 km typically have sigmoidal S shapes indicative of the regional dextral shear component of tectonic deformation. Transfer of strain between the main faults is illustrated later in this chapter in a discussion about a sequence of four northeast-stepping surface ruptures. Patterns of seismic shaking depart from being parallel to the traces of the main faults, which is not surprising given the complex tectonic setting.

The active faults of the alpine fault system are a challenge for paleoseismologists wondering when and where the next earthquake will strike. Previous stratigraphic and forest-disturbance-event studies focused on three recent Alpine fault earthquakes. This chapter uses analyses of the annual growth rings in trees growing in seismically sensitive swamps and steep hillslopes for dating key events, whose patterns of seismic shaking are compared using lichenometry. Then, Hope fault discussions expand the technique further by using lichen-size measurements from fragile outcrops as well as from rockfall blocks. The emphasis is on regional patterns of prehistorical seismic shaking and spatial shifts of a sequence of related earthquakes that are closely spaced in time.

6.2.1.2 Background and Procedures

This introduction starts with characteristics of lichenometry sites, choosing between slow- and fast-growing lichen genera, calibration of lichen growth rates, and assessment of precision and accuracy of this surface-exposure dating method. The Bad Bird site is 39 km from the Alpine fault and is used here to consider the nuances of defining peaks in a nonuniform distribution of lichen sizes.

Lichenometry paleoseismic work is fairly straightforward if rockfall study sites are selected with care. Many erosional processes expose fresh rock substrates to be colonized by lichens and primitive flora. Rockfalls are quite useful for our purposes. They are a very common mass-movement process, are likely to remain unburied for long time spans, and can be caused by distant earthquakes (Keefer, 1984, 1994). Some rockfall blocks travel a kilometer, but a block of rock only needs to tip over, or be split by an incoming bouncing boulder, to expose a fresh substrate for lichens to colonize. First, one needs to consider if lichen sizes on rockfall blocks are only recording aseismic landslide events, rainstorms, and avalanches. Paleoseismology sites should have many blocks that tumbled downslope during earthquakes. Earthquake-generated rockfalls dominate Southern Alps hillslopes to the extent that some sites have not received new rockfall blocks since the Inangahua (Mw

7.4) earthquake of 1968 (Downs, 1995; Bull and Brandon, 1998, Fig. 4). The opposite, and undesirable situation, is an active talus slope where all blocks are buried by incoming detritus within a few years. Unstable landforms such as fragile cliffs, landslides, talus slopes, and young glacial moraines typically are sites where seismic shaking events shift blocks to create fresh rock surfaces that are colonized by lichens. Even subtle block re-orientation may change the microclimate sufficiently to cause re-colonization by a new lichen community.

A sequence of earthquakes generates a series of regionally synchronous coseismic rockfall events in fragile mountain ranges such as New Zealand's Southern Alps and California's Sierra Nevada. The younger regional rockfall events coincide with known times of historical earthquakes (Tibaldi et al., 1995), which facilitates testing of any lichenometry model.

Crustose species of lichens are preferred. They grow as tightly attached symbiotic mixtures of algae and fungi on rock surfaces exposed to sunlight. Slow growing lichens are favored because they date much older events than fast growing lichens – 1,000 instead of 100 years. Yellow-green rhizocarpons, *Rhizocarpon* subgenus *Rhizocarpon*, are the lichen of choice in many arctic-alpine areas. Each millimeter of growth requires about 6.4 years in the Southern Alps of New Zealand and 10.5 years in the Sierra Nevada of California. *Acarospora chlorophana* is excellent for lichenometry in the drier parts of the Sierra Nevada; it needs about 8.4 years to grow 1 mm. I use both slow- and fast-growing lichens in the Sierra Nevada. Measuring sizes of lichens with faster growth rates should improve separation of closely-spaced lichen-size peaks in histograms and improve precision of age estimates. *Lecidea atrobrunnea* needs only 4.3 years to grow 1 mm in the Sierra Nevada.

Landslide, stream terrace, and glacial deposits commonly have enough cobble- and boulder-size blocks to allow a sampling strategy that Bull and Brandon (1998) refer to as the fixed-area largest-lichen (FALL) method. A FALL lichen size is defined as the maximum diameter of the largest thallus, black prothallus rim included, found in a unit sample area. Joint and rockfall blocks rarely are of a uniform size, and yellow rhizocarpons only grow on part of a block, so we typically use a mixture of block sizes within a limited size range. This New Zealand study used individual rockfall blocks ranging in diameter from 0.2 to 1 m, and joint faces of outcrops ranging from 0.05 to 0.5 m².

Many rockfall blocks tumble into stream channels or shady forests, but some accumulate incrementally in sunny repositories. Examples include talus cones, landslide complexes, hillside benches, and as rockfalls scattered along the tread of a stream terrace. Glacial moraines and rock avalanches are examples of bouldery landforms that continue to be modified by seismic shaking events. Active glaciers can transport surficial rocky detritus with young lichens to lateral and terminal moraines. Gravel bars deposited by floods in valley floors are an example of single times of deposition.

The internal consistency of results during a decade of work in the Southern Alps supports several key assumptions. The longest axis of the largest lichen best represents the time since deposition of each rockfall block. It is assumed that this largest lichen was the first to colonize a new rock surface. Measurements from many rockfall blocks provide a statistically more precise estimate of rockfall-event age. I also assume that abundance of rockfalls decreases with distance from an earthquake epicenter, and that earthquake-generated rockfalls dominate our Southern Alps lichenometry datasets (Fig. 6.5). Many species of *Rhizocarpon* subgenus *Rhizocarpon* appear to grow at virtually the same rate in different parts of the study region with diverse climate, altitude, or rock type. Only Section *Alpicola* grows at a different rate (Bull and Brandon, 1998, Fig. 18).

Sizes of elliptical lichen thalli are measured with digital calipers in order to increase precision and reduce operator bias. The long axis of the largest lichen records optimal lichen growth. Largest lichens are easily discerned on blocks smaller than 1 m. The better rockfall blocks have a scattering of isolated lichens (see Figure 6.41 later in this chapter) instead of masses of merged lichens. The thallus of the largest lichen on each block needs to be examined carefully to make sure that it does not consist of several lichens that have grown together, and that the margins are sufficiently sharp that replicate digital caliper measurements are within ±0.1 mm (discussed further in Figure 6.39B later in this chapter). Such replication represents only ±0.6 year of growth in the Southern Alps, so measuring the sizes of lichens is a trivial source of error.

Initial fieldwork found sites where rates of growth for *Rhizocarpon* subgenus *Rhizocarpon* could be calibrated. The goal was to measure lichens at sites of known age that span a century or two in order to get a lichen-growth equation for dating lichen-size

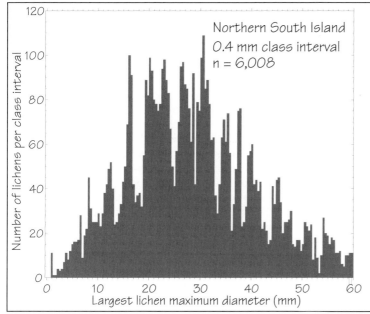

Northern South Island
0.4 mm class interval
n = 6,008

Figure 6.5 Histogram of the largest lichen maximum diameters for 6,008 rockfall blocks. Measurements were combined from 20 lichenometry sites in 20,000 km² of the northern part of the South Island of New Zealand underlain by quartzitic greywacke sandstone, syenite, and argillaceous greywacke sandstone. Site altitudes range from 380 to 1620 m. Altitude and macroclimate do not affect lichen growth rates as shown by the sharply defined lichen-size peaks of coseismic regionally synchronous rockfall events. From Bull, 2003b, Figure 7.

peaks at other sites. Preferred calibration sites are those where the time of exposure of the rock substrate is known to the day or year. Examples include rockfill dams, highway, railroad and trail cuts, landslides that are known to have been created by historical earthquakes, bouldery deposits of specific floods, abandoned quarries, and known times of diversions of streamflow from waterfalls. Historical coseismic landslides are an obvious calibration possibility in earthquake-prone mountain ranges. Tree-ring dated landslides are ideal for extending a chronology into prehistorical times. Use of radiocarbon dated deposits may seem appealing, but radiocarbon dating introduces much larger uncertainties than the more precise tree-ring dating method (Fig. 6.1).

Lichen growth begins with a rapid but exponentially decreasing growth phase for several decades before entering a prolonged uniform-growth phase. Calibration data should document rates for both growth phases and the time that elapses between exposure of the rock substrate and the first appearance of juvenile lichens (colonization time). Lichenometry age estimates are the sum of colonization time, greatgrowth phase, and uniform-phase growth. It is nice to be able to date the exposure event itself, instead of pre- or post-event materials or processes.

A large sample of the lichen-size population is measured instead of the traditional largest five lichens (or single largest lichen) from an entire deposit such

as a glacial moraine (Matthews, 1974; Locke et al., 1979; Birkeland, 1981; Porter, 1981; Rapp, 1981; Innes, 1985; O'Neal and Schoenenberger, 2003). Measuring largest lichen maximum diameters of *Rhizocarpon* subgenus *Rhizocarpon* on 50 to 1,000 rockfall blocks increases data density and allows easy separation of events that are only 4 years apart in the Southern Alps, and 7 years apart in the Sierra Nevada. Fragile landforms in the Southern Alps, such as glacial moraines, and landslide deposits, have multiple lichen-size peaks. Each peak usually represents an earthquake disturbance event (Fig. 6.5). A single landslide or moraine with 400 lichen-size measurements may date and provide estimates of relative seismic shaking intensity for 20 events (Fig. 6.8 and Bull, 2003b). The cost of measuring lichens at one site is a day or two in the field and one does not have to wait months for samples to be dated by a geochemical laboratory.

A basic tenet of surface-exposure dating is that the data collected records a combination of the initial time of exposure plus all subsequent modifications that influence the time-dependent process. Surfaces of sandstone blocks in an unstable glacial moraine record more than the cessation of glacier movement. This applies to a broad range of dating methods including terrestrial cosmogenic nuclide production in surficial rocks, weathering rinds in sandstone and volcanic rocks, and lichenometry.

Post-depositional exposure processes change as a result of splitting and removal of surficial rock by frost or fire, shifting and change of block orientation by earthquakes or random processes, temporary burial by hillslope processes, and growth of bushes and trees. The resulting surface-exposure date may emphasize the time of moraine deposition, but usually documents post-depositional seismic disturbances. Evaluation of results also includes assessment of how much of the surface-exposure dating signal was accumulated before deposition of the moraine.

The growth rate for *Rhizocarpon* subgenus *Rhizocarpon* in the South Island of New Zealand is based on calibration sites that include landslides, flood deposits, and rock walls (Tables 3 and 5 of Bull and Brandon, 1998). The resulting equation continues to be refined as additional calibration data become available. It describes ages of lichens that have entered the uniform-growth phase (larger than 9.4 mm):

$$D = 315.31 - 0.1552\,t \qquad (6.1)$$

Where D is the mean size of a lichen-size peak and t is the substrate-exposure age in years. Lichen sizes from different calibration sites first were normalized to the year 1991 A.D. to facilitate comparison with the large regional dataset collected in that year.

The 95% confidence interval for an estimated lichenometry age depends on the time span of the calibration equation used and the number of lichen-size measurements used to define a lichen-size peak. FALL peaks with 100 to 500 measurements can be dated with equation 6.1 to a precision of ±5 to ±10 years for events as old as 1000 A.D. (Bull and Brandon, 1998, Fig. 20).

Identification of specific lichen-size peaks may use two types of probability density plots. In simple histograms each lichen-size measurement adds another vertical increment to a specific class interval bar. A unit Gaussian represents each observation in the more robust decomposition of Gaussian kernel plots. These plots are constructed by converting each measurement into a unit Gaussian function and summing the densities of the overlapping bell-shaped Gaussians to make a smooth probability density plot. Values for lichen-size peaks are about the same for the two methods, but modeling that involves deconvolution of Gaussian probability density plots identifies peaks that might be hidden in standard histograms. Both probability density plots can be highly polymodal. The degree of smoothing is a function of the kernel standard deviation used in the modeling (Silverman, 1986); the value of 0.5 mm used here in most plots reveals only strong, significant lichen-size peaks. Standard histograms used here generally have class intervals of 0.2 to 0.5 mm, but smaller class intervals can be tested in large datasets. Separation of real peaks from statistical noise is apparent by using progressively smaller class intervals or Gaussian kernel sizes. All lichen-size peaks, small or large, should be tested to determine if they are real.

The choice of an appropriate histogram class interval or Gaussian kernel size involves a trade-off between precision and resolution. Numerous, meaningless, small peaks are obvious noise if the class interval used is too small, so one is tempted to use a large class interval. Resolution is not as good but one gets the impression that the few remaining large peaks are real. This can be misleading. Merger of real peaks creates false peaks with meaningless mean lichen sizes (Bull and Brandon, 1998, Fig. 10).

Lichenometric age estimates are obtained by inserting means of normally distributed lichen-size peaks into equation 6.1. Precision of dating is estimated by comparing ages of lichen-size peaks for a specific regional rockfall event at many sites. I have more than 36,000 lichen-size measurements at 95 sites in the Southern Alps. Accuracy is determined by comparing lichenometric ages of the earthquake-generated rockfalls caused by historical earthquakes with their earthquake dates, and by comparing lichenometry age estimates for deposits whose exact ages have been determined by tree-ring analyses.

Initially, I presumed that each geographic setting might require a separate calibration of lichen growth rate. So I collected 1,400 lichen-size measurements at a low altitude site and 1,000 measurements at a site above timberline. These and intermediate-altitude datasets had a pronounced lichen-size peak at 16 mm (Fig. 6.5) that I soon realized was the result of regional rockfalls caused by the magnitude 7.1 and 7.8 earthquakes in 1929 A.D. I anticipated that this lichen size peak would shift with differences in length of growing season, depth of snow pack, annual precipitation, and temperature. Instead I found that sizes of lichen-size peaks resulting from all earthquake-generated regional rockfall events remained constant at all sites, including sites with different substrate lithologies. Thus equation 6.1 applies to a region. This means that lichen-size peaks remain sharp even when thousands of measurements are combined from diverse sites in a 20,000 km² region. Figure 6.5 shows this distribution of peaks in a standard histogram. For the same data in continuous probability density format see Figure 15B of Bull and Brandon (1998).

The likelihood of describing real peaks improves with density of lichen-size measurements. Density of available measurements affects one's choice of histogram class size. One or two measurements are of little value because one does not know where they would lie in a normal Gaussian distribution for a lichen-size peak recording a single event.

This is why I prefer not to calibrate lichen growth rates with tombstone lichen-size measurements. Comparison of the two calibration methods in Sweden emphasizes the importance of large datasets in lichenometry. Calibration data consisting of single lichen-size measurements from 32 tombstones at Tarnaby, Sweden gave me a reasonable appearing calibration equation,

$$D = 506.50 - 0.25\, t \qquad (6.2)$$

but the correlation coefficient is only 0.47 because of excessive scatter. Calibration data consisting of 23 large datasets in the same Swedish region gave me

$$D = 640.95 - 0.315\, t \qquad (6.3)$$

Confidence in the equation 6.3 analysis is superior because the correlation coefficient is 0.995. Note that *Rhizocarpon* subgenus *Rhizocarpon* grows twice as fast in Sweden as in New Zealand (equation 6.1), but as in New Zealand a single calibration equation can be used in a large region.

The mean size of a lichen-size peak is best defined with 10 to 100 tightly clustered measurements that with luck (relatively low frequency of earthquakes) rise and fall in a distinct peak that is isolated from adjacent lichen-size peaks. Isolation improves with smaller class intervals, but potential for statistical noise also does.

Rockfall lichenometry sites record many events. The central part of a broad plot rises above the left and right sides of the plot with the smallest or largest lichen sizes, which generally have a lower density of lichen-size measurements. The usual approach is to use a class interval that identifies sharp, narrow peaks with minimal noise in the central (high-density) part of the lichen-size plot. However, the trade-off of using a small class interval is possible spurious

lichen-size peaks (noise) in the tails of the distribution where measurement density may be mediocre.

Fortunately the Bull–Brandon method has several ways to distinguish between real peaks and spurious peaks that are mere statistical noise. Real peaks occur at many sites in earthquake-prone mountains because the seismic shaking that causes synchronous coseismic rockfalls is regional. Keefer (1984, 1994) notes that earthquakes cause most synchronous, extensive rockfalls. Rockfall abundance (amplitude and volume of lichen-size peaks) increases towards each known epicenter for historical earthquakes and I presume it did for prehistorical earthquakes. Important, but small, lichen-size peaks resulting from distant earthquakes would be considered as statistical noise were it not for their presence at many sites and systematic change in peak size with distance. Significant peaks can also be defined as those of rise of more than three standard errors above a uniform distribution of lichen sizes defined as lichens growing at a constant rate on blocks in a deposit fed by a continuous trickle of rockfalls (Bull and Brandon, 1998, p. 67, 68).

The typical histogram of lichen sizes consists of a sequence of distinct peaks (Fig. 6.5). The ability of lichenometry to distinguish peaks of specific earthquakes is a function of lichen growth rate and temporal spacing of seismic shaking events. Magnitude Mw >6 earthquakes in the South Island generally occur several years apart and the 1 mm/6 yr growth rate of *Rhizocarpon* subgenus *Rhizocarpon* generally can separate most events. Of course lichenometry generally can't separate multiple events in one year, such as the two Mw >7 earthquakes in 1929 A.D. To do this one needs a regional analysis that reveals two separate areas of seismic shaking; several methods are discussed later (Figures 6.11, 6.12, 6.25, 6.31, 6.32A).

Some readers may not be familiar with the Bull–Brandon (1998) approach to lichenometry, and consequently are surprised by the high precision and accuracy assigned to lichenometric dating in Figure 6.1. Dating precision of lichenometry is excelled only by dendrochronology, so cross-checks of lichenometry age estimates of Alpine fault earthquakes discussed later rely mainly on analyses of the times of suppressions of annual tree-ring growth, and on forest-disturbance events.

Precision of the Bull–Brandon method is tested at every new lichenometry site – does the sequence of lichen-size peaks at this site fit into

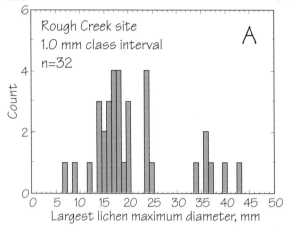

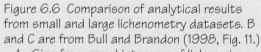

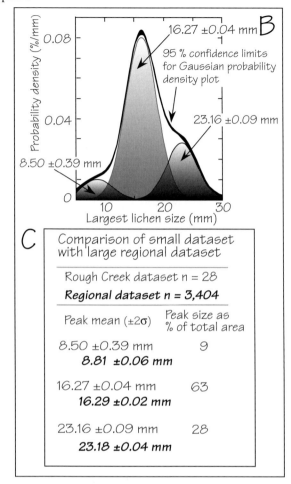

Figure 6.6 Comparison of analytical results from small and large lichenometry datasets. B and C are from Bull and Brandon (1998, Fig. 11.)
  A. Size-frequency histogram of lichen-size measurements at the Rough Creek site; *n* = 32.
  B. PeakFit decomposition of a probability density plot of lichen sizes on historical coseismic rockfalls at the Rough Creek site. Gaussian kernel size is 0.5 mm; *n* = 28.
C. Comparison of analytical results from the small Rough Creek dataset, and a much larger regional dataset.

my regional record of synchronous rockfall events? Initially, I trusted only large datasets (100 to >1,000 lichen-size measurements and large histogram class intervals of 0.5 to 2.0 mm). I now realize that small datasets generally may be sufficiently robust to be trusted and that class intervals of 0.2 mm or smaller can be used if cross-checked with data from other lichenometry sites and verified by precise tree-ring analyses. The example used here compares data from a site in the central Southern Alps with a regional dataset.

   Comparison of datasets affected by the same historical earthquakes provides a rigorous test of the Bull–Brandon lichenometry method. People witnessed rockfalls at Rough Creek near Arthurs Pass during a Mw magnitude 7.1 earthquake in 1929, so this seemed to be a good place to calibrate lichen growth rate. I expected to make hundreds of lichen-size measurements, but plant growth in this extremely humid climate is so rapid that most *Rhizocarpon* subgenus *Rhizocarpon* did not survive. I was only able to find

and measure the longest axis of 32 lichens. Would a sample of only 28 measurements (those smaller than 30 mm) be enough to define the 1929 event? The standard histogram in Figure 6.6A looks quite borderline – many little subpeaks even when a large class interval of 1.0 mm is used.

   The second type of histogram was constructed and PeakFit software was used in an attempt to model the peaks comprising the probability density plot. Each lichen-size measurement was defined as a Gaussian of a chosen width, and these overlapping bell-shaped representatives of the overlapping lichen sizes were stacked to make the probability density plot. Data from the Rough Creek site (Fig. 6.6B) provide an example of decomposition of a probability density plot into component Gaussians. Two shoulders in the plot suggest several components instead of a single lichen-size peak.

The modeling result is three peaks. The large central peak dates to the time of the two 1929 earthquakes. All three peaks are inferred to record regional rockfall events because peaks defining similar lichen sizes are present at other sites in the Marlborough region. Their lichenometry age estimates coincide with the times of historical earthquakes of 1881, 1929, and 1968. The mean of each modeled lichen-size peak was used to estimate rockfall event age. Size (amplitude or area) of a lichen-size peak describes relative abundance of rockfall blocks and is a function of the intensity and duration of seismic shaking and hillslope sensitivity to earthquakes. Peak size provides an index of the intensity of seismic shaking for each event at the Rough Creek site. Peak size for the 1929 earthquakes is the sum of two seismic shaking events, Mw 7.1 in March and Mw 7.8 in June. The distant magnitude Mw 7.8 Murchison earthquake occurred 3 months after the Arthurs Pass earthquake.

Peak size is a function of distance to an earthquake epicenter. The dominant peak has a mean of 16.27 ± 0.04 mm (95% confidence level). It mainly records the 1929 Arthur's Pass earthquake (epicenter 35 km away). The two smaller peaks appear to record coseismic rockfalls associated with the 1881 Hurunui earthquake (epicenter 65 km away) and the 1968 Inangahua earthquake (epicenter 130 km away).

Decomposition of a 20-site regional dataset was analyzed in the same way, and the small Rough Creek sample of data was not included in these 3,404 lichen-size measurements. The tabular comparison of results in Figure 6.6C shows that the mean peak sizes for the three regional rockfall events are the same as those estimated from the small Rough Creek dataset. Rockfalls indeed record the regional nature of seismic shaking – a synchronous tumbling of rocks down the hills of the Southern Alps.

Key conclusions from this test of the method –
1) Modeling of the probability density plot components discerned three peaks in the small dataset whose mean sizes match those in a regional dataset. This is nice precision for a lichen that grows 1 mm per 6 years.
2) Everything has to be just right to obtain the fine comparison shown in Figure 6.6C, including regionally uniform rates of growth for many species of *Rhizocarpon* subgenus *Rhizocarpon*, high-quality normally distributed measurements of lichen thalli measured with digital calipers, correct field and analytical assumptions, and superlative computational algorithms.

3) A major advantage of the Bull–Brandon approach to paleoseismology is that the measurement and modeling of lichen sizes at each new site constitutes a replication test for all assumptions, measurement techniques, and analytical procedures. The work undertaken at each new site in this earthquake-prone mountain range is indeed a prediction that the positions of lichen-size peaks will be similar, and that peak sizes will be larger when the site is close to the fault responsible for a specific earthquake.
4) Precision of dating with the Bull–Brandon method is robust and can be verified by comparing samples from 36,000 lichen-size measurements at 95 sites.

In contrast to the exceptionally limited opportunities to measure only a few lichens at Rough Creek, lichen-size measurements are virtually unlimited at a landslide complex near Barretts Hut in the Seaward Kaikoura Range (Bull, 2003b). Canyon downcutting and favorable geologic structure were conducive for initial landslide formation, which is young enough to have only patchy re-establishment of near-timberline trees (Fig. 6.7). Intermittent unraveling of the prominent headscarp has continued to supply detritus to the landslide surface, which is a good repository for catching falling blocks.

The Barretts site illustrates application of the method at a new site, and allows assessment of accuracy of lichenometry dates. Do the oldest lichens reveal the time of initial landslide formation, and would younger lichen-size peaks record earthquake-induced rockfalls or just random events? The sample of 424 lichen-size measurements is shown in Figure 6.8, using the same histogram class interval as Figure 6.5. Each isolated, sharply defined peak occurs at other sites in the study region. Old lichen sizes begin rather abruptly with a peak at about 45 mm. There are no peaks larger than this, only a few single lichen-size measurements that I presume represent survivors of the initial landslide that, using equation 6.1, occurred at about 1742 A.D. ± 10 years. This date coincides with the time of a regional rockfall event (Fig. 6.5), so the Barretts landslide appears to have been caused by an earthquake.

Comparison of age estimates for nine lichen-size peaks with historical earthquake dates allows assessment of the accuracy of lichenometric dating. The historic record for New Zealand earthquakes extends back only to 1840 A.D., the time of first substantial influx of European colonists into the South Island. Equation 6.1 was used to compute ages of the historical lichen-size peaks. Departures of rock-

Figure 6.7 View of Barretts landslide complex from Mount Stace. Unstable headscarp is partially buried with talus cones. The Hapuku River has removed portions of the toe of the landslide, thus creating younger landslides. Lichen-size measurements were made in the enclosed areas and along the reconnaissance route. Location is number 11 on Figure 6.3. Figure 4 of Bull, 2003b.

fall event lichenometry ages from dates of historical earthquakes range from 0.1 to 5.6 years and average 2.0 years. Accuracy of ±2 years is the same as for other studies in California and New Zealand. My stated uncertainties for a lichenometry ages estimate are routinely rounded up to the nearest decade, ±10 years, but verification by tree-ring analyses of the dates for regional seismic shaking events implies that ±5 years is more appropriate in the following discussions of recent Alpine fault earthquakes. These results support a model that most of the prehistorical lichen-size peaks, those larger than 29 mm (Figs. 6.5, 6.8), also record coseismic regional rockfall events.

I use this new approach to paleoseismology by starting with the Alpine fault because dendrochronological information is available to help decide which of several lichen-size peaks records a particular Alpine fault earthquake. The tree-ring analyses data also test the precision and accuracy of the lichenometric dating of three regional rockfall events. The emphasis is on the magnitude and extent of seismic shaking for three Alpine fault earthquakes that occurred (±10 to ±25 years) at 1580, 1620, and 1717 A.D. as documented by previous paleoseismologists (Wright, 1998; Wright et al., 1998; Yetton, 1998; Wells et al., 1999; Norris et al., 2001; Wells and Yetton, 2004).

These essential studies were a paleoseismology breakthrough because they used more than radiocarbon dating of ruptured stratigraphy. Input included diverse data such as radiocarbon-dated landslides and buried forests, stratigraphy of tectonic sag ponds, and forest disturbance events. Tree-ring analyses helped

narrow age estimates for events whose radiocarbon dates are overlapping. Sequential narrowing of the possible age range for the Crane Creek event of Yetton (2000, Table 6.3) consisted of 4 steps:

Step 1) Trenching and radiocarbon dating of ruptured fault-scarp stratigraphy constrained timing of the event to a broad range of 1480–1645 A.D.

Step 2) Radiocarbon dating of landslides and fluvial aggradation events provided verification of the event age in a broad, general sense, 1488–1640 A.D.

Step 3) Analysis of a forest disturbance event provided a much narrower age range, 1625 A.D. ± 15 years; and

Step 4) Anomalous growth rings in damaged trees improved the age estimate to 1620 A.D. ± 10 years.

Additional work might include:

Step 5) Finding a tree growing in the fault zone that was killed by the 1620 A.D. earthquake, or was strongly tilted to favor growth of distinctive "'reaction wood", or was damaged so severely that growth ceased for a few years. Formal cross dating of sequences of annual growth rings of damaged and undamaged trees might date the earthquake to the year, or perhaps even to the season of a growth year.

Step 6) Cross-checking the tree-ring analysis age estimate by using lichenometry to date the regional rockfall event that occurred as a result of the 1620 earthquake.

Step 7) Mapping the areal intensity of seismic shaking using trends in the abundance of 1620 A.D. rockfalls. Of course the rockfall-abundance seismic shaking index lacks the purity and sensitivity provided

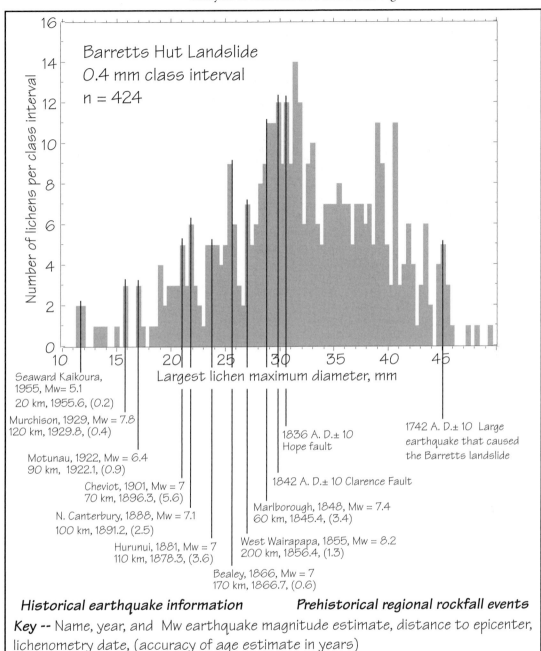

Figure 6.8 Histogram of the largest lichen maximum diameters on 424 rockfall blocks at the Barretts landslide site. Time of creation of the Barretts landslide is assigned to the age of the oldest lichen-size peak. From Bull, 2003b, Figure 6.

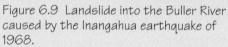

Figure 6.9 Landslide into the Buller River caused by the Inangahua earthquake of 1968.
A. 1969 view. Powerful Buller River has already removed the landslide dam. Photo supplied courtesy of Eric Force of the U.S. Geological Survey.
B. 1981 view. Unstable headscarp at 1 is actively shedding detritus. Location 2 has a mix of damaged and undamaged trees. Location 3 is the tread of a post-1968 aggradation event stream terrace.
C. 2005 view of upper part of the landslide. Cohort of even-aged young trees in the new forest mantles much of the landslide scar. Bedrock collapse in the headscarp on both sides of location 1 continues to add new detritus to the talus slope repository of rockfall blocks with post-1968 lichens.

by seismographs. Systematic regional trends of seismic-shaking index appear to be as useful as Modified Mercalli Intensity maps based on felt reports of New Zealand earthquakes (Cowan, 1991; Downs, 1995). Step 8) Using the seismic shaking index map to assist in reconnaissance of additional trenching sites.

Photos of a 1968 coseismic landslide (Fig. 6.9) illustrate the multidisciplinary approach outlined in the above steps. An impressive volume of material slipped away from the ridgecrest, devastating the forest and damming the Buller River. A typical stream of the Westland district, the large annual unit stream power, quickly redistributed detritus from this and other landslides. The threshold of critical power was crossed during a brief aggradation event, but by 1981 subsequent streambed incision had isolated the stream terrace tread where new trees were beginning to germinate. Exceptionally large boulders remain in the rapids, which appeared unchanged in 2005. The slide surface looks greatly different than in 1981 because of forest regrowth. The headscarp is gradually becoming more stable, but light-toned slashes characterize two recent bedrock collapses (Fig. 6.9C, Location 1).

How might a paleoseismologist study this site 200 years in the future? Numerous tree trunks beneath the landslide and in the aggradation terrace gravels would provide pre-event radiocarbon age estimates that would constrain maximum event age. Tree-ring analyses of the cohort of even-age trees growing on the terrace tread (Fig. 6.9B, Location 3) would constrain the minimum possible event age, perhaps more closely than radiocarbon age estimates from the buried wood. This could be fairly labor-intensive work. Many trees would have to be cored to locate a representative oldest tree. Increment borers rarely pass through the center of annual growth rings, the "pith point", so the number of missing annual growth rings would be estimated (Duncan, 1989). Corrections would also include how long that species of tree usually takes to germinate, and how long it took to grow to the height of the increment borer. A parallel study could be made of the young forest growing on the landslide, which should date as being slightly older by an amount of the time needed to deposit valley-floor alluvium and incise the terrace tread. The edges of the landslide would no longer be obvious (Fig. 6.9B, Location 2), and some of the pre-event trees may no longer be alive.

Search and sampling of many trees older than 200 years in the adjacent forest would define two categories of annual growth rings dating to the time of the coseismic landslide;
1) Trees not affected by either the earthquake or landslide and whose variations in tree-ring width correlate with regional (climate-controlled) variations of annual growth as defined by a formal tree-ring chronology made by crossdating 20 to 50 trees of the same species.
2) Trees with a post-earthquake interval of anomalously slow or fast growth (defined and discussed in Section 6.2.1.4).

Step 6 of this hypothetical paleoseismology investigation would be to measure sizes of lichens on the talus cones shown in Figure 6.9C, working where forest re-growth has not shaded the rockfall blocks. As at the Barretts landslide (Figs. 6.7, 6.8) most lichens will postdate the 1968 landslide and few, if any, lichens will predate the event. The age estimate for the oldest distinct peak in the broad distribution of lichen sizes should match the Step 4

tree-ring analysis age estimate to within ±5 years. The paleoseismologist would conclude that an earthquake caused this landslide if the same lichen-size peak age is present in many datasets in the region.

The following analyses use most of the above steps and make it obvious that lichenometry should be a standard tool to complement trench-and-date stratigraphic studies of prehistorical earthquakes. Three main goals here are to:
1) Improve the precision of the published dates for the three recent Alpine fault earthquakes to ±5 years.
2) Compare the magnitude and extent of seismic shaking generated by the three consecutive earthquakes.
3) Constrain and estimate the surface-rupture length for each event.

6.2.1.3 Diagnostic Lichen-Size Peaks

The Alpine fault story begins by examining lichen-size peaks at a site on the drier east flank of the mountain range. Single-site lichenometry data may not record all prehistorical earthquakes, in part because I use the fairly slow growing *Rhizocarpon* subgenus *Rhizocarpon* instead of faster growing but shorter-lived crustose lichens such as the white *Rhizocarpon candidum*. I can easily distinguish between rockfall events 4 to 6 years apart, but generally not events only 1 or 2 years apart.

Detection of closely spaced events, in a dataset portrayed as a probability density plot that spans centuries of earthquake and landslide activity, starts by using progressively smaller histogram class intervals or Gaussian kernel sizes. We need to demonstrate that many of the resulting small peaks are real events, not mere statistical noise. I do this by showing that 1) the same lichen-size peak occurs at many sites, ± the 2 year resolution that lichenometry is capable of, and 2) that the times of Alpine fault earthquakes inferred by rockfall lichenometry match the dates obtained from tree-ring analyses of disturbed forests.

Use of a single Gaussian kernel size, or class interval such as 0.5 mm, may be too large, ideal, or too small because density of lichen-size measurements (mean data frequency) is variable in all datasets. The Bad Bird dataset illustrates several key points (Fig. 6.10A). The broad lichen-size peaks shown here are compound: they record more than one rockfall event. Combining events into these single broad peaks results from using a conservative 0.5 mm class interval, which seemed a good choice for initial inspec-

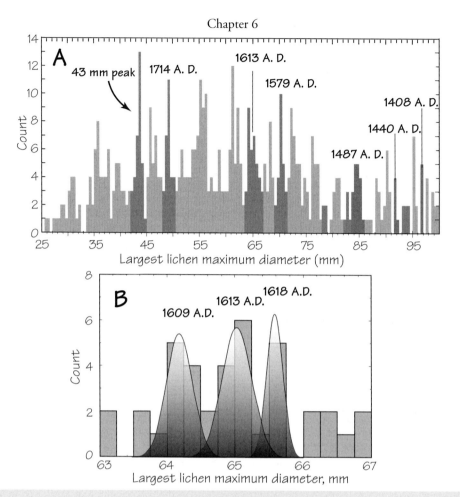

Figure 6.10 Lichen-size peaks that record earthquakes at the Bad Bird talus site in Cameron Range. Location is shown on Figure 6.11.
A. Sizes of *Rhizocarpon* subgenus *Rhizocarpon* in the 25 to 100 mm size range (time span from ~1871 to ~1387 A.D.). Darker shaded peaks are discussed in the text. The six age estimates are for lichen-size peaks that might record Alpine fault earthquakes. These rockfall dates coincide with earthquake dates based on radiocarbon dating of trench stratigraphy and buried forests, and analyses of tree rings in disturbed forests near the fault trace. n = 562. Histogram class interval is 0.50 mm.
B. Broad lichen-size peak at 65 mm becomes three distinct peaks when a smaller class interval of 0.25 mm is used. 63 to 67 mm comprises a time span of ~1626 to ~1600 A.D. n= 39.

tion. Small peaks, such as those at 78.5, 83, and 94 mm would be regarded as statistical noise if Bad Bird was our only lichenometry site, but these peaks also occur at other sites. The 0.5 mm class interval is ideal (symmetrical single, isolated peaks) for displaying these small peaks or other weak parts of a dataset. Most kernel-size choices indeed are compromises.

    Large amplitude peaks, such as at 43, 55, 62, and 73 mm, should also be evaluated as recording local landslide events instead of being the result of regional seismic shaking. Examples of South Island local events include four huge aseismic rock avalanches in the Southern Alps just since 1990 (Chinn et al., 1992, McSaveney, 2002, Hancox et al., 2005). All four Bad Bird site peaks record regional, not local, landslide events. Perhaps the optimal lichenometry site for paleoseismology purposes is one that crumbles a bit each time the earth quakes.

Naming of the Bad Bird site is an interesting digression. A New Zealand falcon flitted from boulder to boulder and kept a wary eye on me as I started measuring lichen sizes on footslope talus blocks. I found larger lichens farther up the slope, but a very angry bird defended them. I threw a rock to ward off the falcon repeatedly diving at my head. In an amazing feat of aerial acrobatics, it changed course by almost 180° and flew along side of the falling rock until it reached the ground. This amazingly agile falcon attacked me yet again until I gave up. Once back on the lower boulder pile, the bird seemed content to return to its nesting area high on the hillside and a truce was maintained for the rest of the day. The appropriate name indeed is the "Bad Bird Site".

A single class interval or Gaussian kernel size is not equally suitable for all parts of a large dataset, so data subsets are analyzed separately (Fig. 6.10B). Reducing the class interval from 0.5 to 0.25 mm for the large 65 mm lichen-size peak clearly displays three lichen-size peaks with calendric ages of about 1609, 1613, and 1618 A.D. But are these three events recorded at many sites within 200 km of the Bad Bird site? How does one decide which of the three, if any, records the ~1620 A.D. Alpine fault earthquake? This is the focus of the following discussions where tree-ring analyses provide essential answers.

The volume of rockfall blocks (amplitude and size of lichen-size peaks) in most of the six peaks labeled as candidates for Alpine fault earthquakes in Figure 6.10A is unexpectedly small compared to the prominent peaks at 56, 62, and 72.5 mm. The Bad Bird lichenometry site is only 39 km from the Alpine fault. Shouldn't all regional rockfall events resulting from Alpine fault earthquakes be recorded by large lichen-size peaks?

Previously, I incorrectly assumed that Alpine fault earthquakes would cause the most intense seismic shaking along the southeastern side of the Southern Alps, and mistakenly concluded (Bull, 1996c) that the large, sharp 43.75 mm lichen-size peak shown in Figure 6.10A was the result of an Alpine fault earthquake whose production of rockfalls varied spatially with topography and lithology (Fig. 6.11). The lichenometry approach to paleoseismology includes the ability to make nice maps of seismic shaking caused by historical and prehistorical earthquakes (see Bull and Brandon, 1998, Figures 19 and 23), but my previous interpretation was wrong.

My present interpretation is that the 43 mm lichen-size peak is the result of two, not one, earthquakes. Mean data frequency between 43 and 44 mm is 20/mm, which allows use of class intervals as low as 0.1 mm (see Bull and Brandon, 1998, p. 67, 69, Fig. 10, and Table 1 for measurements accuracy). The two lichen-size peaks of Figure 6.12A appear to record real events separated by only about 4 years.

Additional verification for seismic shaking events at about 1750 A.D. and 1754 A.D. can be obtained by making similar analyses at other sites, and by narrowing the Gaussian kernel size used in construction of several seismic shaking index maps in order to match mapped earthquake epicenters with their respective lichen-size peaks. The Figure 6.11 map reveals two earthquake epicenters. Pairs of earthquakes on two adjacent fault zones might not be discerned by this method but in this case the epicenters are 300 km apart. The northern epicenter is believed to be the ~1750 A.D. event on the basis of a ~1751 A.D. lichen-size peak at the Barretts landslide lichenometry site (Fig. 6.12C). The Barretts site is 450 km from the southern earthquake epicenter. Matukituki sites (Fig. 6.12B) only have a lichen-size peak dating to ~1753 A.D., thus allowing assignment of the locations of the two sources of seismic shaking recorded at the Bad Bird site. Sites at intermediate locations may date to either event and lichen-size peaks modeled with large kernel sizes may combine the data for both events into a single peak.

This background about a new geomorphic approach to paleoseismology was written with the intent of providing readers with a basis to ask challenging questions. Foremost might be "How can you verify your lichenometry age estimates without resorting to crude radiocarbon dating?" The answer might lie in studies of earthquake-damaged trees because the preeminent method in tree-ring analyses – cross dating with dendrochronology (Fritts, 1976; Sheppard and Jacoby, 1989; Cook and Kairiukstis, 1990) – has the potential of dating an event to a specific year.

6.2.1.4 Tree-Ring Analyses

Precise dating of earthquakes with trees and lichens is a fine response to the plea by Sykes and Nishenko (1984) for improved ways to date surface ruptures on fault zones with short earthquake recurrence intervals. The paleoseismic implications of tree-ring analyses range from suggestive to definitive. Three types of tree-ring analyses were used to fine-tune approximate radiocarbon dates for Alpine fault earthquakes.

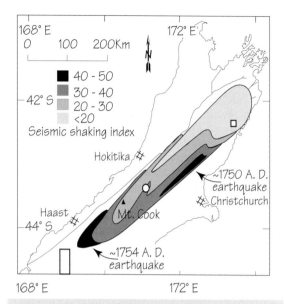

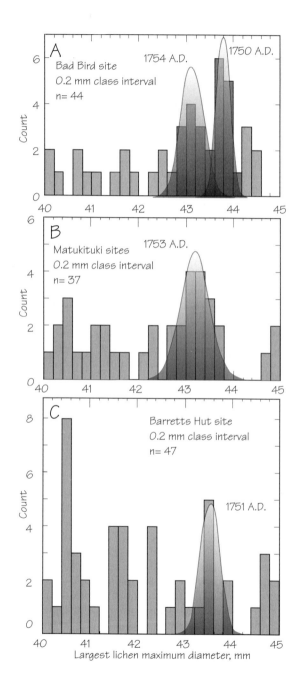

Figure 6.11 Map of regional variations in seismic shaking (rockfall abundance) at the time of the 43 mm lichen-size peak. The broad Gaussian kernel size used in the analysis of each lichenometry site combined two closely-spaced lichen-size peaks into a single peak, but the map suggests two earthquake epicenters. Site locations are shown by a circle for the Bad Bird site, square for the Barretts landslide site, and rectangle for the Matukituki sites. Map is from Figure 13A of Bull (1996c), but current interpretation is much different.

Figure 6.12 Analysis of the 43 mm lichen-size peak along a 500 km long transect.

A. The Bad Bird site dataset discloses two episodes of regional seismic shaking dating to about 1750 and 1754 A.D.
B. Matukituki sites data reveal a single episode of regional seismic shaking dating to about 1753 A.D.
C. Barretts Hut site data disclose a single episode of regional seismic shaking dating to about 1751 A.D.

Tree-ring cross dating is the best because it identifies the calendar year, or even the season of growth, for each annual growth ring in cores from a grove of trees. Ring widths are measured and compared using the bark year date when the sample was collected. Patterns of growth are matched, first between cores from the same tree, and then between trees. If a lack of match is present in a sample, due to growth-ring anomalies (missing or extra rings), it is corrected so each annual growth increment is dated correctly in all trees included in the dataset. This tree-ring chronology, based on exact matches between many trees, is then used to date events in other trees of the same species.

Cross dating is not easy because New Zealand trees have a propensity for wedging – lack of uniform ring width for a given year around the trunk circumference. Initial New Zealand cross dating breakthroughs came with the work of Val LaMarche and Peter Dunwiddie (1979) of the University of Arizona Laboratory of Tree-Ring Research. They developed 21 tree-ring chronologies from seven species of New Zealand indigenous trees.

Simple counting of annual growth rings is used for several New Zealand trees that don't have nice correlations of annual growth rings because of extreme ring-wedging characteristics. Dates should be the same as for cross dating, but how does one identify anomalies? Not knowing the answer to this question makes it difficult to assign uncertainties to age estimates, an ambiguity that may vary with species of tree. Stated uncertainties for ring counting of ±5 or ±10 years seem reasonable for ring counting age estimates of 300 to 500 years and might even be conservative where growth rings are wide, of uniform width, and are measured on slabs instead of in cores.

I found three types of anomalies in the ring series of New Zealand cedar, *Libocedrus bidwillii*. A growth ring may be missing in an increment core for a particular year. Polished slabs are better than cores when they show a growth ring that narrows to the point where the latewood of two consecutive years merges. Ring counters working with cores will register only one year and thus miss a year. Two increments of late-wood growth within a single growing season results in an additional or "false ring", and an incorrect width measurement for annual growth. Ring counters will register an extra year and ring widths that are too narrow. The third type has not been widely recognized. It forms when latewood cells fail to develop. The result is a "double ring" that is anomalously wide because it is the product of two growing seasons. Ring counters will register one year instead of two years.

Forest-disturbance-event analysis is the third type of tree ring analysis and was important in estimating the time of the 1620 A.D. earthquake (Wells et al., 1999). Episodes of destruction of the rainforest near the trace of the Alpine fault attests to the seismic character of slip on the Alpine fault (Holloway, 1957; Wardle, 1980; Stewart and Veblen, 1982; Wells, 1998). Andrew Wells (1998) searched for the oldest tree in a post-earthquake grove. Estimating ages of post-earthquake regeneration of forests has the uncertainties of finding the oldest tree, post-earthquake germination time, and time span needed to grow to increment borer height. Estimated times of forest destruction have summed uncertainties of ±20 years or more, which is better than the sum of analytical and field uncertainties for stratigraphic radiocarbon dates of earthquakes.

The key to successful dendro-paleoseismology is recognition of earthquake-induced departures from normal growth, and distinguishing earthquake events from storm damage or droughts. Applications of dendrochronology to paleoseismology (Sheppard and Jacoby, 1989: Jacoby et al., 1997) answered questions about the Cascadia earthquake of 1700 A.D., and the San Andreas fault earthquake of 1812 A.D., and described how trees responded to damage caused by the 1964 Alaska earthquake. Dendro-paleoseismology entails description of departures from normal sequences of annual growth rings. Only some of the trees in a grove may record earthquake damage, and then perhaps only in one core from the same trunk (Sheppard and Jacoby, 1989, Figs. 1 and 2).

Tree-ring analyses were fundamental in fine-tuning the initial dating of Alpine fault earthquakes by stratigraphic radiocarbon analyses (Wells et al., 1999). Post-earthquake growth of even-aged stands of trees growing near the fault provided better estimates of the times of Alpine fault earthquakes than was possible with many radiocarbon dates (Yetton et al., 1998; Wells et al., 1999). Severe earthquake damage to cedar trees (*Libocedrus bidwillii*) was cross dated by Norton (1983a) thus providing paleoseismologists a dendrochronologic age estimate of 1717 A.D. for the most recent Alpine fault earthquake.

Most paleoseismic tree-ring analyses used simple ring-counting procedures. Rimu (*Dacrydium cupressinum*), matai (*Prumnopitys taxifolia*), and kamahi (*Weinmannia racemosa*) have annual growth

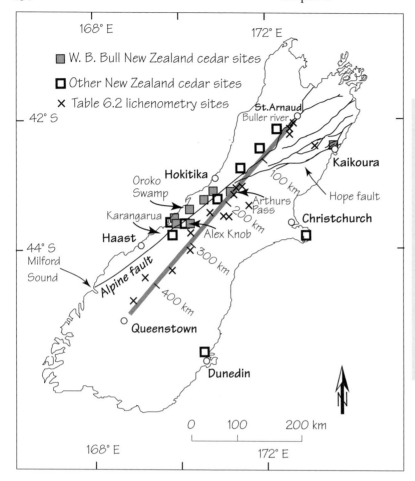

Figure 6.13 Map of some of the dendrochronology sites for New Zealand cedar, *Libocedrus bidwillii*, in the South Island. Lichenometry is used to assess regional changes in the intensity of seismic shaking caused by Alpine fault earthquakes along a transect between St. Arnaud and Queenstown. The transect line departs from the Alpine fault where the climate is too wet for the lichens used in this study.

rings but attempts to cross date sequences of rings between adjacent trees have failed. Samples have been collected from many cedar groves (Fig. 6.13), but exact measurement of all annual ring widths and generation of statistically valid chronologies requires a funded research project. The results reported here are mainly based on cross dating of annual growth rings using skeleton plots, which correlates growth patterns between trees on the basis of anomalous years, and can identify absent and false rings (see how this is done at http://tree.ltrr.arizona.edu/skeleton-plot/introcrossdate.htm).

Tree-ring analyses of several species were essential for matching lichen-size-peak ages to times of Alpine-fault-earthquake forest disturbance events. New Zealand cedar (*Libocedrus bidwillii*) and silver pine (*Lagarostrobos colensoi*) are the best trees for dendrochronologic dating in the South Island (Xiong,

1995; Xiong and Palmer, 2000; Hogg et al., 2002). Wells and Yetton (2004) favor using silver beech (*Nothofagus menziesii*) because of its responsiveness to seismic shaking (Vittoz et al., 2001) and ability to grow in diverse topographic settings. Alas, silver beech does not grow along the central Alpine fault. Although not nearly as widespread as rimu, cedar grows both near alpine treelines and in lowland valleys and boggy sites. Its variations in annual growth permit future development of cross-dated chronologies at many localities (Fig. 6.13), and a single chronology may be developed for the South Island. The cross-dated cedar chronology of Xiong and Palmer extends back to 1140 A.D.

Site locations straddling the Alpine fault (Fig. 6.13) are particularly useful for paleoseismology studies. Oroko Swamp, 9 km from the Alpine fault, has many tilted and toppled cedar and silver pine trees indicative of liquefaction of the saturated swamp deposits during times of intense seismic shaking. Swamps are good sites for trees to record earthquakes because of high potential for liquefaction events, minimal climatic regional growth suppressions resulting from droughts, and less chance of windstorm damage in topographic hollows. I picked this swamp because it was where Jonathan Palmer was cross dating silver pine (*Lagarostrobos colensoi*). Work in progress includes COFECHA cross dating of cedar at Oroko Swamp. I note a good statistical correlation back to 1600 A.D. The older time span used here is based on skeleton-plot cross dating of tree rings. Alex Knob cedar trees grow near alpine treeline on a steep, crumbly mountainside, and on the adjacent ridgecrest where seismic energy tends to be focused. Dunwiddie (1979) notes that ring wedging in cedar is worst at stressed subalpine sites, but many trees nicely fit my skeleton plot master. Alex Knob cedars are only 2 km from the trace of the Alpine fault, which dips under the site.

The examples used here emphasize suppressions of tree growth but accelerations of tree growth also can be important near earthquake epicenters. Consider the changes to the rainforest with increasing distance from an Alpine fault surface rupture. Earthquakes devastate the forest near the fault trace. Heavy, stiff trees are snapped or knocked down by the intense seismic shaking. Only a few older growth trees remain alive to record earthquake damage in the form of suppression of annual growth rings. Limber saplings not only survive, but many accelerate their rate of growth because of earthquake-induced increases in available sunlight and nutrients. Devastation drops off rapidly within 2 km of the fault trace, as does the opportunity for prolonged acceleration of tree growth in surviving trees. Growth suppressions become relatively more important with increasing distance from the fault trace. The growth acceleration/growth suppression ratio in the trees of a forest should be a good way to map intensity of seismic shaking for specific prehistorical earthquakes in humid regions. Damage caused by landslides (Fig. 6.9) should be classed differently because mass movement has occurred in addition to the seismic shaking. Accelerations, although quite valuable, may be recognized only in a small linear area along the trace of a fault. Suppressions are more likely to be found in two or three orders of magnitude larger study areas, but need to be identified as being the result of a specific earthquake rather than being part of the background

of ubiquitous random variations (Fig. 6.14C) found in any grove of trees. The key to success at sites like Oroko Swamp includes noting the relative intensity of a particular growth suppression, and matching the suppression date with a time of regional seismic shaking as precisely dated by lichenometry.

Annual growth rings in cedar trees at Oroko Swamp and Alex Knob may also record damage from earthquakes or windstorms. Suppressions of widths of annual growth rings may range from slight to extreme to the point that growth stops, even in cores from the same tree. Neighboring trees may have a concurrent acceleration of annual growth when earthquake toppling of trees reduces competition for sunlight and soil nutrients. *Libocedrus bidwillii* sites on the east coast of the South Island (Fig. 6.13) are valuable because suppressions in annual growth are not the result of Alpine fault earthquakes. Climatic signals predominate at sites near Dunedin and Christchurch, but both local Hope fault earthquakes and climatic signals may be recorded in cedar trees at the Kaikoura site.

Suppression of annual growth may be temporary or permanent (Jacoby et al., 1988; Sheppard and Jacoby, 1989; Jacoby, et al. 1995). Unusual and extreme reductions in growth, such as the 1408–1415 A.D. example illustrated in Figure 6.14A, B are unlikely to be the result of climatic factors in this extremely humid climate. Mean growth rate between 1387 and 1409 was 800% faster than during the 1411–1412 A.D. time span of maximum suppression.

In what year did an earthquake damage this tree? The 1412 A.D. year of maximum suppression might seem to be the obvious answer, until one realizes that the year with narrowest rings in a suppression may be delayed as the tree uses its stored nutrients (Kramer and Kozlowski, 1979). Annual ring widths were increasing from 1404 to 1408 in Figure 6.14B. Would this upward trend have continued if the tree-damaging event had not occurred? The year of damage might be 1408, or even 1407. The point here, as noted by Sheppard and Jacoby (1989), Jacoby, (1997), and Jacoby et al. (1997) is that some studies need to assign several years of uncertainty to allow for the possibility that an earthquake date is 1 to 5 years older than the maximum-suppression year.

A nice analysis by Wells and Yetton (2004) of growth changes in several New Zealand species, including beech and cedar, illustrates substantial delays before onset of earthquake-induced suppressed growth. Growing seasons for Southern Hemisphere

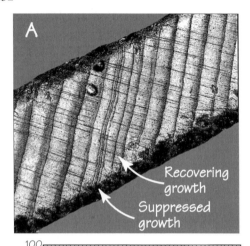

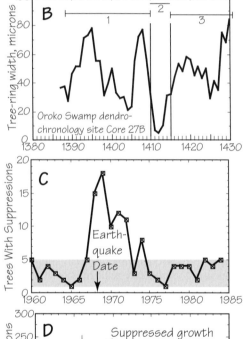

Oroko Swamp dendro-chronology site Core 27B

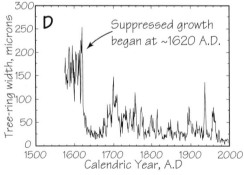

Figure 6.14 Temporary and permanent suppression of annual growth of trees growing at sites susceptible to seismic shaking.

A. Anomalous suppression of annual growth rings in 4 mm wide increment borer core of New Zealand cedar (*Libocedrus bidwillii*) from Oroko Swamp, 9 km northwest of the Alpine fault. Darker, smaller cells at right side of each ring mark the end of the growing season for that year.

B. Variations in annual growth of cedar shown in Figure 6.14A. Marked suppression of ring width dates began in 1408 A.D. and continued until 1415 A.D. Mean sizes of annual growth rings: 48 microns during the 1387 to 1409 A.D. interval, 19 microns during the 1410 to 1415 A.D. interval (6 microns for 1411–1412 A.D.), and 52 microns during the 1415 to 1430 A.D. interval.

C. Delays in the onset of suppressed growth in several species of trees in the Buller River watershed after the magnitude Mw 7.4 Inangahua earthquake of 24 May 1968. Normal range of 1 to 5 suppressed trees per year is shown by the gray band. Pattern does not return to normal until 1975. Data is from Table 6 of Wells and Yetton (2004).

D. Permanent suppression of growth of a matai tree (*Prumnopitys taxifolia*) that lost many limbs at about 1620 A.D. Rings widths of this Karangarua River watershed tree were counted by Andrew Wells (1998).

trees span two calendar years, so the convention used for an annual growth ring is to assign the calendric year in which growth started. September 1968 was the first season after the 23 May 1968 Inangahua earthquake and 15 of the trees sampled had markedly suppressed growth during the September 1968 to May 1969 growing season. No delay here – response to earthquake-induced damage was immediate. The time of maximum numbers of trees with growth suppressions – 18 – was delayed until the 1969 growth year. Onset of suppressed growth was still more than twice the background level five growth years after the earthquake (Fig. 6.14C).

Damage can be so severe that the tree never fully recovers. A Karangarua watershed matai tree (Fig. 6.14D) is a good example of permanent suppression dating back to about 1620. This matai tree resumed normal growth fluctuations after 60 years of suppressed growth, but rapid annual growth never resumed. Damage of this sort is not the result of distant earthquakes (unless a secondary event such as a landslide occurs), so this type of earthquake signal could be important to paleoseismologists (Wells and Yetton, 2004).

The ultimate category of seismic damage is to kill trees. The floors of swamps near the Alpine fault contain many fallen trunks. Cedar decays slowly and silver pine seems immune to rot.This treasure trove of datable wood may be especially useful to paleoseismologists working at Oroko Swamp because tree-ring chronologies have been made for both cedar and silver pine. The outermost rings can be dated where preserved beneath remnants of bark. Even here one has to allow for the possibility that tree growth might continue for a year or two. But the severe ring-width suppression for such a toppled tree surely will date to the year of the earthquake. I assume that the surface rupture was in the immediate vicinity of a devastated forest in order to generate a Mercalli seismic shaking intensity of 9 or 10 (Wells and Yetton, 2004). For example, finding a toppled tree dating to 1408 A.D. would be further evidence for a catastrophic level of damage to support the severe level of damage indicated by Figures 6.14A, B.

Several cross-checks should be used to verify earthquake-caused suppressions. Earthquakes and wind cause severe suppressions of annual growth and toppling events in swamps, and landslides are an additional causal factor on steep hillslopes where windstorms are more likely too. The damage event should be obvious in several trees at a site(s) where the substrate they grow in is susceptible to seismic shaking (Table 6.1). One other suppression maximum, also dating to 1412 A.D., was noted at Oroko Swamp where very few of the trees cored are this old. Three suppression maxima were noted in the cedar tree cores from Alex Knob steep hillsides, two dating to 1413 A.D. and one dating to 1412 A.D.

Using lichenometry as a cross-check has the advantage of dating the time of the event, thus avoiding the possibility of several years of delay of maximum suppression of tree-ring widths. Of course lichenometry cannot match the exact precision possible with the best tree-ring dating (Fig. 6.1). A strong

regional rockfall event occurred several years before 1412 A.D. Intense, seismic shaking disrupted glacial moraines at the Mt. Cook and Cameron sites, which are 25–40 km from the Alpine fault. The mean age of the lichen-size peaks shown in Figures 6.15A and B is 1408 A.D. The 96.71 mm lichen-size peak at the Bad Bird talus site (Fig. 6.10A) also has an estimated age of 1408 A.D. It appears that an earthquake dating to about 1408 A.D. damaged cedar trees growing at seismically sensitive sites straddling the Alpine fault, and was strong enough to cause a major regional rockfall event. Paired lichenometry and tree-ring analyses are a potentially powerful tool in paleoseismology because the precision of both dating methods is much better than for stratigraphic or other surface-exposure dating procedures.

One example of this multidisciplinary approach does not suffice, so let us compare paired lichenometry and tree-ring analyses for the three most recent prehistorical Alpine fault earthquakes (Tables 6.1 and 6.2). The most recent event dates to 1716.5 A.D. for the tree-ring analyses, and 1 year older for lichenometry analytical results, with a 2σ uncertainty of only 2.8 years for both methods. The case for a second event dating to 1717 A.D. in Table 6.2 is discussed later. The penultimate earthquake dates to 1615 A.D. using tree-ring analyses and to 1613 A.D. using lichenometry, and the next oldest dates to 1578 A.D. using tree-ring analyses and 1579 A.D. using lichenometry. Not having a dendrochronology level of precision (might date to the exact year in some cases), we do not know the exact lichenometry error in this specific age range. However, the crosscheck with tree-ring analyses age estimates indicates that lichenometry calibration equation 6.1 is robust, and that precision is ±5 years.

I prefer to round off these age estimates to the nearest 5 years. The names are those used by previous workers. The three Alpine fault earthquakes are the Toaroha event of 1715 A.D. ± 5 years, the Crane Creek event of 1615 A.D. ± 5 years, and the Waitaha event of 1580 A.D. ± 5 years.

A University of Otago team studied the interesting Waitaha site. A short left step in right-lateral Alpine fault displacements resulted in an uphill-facing fault scarp with tilted trees growing on it. Adjacent sag-pond stratigraphy records episodes of deposition after surface-rupture uplift of the fault-scarp dam. Radiocarbon dating of samples from trenches and cores of sag-pond deposits constrain times of local surface ruptures.

Tree ring study site	Age for 1715 A.D. event	Age for 1615 A.D. event	Age for 1580 A.D. event
Oroko Swamp	1714	1616	1578
cedar W. B. Bull	1717	1616	1579
	1716	1614	1578
Alex Knob cedar-	1715	1613	1579
W. B. Bull	1715	1614	1576
	1714	1614	1580
	1715	1617	1576
	1714	1615	1580
	1718	1614	
	1714		
	1716		
	1717		
Waitaha rimu C. A. Wright			1578
Mean age, 2 standard deviations	1716 A.D. ± 2.78 years	1615 A.D. ± 2.58 years	1578 A.D. ± 3.12 years

Table 6.1 Initial year of marked suppression of annual growth rings in New Zealand cedar (*Libocedrus bidwillii*) and rimu (*Dacrydium cupressinum*) growing on or near the Alpine fault. These dendrochronology data can be compared with lichenometry age estimates for the same seismic shaking events in Table 6.2.

Lichenometry site	Age for 1715 A.D. event	Age for 1715 A.D. event	Age for 1615 A.D. event	Age for 1580 A.D. event
Robert Mountain	1719	·	1614	·
Rainbow-RosePatch	1715	·	1614	·
Zig Zag	1717	·	1613	1580
Deaths Corner	·	·	1613	·
Arthurs Pass	1715	·	1615	1580
Craigieburn	1718	·	1614	1579
Cameron	1716	·	1613	1579
Bad Bird	1716	·	1613	1579
Clyde	1718	·	1612	1581
Mt. Cook	1717	·	1612	1578
Ohau	1715	·	·	1578
Lindis	·	1718	·	·
Matukituki	·	1716	·	·
Crown Pass	·	1718	·	·
Mean age, 2 standard deviations	1716 A.D. ± 2.80 years	1717 A.D. ± 1.54 years	1613 A.D. ± 1.78 years	1579 A.D. ± 2.08 years

Table 6.2 Age estimates for lichen-size peaks for *Rhizocarpon* subgenus *Rhizocarpon*. Site locations are shown in Figure 6.13, and the sizes of the lichen-size peaks are used as an index of seismic shaking intensity for the three earthquake events (see Figures 6.22, 6.23). These lichenometry data can be compared with tree-ring analyses for the same seismic shaking events in Table 6.1.

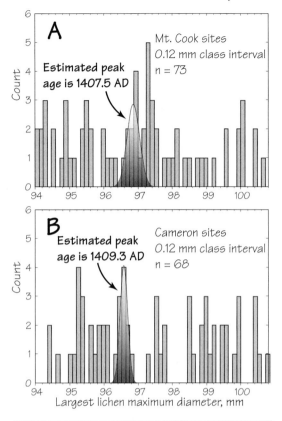

Figure 6.15 Comparison of the beginning of the ~1408 A.D. suppression of growth in an Oroko Swamp cedar tree with estimated ages of a synchronous rockfall event dated by lichenometry.

A. Estimated age of ~ 1408 A.D. for a lichen-size peak in the Mt. Cook area about 26 km from the Alpine fault.
B. Estimated age of ~ 1409 A.D. for a lichen-size peak in the Cameron Range area about 34 km from the Alpine fault.

Craig Wright (1998) estimated the times of Alpine fault surface ruptures at the Waitaha site by counting annual growth rings in 26 polished slabs cut from stumps of logged rimu and miro trees. Strong growth suppressions were found in 7 out of 20 mature trees during the decade beginning in 1715 A.D., and in 12 out of 16 trees in the decade beginning in 1575 A.D. Smaller disturbance modes include the decades beginning in 1415 A.D. and 1615 A.D.

Stratigraphic radiocarbon dating of Alpine fault surface rupture events, refined by annual ring counting of the times of suppression and acceleration of tree growth, identified local earthquakes as occurring in 1720 A.D. ± 10 years, 1580 A.D. ± 10 years, and 1440 A.D. ± 15 years (Wright, 1998; Norris et al., 2001). Wright's Figure 6.15 shows a moderately strong suppression of annual growth beginning in the decade starting in 1615 A.D. that was followed by three decades of accelerated growth in some trees. This suggests that the 1615 A.D. Crane Creek event also ruptured through the Waitaha site.

The Alpine fault perhaps is characterized by intervals of frequent earthquakes of moderate size. Clustered temporal behavior of Alpine fault earthquakes is obvious for the recent three events that occurred during a 135 year time span – 1580, 1615, and 1715 A.D.

Did another possible cluster of three events occur during the 15th century? (Goff and McFadgen, 2002). Evidence is adequate but not as abundant for older seismic shaking events. Trees have a limited lifespan and old rockfall blocks are more likely to have been smashed or buried by younger rockfalls, or have their lichens shaded by growth of trees. Availability of stratigraphic evidence from trench excavations drops off rapidly with increasing age. Even if they have not been removed by erosion, the old stratigraphic sections may be below the ground-water table or buried beyond the reach of most excavation projects.

The three main candidates for 15th century earthquakes are the times of regional seismic shaking recorded by the 84.5, 92, and 96 mm lichen-size peaks. Figure 6.10A is typical of these lichen-size peaks with the 84 mm peak being the largest. From youngest to oldest, let us consider whether or not these regional seismic shaking events resulted from surface ruptures on the Alpine fault.

The consistently larger size of the 84.5 mm lichen-size peak implies that the main earthquake of the cluster was the ~1490 A.D. event. The seismic shaking index map (Fig. 6.16) suggests rupture of all of the central section of the Alpine fault at this time, and substantial parts of adjacent sections. Tree-ring analyses provide strong support for a major surface rupture at this time (Table 6.3). Marked suppressions in annual growth were noted in three of the Oroko Swamp trees and in five trees near Alex Knob, where more cedar trees were sampled. Ages of lichen-size peaks typically are within 2 years of the growth suppressions noted in the cedar trees.

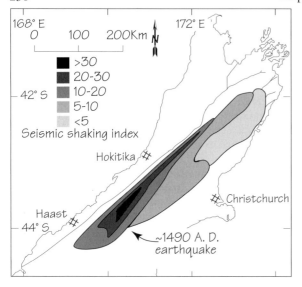

Tree-ring study site	Age for ~1490 A.D. event	Age for ~1410 A.D. event
Oroko Swamp		
OKO 28	1488	
OKO 16	1486	1412
OKO 27	1489	1412
Alex Knob		
ALX 11	1489	
ALX 13	1487	1413
ALX 14	1486	1419
ALX 17	1487	1413
ALX 18	1486	1412
ALX 40		
Mean age, 2 standard deviations	1487.25 A.D. ± 2.56 years	1413.50 A.D. ± 5.48 years **

** Without 1419 the mean is 1412.40 ± 1.10 years.

Table 6.3 Initial year of marked suppression of annual growth rings in New Zealand cedar (*Libocedrus bidwillii*) for two 15th century events at two sites straddling the Alpine fault.

A study by Wells and Goff (2006) presents a novel and much different line of evidence. They estimated the germination ages of sequences of forests on progressively older sand dunes (Fig. 6.17) at the mouth of the Haast and Okuru Rivers in south Westland. Each new inter-dune forest germinated about 20 to 40 years after Alpine fault earthquakes of the past millennium. For example, one cohort of even-aged oldest trees between two of the older sand dunes germinated at about 1520 A.D., about ~30 years after the implied Alpine fault earthquake of ~1490 A.D.

This process-response model involves massive earthquake-induced sediment yield increases that are quickly transported to the coast where sand is reworked by powerful west-coast surf and wind to become a new dune ridge in the prograding shoreline. Frequent mass movements within these steep watersheds (Korup, 2005) contribute significantly to the present extraordinarily high sediment yield of 12,700 tonnes km²/year (Griffiths, 1981; Griffiths and Glasby, 1985). Prehistorical pulses of extreme sediment yields were much higher during times of fluvial-system aggradation events caused by Alpine fault earthquakes. Variation in dune size is an index of size of watershed sediment-yield and of seismic shaking intensity (Table 6.4).

Figure 6.17 Aerial photo of coastal sand-dune ridges, Arawata River, South Westland, New Zealand. Two-lane road along coast for scale. Photograph courtesy of Land Information New Zealand. Crown copyright reserved.

Table 6.4 Comparison of the intensity of seismic shaking by sizes of Alpine fault earthquakes using regional rockfall events and pulses of sediment-yield increase that result in coastal sand dunes.

Lichenometry analyses		Coastal dune-ridge analyses
Earthquake date ±20 yrs	Relative size of earthquake	Characteristics *
960	Great	
1220	Great	Much larger than other dune ridges
1490	Large	Prominent large dune
1615	Large	Large at Haast, smaller at Okuru
1410	Noteworthy	
1715	Noteworthy	Large and continuous
1440	Small	Discontinuous and small
1580	Small	
1826	Small	Small but continuous

* Coastal dune-ridge analyses are by Wells and Goff (2006)

The bursts of earthquake generated sediment-yield increase that made coastal sand dunes are so impressive that it is worthwhile to assess geomorphic responses to forest disturbance events in the drainage basins of this part of the Southern Alps. The watershed of the Karangarua River served this purpose admirably for Wells et al. (2001). Only 14% of the 1412 ha study area has trees old enough for an all-aged forest where individual tree senescence provides the gaps for replacement seedlings. Instead, even-aged stands of trees – cohorts – are typical of this forest. Of the 51 recognized cohorts, 47 of them

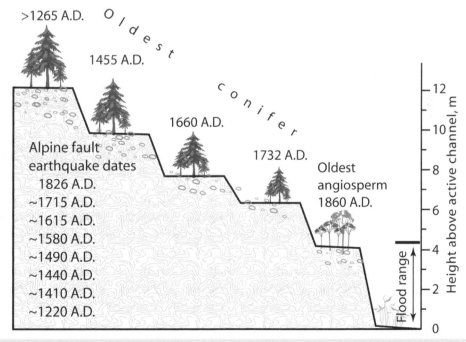

Figure 6.18 Diagrammatic section of the Karangarua River valley showing fill-terrace heights. Each brief aggradation event records a time of hillslope disturbance. Age estimates for stream-terrace treads include a colonization delay of 28 years for the oldest trees. Figure 6 of Wells et al., 2001.

germinated as a result of the 1615, 1715, and 1826 earthquakes.

Forest regeneration is dominated by earthquake-generated destruction of swaths of forest. Active channels and avalanche paths and bare rocky hillslopes comprise 10% of the watershed and remain unforested. Extreme rainfall events cause landslides, but it is earthquakes that restructure this forest. Each Alpine fault seismic shaking event creates landslide scars over much of the hillslopes. The resulting footslope debris blanket kills more forest. Debris avalanches are cataclysmic, but slower processes are just as effective. These include pulses of alluviation that create small, new alluvial fans and valley floor aggradation events that create fill terraces. Each fan or terrace tread becomes the site of a new forest that germinates after aggradation ceases and stream channels start to re-entrench. Andrew Wells searched for the oldest tree on each distinctive terrace tread, obtained an increment-borer core sample, and counted the rings to estimate the age of the tree.

The sensitivity of geomorphic processes in the extremely humid Southern Alps to seismic shaking perturbations is remarkable. Each of the five fill terraces of Figure 6.18 appears to be the result of an earthquake. Uplift along the Alpine fault increases relief and stream gradients. But this potential increase in stream power is overwhelmed by landslide-induced increases in sediment yield, which quickly aggrade the trunk valley floor. Hillslopes eventually become more stable and the threshold of critical power is crossed when stream-channel downcutting is renewed.

A new forest germinates with minimal delay. If we assume that the causative earthquakes occurred in about 1220, 1440, 1615, 1715, and 1826 A.D., then the response times are 50, 20, 50 22, and 39 years. If the second oldest terrace resulted from the 1410 Alpine fault earthquake, the response time would be 50 years instead of 20 years.

Not all recent Alpine fault earthquakes are represented. The small 1580 A.D. event did not rupture this far south. The 1490 A.D. event was the strongest of the 15th century and should be represented. Perhaps its fill terrace was buried by the 1615 A.D. event.

Meager evidence for a 1440 A.D. Alpine fault earthquake in the Haast region sand-dune record consisted of small, discontinuous dunes. This event may not be represented in the Oroko Swamp-Alex Knob cedar trees. A single growth suppression in an Alex Knob cedar tree dates to 1446 A.D.

Studies by other workers date this earthquake. The 1440 A.D. regional rockfall event corresponds to the "Geologist Creek" Alpine fault earthquake studied by Yetton (2000).

One rimu tree growing on a 2 m high scarp of the Alpine fault at the Waitaha site was tilted at this time, resulting in the formation of reaction wood typical of tree trunks tilted by any process. Either complete cessation of growth of new wood cells or reaction to strong tilting of a tree trunk may be better dendrochronology indices than temporary suppressions as indicators of the exact year, or season of a year, in which an earthquake occurred.

Reaction wood (Shroder, 1980) is a gravity-controlled process where asymmetric growth of annual rings on opposite sides of a tilted trunk returns the upper portion of a tilted trunk to a vertical stance. Microscopic examination of reaction wood shows thick-walled, denser cells forming wider rings. The start of growth of the reaction wood, and the beginning of a marked suppression of annual growth in this Waitaha rimu tree, date to 1440 A.D. Damage to trees not growing in the fault trace requires consideration of several possible causes. But a tree like this Waitaha site rimu is as definitive evidence as liquefied and ruptured strata in a trench exposure for abrupt surface rupture and unequivocal identification of the fault zone responsible for the earthquake. Many other trees at Waitaha have suppressions of annual growth rings at this time.

The third 15th century Alpine fault earthquake is the ~1408 A.D. event whose consequences are described in Figures 6.14A, B, and 6.15. It had a marked affect on cedar trees growing at Oroko Swamp and Alex Knob (Table 6.3), so the postulated surface rupture includes that portion of the Alpine fault.

It seems that moderate to small size Alpine fault surface ruptures occurred in clusters: between ~1410 and ~1490 A.D., and again between ~1580 and ~1715 A.D., with intervening time spans of tectonic quiescence between 1490 and 1580 A.D., and since 1715 A.D. All six seismic shaking events, in only 305 years, caused rockfalls at the Bad Bird (Fig. 6.10A) and at many other lichenometry sites.

After an apparent 290 years of quiescence since 1715 A.D. it might appear that an Alpine fault earthquake is overdue. Have we missed a recent earthquake? Indeed, the 1826 Fiordland "sealers" earthquake may have extended far to the northeast. It has yet to be defined in trenches excavated across the fault scarp, leading to the perception that its surface rupture might have occurred on an offshore fault zone. Paleoseismologists need to consider alternatives when trench-and-date approaches omit a surface rupture that occurred during a time of nondeposition.

Richard Taylor's 1870 book gives a graphic account by Europeans on a sealing expedition in Dusky Sound, Fiordland in the period 1823-1827.

"… from 1826 to 1827 there was an almost constant succession of earthquakes … sufficiently violent to throw men down … such was the flux and reflux of the ocean, that they were in the greatest danger of being swamped. … small cove called the jail … was completely altered; the sea had so retired from the cove that it was dry land. Beyond Cascade Point (260 km NE of Dusky Sound) the whole coast presented a most shattered appearance, so much so that its former state could scarcely be recognized; large masses of the mountains had fallen, and in many places the trees might be seen under the water."

Obviously this was a substantial earthquake and the tsunami descriptions suggest an offshore earthquake epicenter on the Fiordland complex of the submarine Alpine fault zone carefully described by Barnes et al. (2005). But did the surface rupture(s) include an onshore portion of the Alpine fault? Was there a propagating sequence of large earthquakes to the northeast during this two-year time span? Trees hold the answer to this earthquake riddle.

Earth scientists have been slow to accept forest disturbance events as a valuable paleoseismology tool. This attitude has resulted in part because the geophysical community maintains that Alpine fault movements are gradual and aseismic for a variety of reasons discussed by many authors (Adams, 1980b; Holm et al., 1989; Anderson and Webb, 1994; Upton et al., 1995; Vry et al., 2001; Warr and Cox, 2001; Okaya et al., 2003). In brief, the Alpine fault plate boundary has suppressed levels of modern seismicity – a seismic gap, which suggests either long elapse times for large earthquakes or ductile creep and aseismic slip. Historical Alpine fault seismic quiescence allows several contrasting, models.
1) Aseismic slip on fundamentally weak plate boundary. Factors that contribute to the low rock mass strength of the Alpine fault zone include a high geothermal gradient caused by extremely high erosion rates, fluid penetration from meteoric sources, fluid enhanced metamorphism and clay mineral transfor-

mations, and elevated fluid pressures in a fault zone with an upward constriction.
2) Episodic large magnitude rupture events on a fault plane characterized by stick-slip seismic behavior.

Two diverse observations dispel several largely hypothetical considerations of model 1. Melted rock formed by frictional heating during an earthquake-rupture event creates melted rock that seeps into fault-zone fractures to create mylonitic pseudotachylite. Linear groves of trees of similar age that coincide with the trace of the Alpine fault record abrupt seismic devastation instead of gradual creep.

I conclude that the geophysical arguments are sufficiently valid to allow for a third model.
3) Intervals between earthquakes would be longer, surface ruptures longer, and earthquake magnitudes would be greater were it not for weakened rock mass strength along this plate-boundary fault zone.

Foresters were the first to describe local destruction of trees. Wardle (1980) ascribed forest devastation that coincided with the trace of the

Figure 6.19 Tourists walk under tilted kamihi tree at the Minnehaha track near the township of Fox Glacier (between Karangarua and Alex Knob on Figure 6.12). Tilted trees and snapped tree trunks are progressively more common closer to a 3-m high scarp that coincides with the trace of the Alpine fault. Surrounding forest is a young cohort of even-aged kamihi trees that sprung up after the seismic devastation caused by the 1826 A.D. earthquake.

Alpine fault as being caused by a recent earthquake. Stewart and Veblen (1982) studied two trees that are the dominant species in severely disturbed land, southern rata (*Metrosideros umbellata*) and kamahi (*Weinmannia racemosa*). They ascribed local forest damage in central Westland to violent windstorms, but many of their sites are along the Alpine fault. Norton (1983b) noted similar damage near the Alpine fault as far north as the Cropp drainage basin near Hokitika.

In 1996, I cored silver beech (*Nothofagus menziesii*) growing on a low strike-slip fault scarp of the Alpine fault southwest of the Arawata River in south Westland. The site is only 28 km east of the Cascade Point earthquake devastation described by Taylor (1870). Trunks of the older trees had been snapped by a recent earthquake, which I concluded was the 1826 event on the basis of the time span since the snapped trunks started to grow again.

Snapped trunks of kamahi trees are present in the immediate vicinity of the fault trace 140 km to the northeast near the townships of Fox Glacier and Franz Josef. Earthquake-damaged trees are surrounded by young kamahi forests (Fig. 6.19) that I believe postdate the 1826 Alpine fault earthquake.

Paleoseismologists may tend to emphasize earthquake-induced damage that causes suppressions of tree growth, but releases (accelerated growth) of the surviving trees can be useful too. Forest devastation in the immediate vicinity of a surface rupture

(Fig. 6.19) tends to be a more local consequence of an earthquake. Times of maximal releases and suppressions in cedar trees (Fig. 6.20) at the Karangarua study area coincide with the times of the three most recent Alpine fault earthquakes (Wells et al., 2001). Suppressions are more common than releases, which makes sense for a site that is 12 km from the Alpine fault. Forests only 3 to 8 km from the fault may have an opposite relation. Mapping this ratio requires data from many sites but should easily identify the responsible fault zone for a seismic shaking event and may distinguish the limits of a specific surface rupture.

Four other approaches support the concept of an 1826 Alpine fault earthquake. The Goff et al. (2004) study of the sediments and shoreline trees at Okarito Lagoon (north of Oroko Swamp) concluded that inundation by a tsunami occurred in about 1826. If this tsunami was the result of an Alpine fault earthquake, an undersea landslide triggered by an onshore fault rupture would have caused it. Second, I found many suppressions of annual growth rings dating to about 1826 in cedar trees at the Oroko Swamp and Alex Knob sites. Third, the youngest forest germination event in the sand dunes adjacent to the mouth of the Haast River dates to shortly after 1826 A.D. (Wells and Goff, 2006). Fourth, I note a young regional rockfall event dating to 1826 at sites ranging from Crown Pass near Queenstown in the south to Arthurs Pass in the north. The regional extent is large but lichen-size peaks are small, suggesting that more

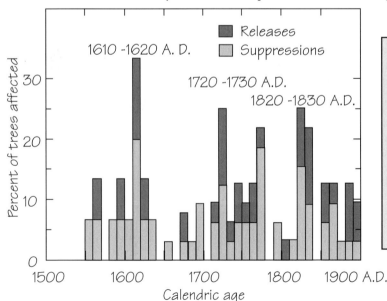

Figure 6.20 Frequency distributions of growth releases and suppressions in 33 cedar trees, *Libocedrus bidwillii*, at Welcome Flat, south Westland, New Zealand. A suppression or release was defined as > 150% difference in mean growth rate between consecutive 5-year means. From Figure 4 of Wells et al., 2001.

than one earthquake occurred in the 1826–1827 time span as described by Taylor (1870).

Visual first impressions of lichens and tree rings can be deceiving. Abrupt annual growth rings in trees look so flawless, but dendrochronologists are pleased to get a statistical significance of $r = 0.35$ to 0.40 when correlating between the trees in a grove of same species. Being able to date to the year is noteworthy, but the data carefully selected for a chronology are so messy that it becomes delicate, painstaking work. In contrast, lichen thalli rarely appear flawless. They are not circular, and the margins at measuring points generally do not seem ideal. *Rhizocarpon* subgenus *Rhizocarpon* thalli looked so erratic that I started entering a quality-assessment number for each lichen-size measurement included in a dataset. In spite of this deceptive appearance, largest lichen maximum diameters on multiple rockfall blocks can date known events with an accuracy of ±2 years. This requires consistently high-quality data.

6.2.1.5 Alpine Fault Earthquakes

We complete this story by evaluating seismic shaking caused by several recent Alpine fault earthquakes. What was the extent and magnitude of seismic shak-ing and extent of surface rupture during the earthquakes of 1580, 1615, and 1715 A.D.? Regional variations in rockfall abundance provide insight for each event.

The number of lichen-size measurements varies from site to site so this analysis calculates the size of a lichen-size peak as the percentage of measurements in a 10 mm size range, ±5mm from a peak.

Normalization also needs to account for variation in density of lichen-size measurements in some datasets. The combined dataset of lichenometry sites near Mt. Cook is used to illustrate normalization procedures. The 1448 lichen-size measurements range from 9 to 226 mm. They were measured primarily on unstable blocks of young glacial moraines that are only 24 to 29 km from the trace of the central Alpine fault. Examples of situations where a normalization correction is, and is not, needed are shown in Figure 6.21.

The peak corresponding to the 1715 A.D. earthquake is assigned a value of 100% in the following analyses of relative sizes of the three lichen-size peaks. Peak sizes for the 1615 and 1580 events are increased proportionally but only where the trendline slopes down with increase of lichen size in the 40 to 70 mm size range.

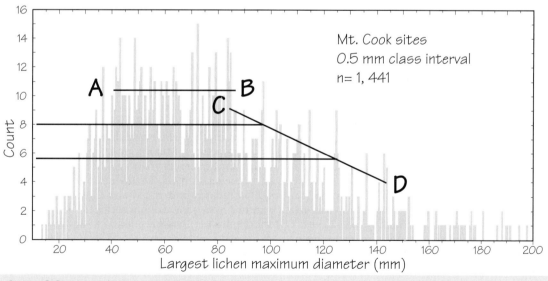

Figure 6.21 Normalizing variations in density of lichen-size measurements. Adjustment would not be needed for an analysis comparing lichen-size peaks in the 40 to 90 mm range: A to B. Peak sizes along the C–D trend need to be normalized. Intersection of horizontal lines with the Y axis provides numerical values of change for specific peaks. Using the 97 mm peak as the norm (100%), the 125 mm peak size would need to be multiplied by 143% (8/5.6 × 100=143).

Lichen-size peaks in 10-year time spans bracketing the times of three recent Alpine fault earthquakes were defined and dated. Lichen-size measurements are sufficiently abundant near key sizes of 49, 65, and 70 mm to define six to eight prominent lichen-size peaks in a ±3 mm size range (26 to 28 year range in peak ages) from each of the three central positions (Fig. 6.22).

The oldest event at 1580 A.D. ± 5 years has only two lichen-size peaks near 1580 A.D. I chose the 1578 A.D. peak because its age estimate matches the corresponding time of tree-growth suppression (Table 6.1).

The choice for the 65 mm lichen-size peak is not as easy because many peaks are present and three are possible candidates – 1617, 1612, and 1609 A.D. This 1612 peak is the oldest in a dataset of 10 sites, the youngest being 1615 A.D. The average lichenometry event time of 1613 A.D. (Table 6.2) does indeed precede the tree-growth suppression time of 1615 A.D. as estimated by tree-ring analyses (Table 6.1) by the ideal 1 to 5 years. But then all the event comparisons between these two tables are within the 2σ overlap uncertainties. The 1617 and 1609 suites of lichen-size peaks also should be considered carefully.

So, that is what I did, analyze each one separately. Only the 1613 A.D. dataset has strongest seismic shaking near the Haupiri River in the northern part of the transect, an essential criterion that is needed to match that of the Crane Creek event of Yetton (1998) and Wells et al. (1999). I decided to use the 1613 A.D. event despite its relatively small peak size in Figure 6.22B.

The youngest event was much easier to pick because lichen-size peaks are few and the large peak dating to 1717 A.D. matches the 1717 A.D. tree-ring dating of the event. Wells et al. (1999) are specific – "The most probable date for the earthquake is therefore when chronologies first show major growth suppression: some time after the 1716 growing season but before the end of the 1717 season" – a nearly perfect match with Tables 6.1, 6.2, Fig. 6.22C.

Normalized peak sizes for the three events were compared at sites projected to a transect line between the towns of St. Arnaud and Queenstown (Fig. 6.12). Despite appearing fairly rigorous, this analysis has some deficiencies.

My comparison of magnitudes of normalized lichen-size peaks does not take into account:
1) Variations in site characteristics that influence ability to record earthquakes. Some sites produce rockfall blocks only at times of seismic shaking (Figs. 6.25 B, C shown in the next section) whereas others may also record storm runoff and snow avalanches, even though site selection minimized nonseismic processes (Fig. 6.25A).
2) Variation in the landslide sensitivity of a hillslope to seismic energy coming in from different directions, a topic deferred to Sections 6.2.2.3 and 6.2.2.4.
3) Directivity effects of the release of seismic energy resulting from propagation direction and surface rupture termination for each prehistorical earthquake.
4) The number of lichen sizes in the interval including the lichen-size peak of interest is not constant. Adjacent lichen-size peaks may increase or decrease in relative importance between the towns of St. Arnaud and Queenstown.
5) Although the lichenometry sites are on both sides of the transect line, it is farther from the Alpine fault towards the south.
6) Distance of a lichenometry site from the epicenter of a specific earthquake is an unknown. Creation of a complete set of seismic shaking index maps for all major regional prehistorical rockfall events since 1200 A.D. would resolve this important aspect.

My seismic shaking assessments generally support the interpretations of previous workers. The 1580 A.D. surface rupture was not described in trench studies other than at the Waitaha site, so it is no surprise that both the magnitude and extent of the event are relatively small. The symmetrical peak of Figure 6.23A describes changes as the line of transect cuts across what would be a lobe of isopleths on a seismic shaking index map.

The 1615 A.D. Crane Creek earthquake has the largest amplitude of seismic shaking index and overlaps the Waitaha event that occurred only about 35 years before. Maximum seismic shaking extends much farther north than the 1715 A.D. event. A more recent study by Mark Yetton (email of 25 May 2005) radiocarbon dated the Crane Creek surface rupture in a trench opposite the 15 km position of the Figure 6.13 transect line. A study of historical earthquake impacts on trees (Wells and Yetton, 2004) found numerous post-1615 A.D. suppressions of annual growth in trees located 40 km from the Alpine fault in a study area opposite the 20 km position on my transect line. The percentage of trees damaged by the 1615 A.D. earthquake was similar to that of the magnitude Mw 7.8 and 7.4 earthquakes of 1929 and 1968. They concluded that the 1615 A.D. event had a Modified Mercalli shaking intensity of about MM

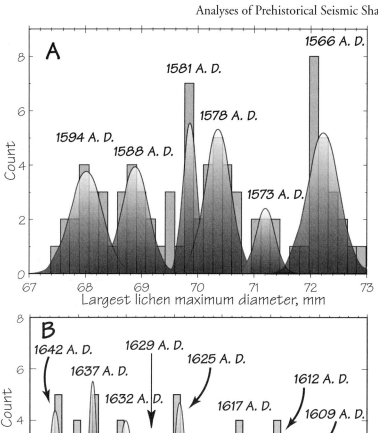

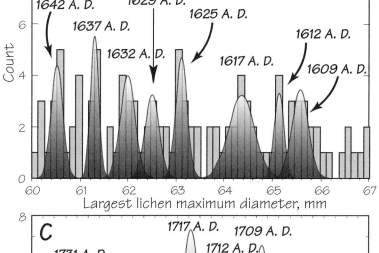

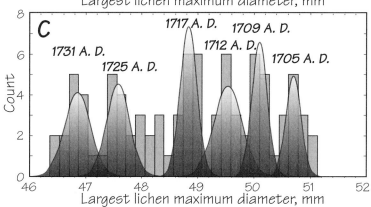

Figure 6.22 Lichen-size peaks in the Mt. Cook dataset dating to the times of the three most recent prehistorical Alpine fault earthquakes.
A. 1580 A.D. regional rockfall event. 0.17 mm class interval. n=86.
B. 1615 A.D. regional rockfall event. 0.13 mm class interval. n=112.
C. 1715 A.D. regional rockfall event. 0.17 mm class interval. n=89

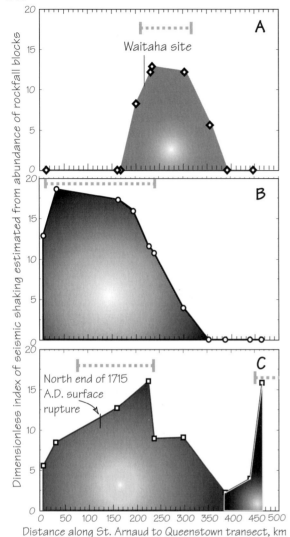

Figure 6.23 Trends in seismic shaking index based on spatial variations in the abundance of rockfall blocks dating to times of the three most recent prehistorical Alpine fault earthquakes. Dashed gray lines (using an index value of 10) depict extents of most intense seismic shaking, approximately a Mercalli seismic shaking intensity of 7 or greater.
A. Alpine fault earthquake of 1580 A.D.
B. Alpine fault earthquake of 1615 A.D.
C. Alpine fault earthquake of 1715 A.D.

8, which fits the intensity suggested by the lichenometry analysis (Fig. 6.23B). John Adams (1981) suggested that a major earthquake dating to roughly 1600 A.D. caused landslide dams in valleys along this part of the Alpine fault. My large lichenometry dataset at nearby Mt. Roberts has a large, distinct lichen-size peak dating to 1615 A.D., but has even larger peaks in the 30-year time spans before and after this Alpine fault earthquake. This part of the South Island has a long-term history of large earthquakes, with only few of them occurring on the Alpine fault.

This level of damage to trees that are 40 km from the Alpine fault might seem surprising, but can be explained in part by the range of sensitivity of their sample sites to seismic shaking. Trees growing on flat treads of stream terraces and gentle hillslopes will record earthquake-induced suppressions of annual growth only when the surface rupture is nearby, perhaps <10 km (Wells and Yetton, 2004). Swamps and steep hillslopes seem ideal for recording mainly local seismic shaking events. Steep alluvial fans and talus slopes can have influxes of detritus as a result of either local or distant earthquakes. Sudden deposition of boulders affects many trees. Suppressions and accelerations of tree growth can be numerous, and may be either temporary or permanent. Successful identification of the source of the seismic shaking can be tricky when an earthquake triggers landslides more than 300 km away from the epicenter (Keefer, 1984).

Dendroseismologists may prefer to include a few sites that are sensitive to distant seismic shaking. A site that is subject to snow avalanches is used as an example here. Peter Dunwiddie was the first to study *Phyllocladus alpinus* (Mountain toatoa) at my timberline Pegleg Creek site near Arthurs Pass. It crossdates well and suppression of tree-ring widths after times of three historical earthquakes indicates that these alpine trees record seismic shaking events (Fig. 6.24). Damaged trees occur on slopes with talus but no outcrops. Blocks of graywacke sandstone stopped against the upslope-facing sides of other trees, especially in depressions and swales. Many trees have impact scars several meters above ground level. Snow avalanches may be the main process damaging trees on these slopes. If so, both distant or nearby wintertime earthquakes may be recorded in the tree rings. The north Canterbury earthquake of 1 September 1888 occurred in late winter (epicenter about 75 km away). The local Arthurs Pass earthquake of 9 March 1929 occurred during late summer with an epicenter

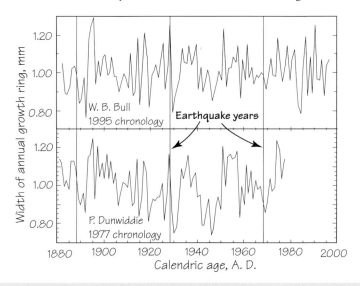

Figure 6.24 Growth suppressions in *Phyllocladus alpinus* (Mountain toatoa) after times of three historical earthquakes shown by vertical lines. Rockfall events for these three times are shown in Figure 6.5. Suites of trees were sampled in 1977 and 1995 from a timberline forest at Pegleg Creek, near Arthurs Pass.

that may have been 45 km away. It was followed by the magnitude Mw 7.8 Murchison earthquake of 17 June 1929 whose epicenter was about 140 km away. The smallest of the three mass-movement suppression events was after the Inangahua earthquake (epicenter 130 km away) that occurred on 23 May 1968—before the time of deep snowpack. As noted for many previous examples, the times of maximum suppression of annual growth rings seem to occur several years after each earthquake.

Yetton's studies in several trenches constrain the northern termination of the 1717 A.D. surface rupture of the Toaroha earthquake (Fig. 6.23C). Seismic shaking continued on to the north from the surface-rupture termination, at decreasing intensity, but strong enough to cause rockfalls for at least another 150 km. This suggests a surface rupture propagating from the south-southwest that directed seismic energy towards the north-northeast as the rupture energy was arrested. Maximum seismic shaking appears to be less than for the 1615 A.D. event, as does the extent of most severe seismic shaking.

The seismic shaking index dating to 1717 A.D. decreases to near zero southwest of Mt. Cook. This agrees with Wells et al. (1999, p. 3) statement

"On the basis of forest ages alone, the most recent earthquake can be traced from its northern limit at Haupiri to Paringa, 220 km to the south." But the Figure 6.23C plot then rises to large values again.

This double-peaked pattern is best interpreted as two earthquakes that were closely spaced in time. Wright (1998, p. 130) notes a possibility of two earthquakes for the 1715 A.D. event. In regard to the southern section of the Alpine fault, Berryman et al. (1998) say that "A 560 year old Matai tree close to the trace at Okuru shows a major cessation in growth in the 5 years following 1718 ± 5 AD". I suspect that this statement was the basis for Wells et al. (1999) saying "The most recent earthquake occurred in A.D. 1717 and the rupture extended over a section of fault between Milford and the Haupiri River, a distance of 375 km. Alternatively, the Alpine fault may have ruptured in two closely spaced events, at about 1717–1718 A.D. (Table 6.2, and Berryman et al., 1998).

A few summary thoughts and key conclusions – my previous interpretation of widespread intense seismic shaking at about 1750 A.D. now appears to have been the result of ~1749 and ~1754 earthquakes on widely separated faults; neither was an Alpine fault event (Figs. 6.11, 6.12). Sequences of Alpine fault earthquakes seem to be clustered in time. The clustered temporal behavior and small size of the Alpine fault earthquake peaks in Figure 6.10 raises the possibility of magnitude Mw 7 instead of the Mw 8 earthquakes suggested by the apparent 8 m dextral offsets on the Alpine fault near Milford Sound (Sutherland and Norris, 1995; Norris et al., 2001). One elusive possibility remains untested – 150–300 km of the Alpine fault may have ruptured in a sequence of prehistorical earthquakes so closely spaced in time that their dates cannot be separated by either lichenometry or tree-ring analyses. It is improbable that this plate-boundary fault with a minimum long-term dextral slip rate of 27 ± 4 mm/yr (Sutherland, 1994) would have elapsed times of ~300 years between earthquakes. The 1826 A.D. "sealers" earthquake divides the 290-year time span since 1715 A.D. into more realistic earthquake recurrence intervals. Important future work should include making seismic shaking index maps for all historical earthquakes and for all magnitude Mw >7 earthquakes in the South Island since 1200 A.D.

Such maps are a major topic of the next section, for lichenometry sites having different sensitivities to seismic shaking.

6.2.1.6 Recent Marlborough Earthquakes

Both rockfall and bedrock lichenometry analyses were used in a study of recent earthquakes on the Hope, Clarence, Awatere, and West Wairapapa faults.

This study tested hypotheses that emerged during analyses of data at lichenometry sites near the Conway segment of the Hope fault (Figs. 6.2, 6.3):
1) A prominent 30 mm lichen-size peak in the rockfall-block data suggested that a recent local earthquake had created many fresh rock surfaces.
2) This rockfall event is regional as indicated by the presence of 30 mm lichen-size peaks at 47 rockfall sites, some nearby and some distant.
3) Measuring the sizes of lichens on outcrops with insufficient strength to resist disruption by intense local seismic shaking might identify the Conway segment as the source of the seismic shaking.
4) Lichenometry might be capable of describing three coseismic rockfall events during a 6 year interval.
5) Regional variations in rockfall abundance should be used to make seismic shaking index maps to identify the faults responsible for four recent earthquakes.

The rockfall block sites used here are 0.5 to >50 km from the trace of the Conway segment. Blocks of rock on 26° to 30° hillsides may be moved by seismic shaking from either local or distant earthquakes. Fresh blocks may be derived from crumbly outcrops. Newly arrived blocks, or recently smashed or overturned blocks, provide fresh rock substrates for colonization by lichens. Such landscape responsiveness to both nearby and distant earthquakes results in frequency distributions with many prominent lichen-size peaks Figs. 6.8, 6.10, 6.22B).

One drawback of this sensitivity is inability to use data from a single rockfall site to identify which fault was the source of seismic shaking represented by a prominent lichen-size peak. Furthermore, a small peak might be the result of a small local earthquake, a large distant earthquake, or just statistical noise. The way out of this quandary is to measure lichens at many rockfall block sites in a region, and to measure lichen sizes on joint faces of weak to strong outcrops near tectonically active fault traces.

The outcrop method was used to identify which lichen-size peak records the most recent earthquake generated by the Conway segment. Outcrops provide an array of sensitivity to seismic shaking ranging from durable to fragile. My strategy was to find outcrops that were disrupted by a local, strong earthquake.

Outcrop rock types vary in their ability to favor growth of high-quality, large lichens that record prehistorical earthquakes. The ideal lithology:
1) Has smooth planar joint faces, such as quartzitic sandstone, so that the digital caliper measurement of largest lichen maximum diameter matches the true sum of lichen incremental growth. Surfaces of rough joints, such as in porphyritic granite, blocky limestone, or andesitic lava, can be so irregular as to introduce measurement errors of a millimeter or more. Large lichens are more likely to grow onto irregularities and thus be excluded from a dataset.
2) Has sufficiently large joint surfaces to improve the chances of having several thalli close to the largest lichen maximum diameter, which enhances one's confidence in that measurement. Favorable rock types include granitic blocks and sandstone slabs. Unfavorable lithologies include argillite chips and breccia fragments.
3) Weathers slowly into landforms that are disrupted by earthquakes but not by rainstorms and avalanches. Favorable rock types include phyllitic schist and columnar basalt in tor landforms. Unfavorable

lithologies include fractured mudstone, and unconsolidated bouldery alluvium in a stream terrace riser. Depositional cones are distinctive landforms created by mass movements and streamflow. Planar sheets of talus beneath linear outcrops that have yet to develop small watersheds are less likely to record floods and avalanches, so most of the blocks are deposited during earthquakes, or by aseismic landsliding.

Sensitivity of outcrops to seismic shaking is a function of rock mass strength (Selby, 1982a; Augustinus, 1992, 1995; Brook et al., 2004). Outcrops are useful if they are not so strong as to remain unchanged after a strong, local earthquake. Two classes of outcrops were avoided: those that are too sensitive (crumbly with only small lichens), and those that are too insensitive to seismic shaking (solid, massive rock with continuous lichen cover or with large thalli that may not even be first-generation lichens). Moderately sensitive outcrops generally have a variety of largest lichen sizes on joint faces, which reveal multiple bedrock failures. The fractured graywacke sandstone outcrops used in this Hope fault study were only 1 to 5 m high; they lack the strength to persist as high cliffs.

The typical outcrop histogram of lichen-size measurements is not as strongly polymodal as for rockfall histogram datasets. Many processes create new joint faces on outcrops including rainstorms, frost splitting, passage of animals, avalanches, wedging by roots and soil, and earthquakes. The result is a relatively greater abundance of background lichen sizes. Therefore the earthquake-related lichen-size peaks are not as isolated or sharp as in rockfall block datasets.

The relative importance of earthquakes in rockfall histograms is also a function of rock type. FALL (fixed area largest lichen) unit area is much smaller than for rockfall blocks because of the characteristic crumbly, fractured nature of graywacke in the Seaward Kaikoura Range–this lithologic drawback could not be avoided in the Hope fault study. Larger joint blocks that are less likely to be moved by nonseismic processes characterize sites in Otago (Figs. 6.25B, C) or nearer the Alpine fault where the ubiquitous graywacke has been slightly metamorphosed to phyllite or low-grade schist. The FALL unit area is larger (better) and joint planes smoother. The result is better quality lichen-size measurements and a wider range of lichen sizes from which to select the largest lichen on each block. This raises one's confidence that the largest lichen dates the true time

of substrate exposure. At the other extreme, outcrops that shed small rock fragments to scree slopes during rainstorms or snow avalanches (Figs. 6.25A, D) are not appropriate for paleoseismology studies because seismic shaking signals may be masked by too many nonseismic background measurements.

An unsurpassed outcrop for paleoseismology purposes has an ideal lithology, maximal exposure to sunlight, and many nice yellow rhizocarpons. The rockfall block repository incrementally accumulates 0.2–1 m long rockfall blocks shed from the outcrop and won't be eroded by streamflow or buried by mass-movement processes in the future.

Lichen sizes can be measured on both the outcrop joint faces and on rockfall block surfaces at the Otago site shown in Figures 6.25 B, C, which appears fragile and ready to collapse again during the next earthquake. The planar joints in the phyllite improve the precision of lichen-size measurements and reduce the chances of rejecting otherwise nice lichens because the surface between the longest axis measuring points is too irregular. This fragile-appearing outcrop is small, perhaps because of many disruption episodes. A major collapse event deposited the lobe of blocks at the base of the hill. Lichen quality is optimal at this altitude where *Rhizocarpon* subgenus *Rhizocarpon* is the dominant lichen. Some thalli are more than 800 years old. Nearby tors invite one to collect replicate outcrop lichen-size measurements. Tors on steep hillslopes have adjacent talus slope datasets in less stable repositories, which expands the range of lichen sizes to include more younger events.

Geomorphically diverse sites such as a rugged valley with active glaciers (Fig. 6.25D) may require more reconnaissance time. The best quality, oldest lichens here may be on rockfall blocks partially hidden amongst that part of the hillslope that is stable enough to allow scrubby bushes to grow. Plants also provide partial shelter from strong winds that blast lichens with snow and sand.

Two lichenometry sites in the Goat Hills 4.3 and 4.6 km south of the Hope fault appear to record the time of the most recent Conway segment earthquake. One dataset is highly sensitive. Largest lichen maximum diameters on rockfall blocks plot as a highly polymodal histogram. Only the 30 mm peak is modeled in Figure 6.26A. The other prominent peaks occur in the datasets of many sites in the surrounding region. The largest peak is at 30.5 mm, which suggests a powerful and perhaps nearby earthquake.

Figure 6.25 Types of lichenometry sites.

A. An unstable hillslope that produces rockfalls during earthquakes and during times of active slope processes. This site has variable quality, so a careful reconnaissance is needed. The highly active part of the slope with small rock fragments should be avoided. The area of gradual accumulation of rockfall blocks between 1 and 2 is good. West facing ridge of Mt. Lyford in the Amuri Range. Altitude ranges from 1,300 to 1,600 m. Site location is S42° 27.783' E172° 08.483'.

B. Ideal site that produces rockfalls only during earthquakes. Fragile tor outcrop has joint blocks that would be easily dislodged by intense seismic shaking. Rock type is greywacke sandstone that has been metamorphosed to phyllite or low-grade schist, which weathers to blocks instead of chips of rock. Other tors are shown at the upper left. Flat summit of the St. Marys Range, northeastern Otago is at upper right. Hat, vehicle, and single-track road for scales. The altitude is 1,350 m and the site location is S44° 40.885' E175° 15.834'.

C. Ideal site rockfall repository consists of blocks and minimal fine detritus deposited on a flat surface without stream channels. Collapse of the outcrop at right side of view resulted in deposition of the lobe of blocks, and has continued to shed younger blocks during earthquakes.

D. Variety of sites in valley of the Hooker Glacier near Mt. Cook. Small lichens on active debris cones (1) record many recent avalanches. Youngest unstable moraines (2) are highly sensitive to recent seismic shaking. Older moraines (3) continue to be modified and they catch recent blocks tumbling down from the high cliffs. Bushy slopes between cones (4) gradually accumulate blocks over long times; could be ideal paleoseismology sites.

The other dataset reveals a less sensitive record of seismic shaking. It consists of largest lichen maximum diameters on outcrop joint faces. By "less sensitive" I mean that lichen-size peaks are not as isolated as for the rockfall block dataset. This is largely due to a mediocre lithology. Fractured graywacke typically has small joint surfaces so the FALL unit area is smaller than optimal, which means having smaller populations from which to select the largest (oldest) lichen. Deconvolution of the probability density plot reveals several outcrop-detachment events. Peaks at about 37, 40, and 45 mm also record regional events (see Fig. 6.5) so perhaps this outcrop dataset is more sensitive than we would have preferred. The most prominent peak is shown in Figure 6.26B and has a mean size of 31.1 mm. The presence of 30.5 and 31.1 mm peaks in both the rockfall and outcrop datasets suggests that a strong, local earthquake occurred recently in the Goat Hills. Peak sizes are not identical, with the bedrock lichen-size peak appearing to be 4 years older.

Other rockfall sites support the hypothesis for a recent 30 mm regional seismic shaking event caused by a surface rupture on the Conway segment (Fig. 6.27). Note how the 30 mm peak dominates the plot for the Stone Jug rockfall site, which is directly above the steeply dipping Hope fault whose trace is only 1.4 km from this site.

The advantage of having many sites within this study region allows us to check out possible compound lichen-size peaks resulting from using too large a class interval or Gaussian kernel size. The lichen-size peak at about 40 mm is an example. Both 39 mm and 40 mm peaks commonly are present at many lichenometry sites. Note the distinctly separate peaks at 39.2 and 40.6 mm on Figure 6.8. These lichen-size peaks record separate seismic shaking events only about 9 years apart. The breadth of the 40 mm lichen-size peak in Figure 6.5 suggests that more than one event might be represented. I don't know the location of the source for the 39 mm seismic shaking event but the 40 mm event occurred close to the range-bounding trace of the Conway segment of the Hope fault.

The 37 mm peak is another example. The sharp isolated peak in Figure 6.5 strongly suggests a single event. That supposition can be tested by an analysis that uses modeling results from many sites. Rather than trust the results from one dataset, let's see what 67 modeled lichen-size peaks from 24 sites tells us about lichen-size peaks in the 36 to 39 mm size range. The result is an unequivocal (isolated and sharp) pair of lichen-size peaks (Fig. 6.28) recording regional rockfall events that are only 3 or 4 years apart. This approach has potential for separating events that are only 2 years apart.

Perhaps lichens from bedrock sites will provide additional insight regarding the 31 mm lichen-size event. Outcrop method plots have prominent 30 mm peaks, but only for sites that are close to the Hope fault (Fig. 6.29). Intense seismic shaking accompanying the 30 mm rockfall event seems to have extended only 3 to 5 km into the hanging-wall block, and 8 to 10 km into the footwall block. Cowan (1991) noted a similar pattern of asymmetric seismic shaking during the 1888 earthquake on the adjacent Hope River segment of the Hope fault. Footwall

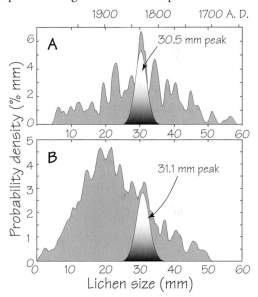

Figure 6.26 Frequency distributions of largest lichen maximum diameters, Goat Hills site on the footwall block of the Hope oblique-thrust fault. Lichen-size peaks record times of large rockfall events or outcrop-disruption events. 0.5 mm Gaussian kernel size.
A. Lichen-size peaks for a sensitive rockfall block site 4.6 km from the Hope fault records both nearby and distant earthquakes. n = 402.
B. Plot for less sensitive outcrop site 4.3 km from the Hope fault has fewer obvious peaks. n = 1,194.

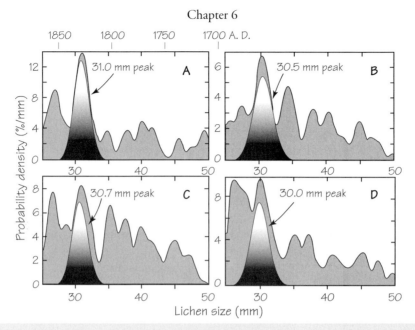

Figure 6.27 Frequency distributions of lichen sizes on rockfalls at sites 1.4 to 13 km from the Conway segment of the Hope fault. 0.5 mm Gaussian kernel size.
A. Stone Jug site above dipping fault plane, 1.4 km from the fault trace; n=84.
B. Goat hills site, 4.6 km from the fault trace; n = 256.
C. Cattle Gully site, 8.2 km from the fault trace; n = 690.
D. Stonehenge site, 13 km from the fault trace; n = 238.

block buildings were severely damaged but buildings on the hanging-wall block were hardly damaged. The strong bcdrock seismic shaking recorded by lichens at the Fenceline Crags site may reflect ridgecrest focusing of seismic waves, or it may reflect displacement along a blind thrust fault several kilometers to the north of the range-bounding oblique-slip fault. The Stone Jug site is well situated to record the 30 mm event. The Cattle Gully site is 8 km from the trace of the Hope fault, but being on the footwall block still has a fairly large 31 mm lichen-size peak.

Age estimates for the 30 mm lichen-size event from seven bedrock sites are slightly older than those from seven local rockfall sites: 1833 as compared to 1836 (1.8% older). I attribute these discrepancies to minor differences in the growth rate of *Rhizocarpon* subgenus *Rhizocarpon* in slightly different microclimatic settings. Lichens that are sheltered from sun and strong wind grow faster (Benedict, 1967). Measuring only those lichens on the tops of rockfall blocks generally can minimize this complication. However, inclusion of some partially sheltered lichens could not be avoided on highly varied and generally steep outcrop joint faces. Adjacent outcrop joint faces also tend to be wetter during rain or snowmelt than isolated tops of rockfall blocks that do not receive runoff. Outcrop lichen sizes should be multiplied by 0.98 when combined with rockfall lichen-size measurements.

Rockfall and outcrop data were combined for two sites 2 to 4 km from the trace of the Conway segment of the Hope fault in order to get a mean frequency of data of more than 100/mm. Use of a 0.5 mm Gaussian kernel size modeled two lichen-size peaks, but a third, much smaller, peak emerged when modeled using a 0.1 mm Gaussian kernel size (Fig. 6.30). Using equation 6.1, the 28.70 mm peak dates to about 1847, so it probably records the oldest historical earthquake. The Marlborough earthquake of 1848 had an epicenter approximately 110 km northeast of Goat Hills. A 105 km surface rupture and mean dextral displacement of 6 m (Grapes et al., 1998; Benson et al., 2001) suggests an earthquake magnitude similar to that on the Conway segment of the Hope fault; about Mw 7.4. The 29.43 and 30.36 mm peaks are only slightly older; these events

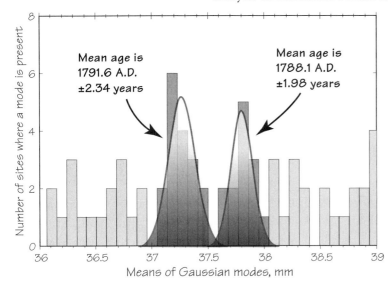

Figure 6.28 Plot of the means of 67 modeled lichen-size peaks from 24 lichenometry sites. 30 of the peak means are between 37 and 38 mm. The double peak supports the premise of two closely-spaced regional rockfall events at this time. Class interval of 0.09 mm. 2σ uncertainties.

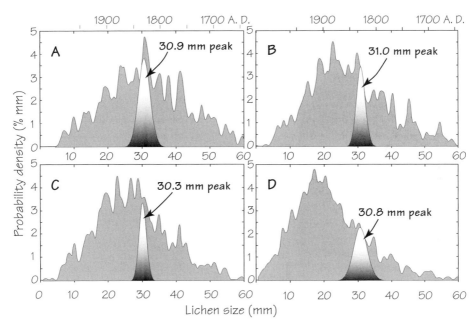

Figure 6.29 Frequency distributions of lichen sizes on outcrops at sites 1.6 to 8.2 km from the Conway segment of the Hope fault. A, B, and C are on the hanging-wall block of the fault and D is on the footwall block. 0.5 mm Gaussian kernel size.
A. Fenceline Crags site on sharp ridgecrest, 2.7 km from the fault trace; $n = 705$.
B. Stone Jug site above dipping fault plane, 1.6 km from the fault trace; $n = 785$.
C. Dog Hill site, above dipping fault plane, 3.7 km from the fault trace; $n = 704$.
D. Cattle Gully site, 8.2 km from the fault trace; $n = 1,462$.

record a sequence of earthquakes in the Marlborough district. The size of the 30.36 mm peak in Figure 6.29 implies a nearby, large earthquake – the most recent Conway segment event. The diminutive 29.43 mm peak might suggest a more distant event. Its small size might also be the consequence of exceptionally severe local seismic shaking only 6 years earlier, which shifted most of the unstable blocks and left hillslopes near the trace of the Conway segment temporarily less sensitive to seismic shaking. These several possibilities can be evaluated by analyzing data from additional lichenometry sites in the region, and by making seismic shaking index maps.

The 29 and 30 mm seismic events are recorded at 47 sites; mean peak sizes are 29.5 mm and 30.4 mm. Decomposition of the density plot of estimated ages (Fig. 6.31) into its component Gaussian peaks reveals still another but smaller regional peak that was not apparent at the Hope fault sites. Estimated calendric ages for these three regional rockfall events are about 1842, 1838, and 1836 A.D. Can we really separate seismic shaking events only two years apart? Do these closely spaced peaks describe seismic shaking events, or is this merely noisy data?

Seismic-shaking index maps provide unequivocal answers to these essential questions. They describe the spatial variations in seismic shaking for three recent Marlborough earthquakes, and the 1838 A.D. earthquake too. Such maps have additional potential for 1) comparing of relative shaking intensities on footwall and hanging-wall blocks of fault zones, 2) describing intense seismic shaking well beyond the terminations of a surface-rupture event to ascertain the propagation direction of a surface rupture, and 3) evaluating the effects of topography on local seismic shaking.

The hypothesis of three regional coseismic rockfall events during a 6-year time span between about 1836 and 1842 A.D. can be tested with seismic shaking index maps to locate their epicentral regions, and to compare the sizes of areas of most intense seismic shaking. The peak-size pattern for the 30 mm event (Fig. 6.32A) is suggestive of a composite pattern of seismic shaking resulting from two earthquakes. The smaller event occurred in about 1838 A.D. near the western edge of the study area. The 1836 A.D. event occurred along the Conway segment of the Hope fault, as is suggested by using both the Figure 6.32A peak-size map, and analysis of lichen-size measurements on disrupted outcrops close to the fault trace (Fig. 6.29). The linear area of most intense seis-

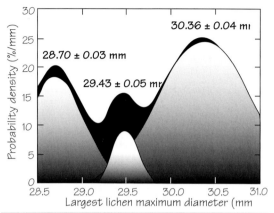

Figure 6.30 Three modeled lichen-size peaks for a Gaussian probability density plot (in black) for combined rockfall block and outcrop data from the Stone Jug and Goat Hills sites. Gaussian kernel size is 0.1 mm; n = 297. From Figure 21 of Bull and Brandon, 1998.

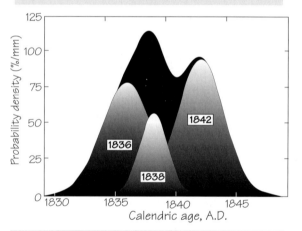

Figure 6.31 Decomposition of a single Gaussian probability density plot (in black) of times of regional rockfall events at 47 lichenometry sites suggests that three coseismic events occurred in a 6-year time span. Gaussian kernel size is 1.0 yr. From Figure 22 of Bull and Brandon, 1998.

mic shaking parallels the Conway segment fault trace, but continues on to the southwest for another 40 km including the northwest side of the Culverden basin. This apparent extension of seismic shaking may have resulted as a directivity effect of a fault rupture that started near Mt Fyffe and had its propagation energy absorbed at the southwest terminus. Another possibility is that both the Conway segment and a cross fault (Fig. 6.4) ruptured during this earthquake.

The simpler pattern associated with the 29 mm seismic shaking index event (Fig. 6.32B) clearly suggests seismic shaking associated with a single large ~1842 A.D. earthquake. The northern part of the area of most intense seismic shaking coincides with a large diamond-shaped block bounded by the Clarence and Elliott faults. It also extends southwest into Hanmer basin. This extension follows prominent cross faults of the Marlborough fault zone at the regional hinge line discussed in Figures 6.3 and 6.4. Likely involvement of cross faults within the main fault-bounded blocks of Marlborough in ~1836 and ~1842 A.D. is reasonable because the long-term rotation of the primary fault blocks should have reached the maximum of ~40° (Nur et al., 1986). The earthquake epicenter, if near the north end, was only about 30 km southwest of the historic 1848 A.D. surface rupture on the Awatere fault. Peak sizes for the A.D. ~1842 event are anomalously low near the Conway segment of the Hope fault, most likely because the A.D. ~1836 earthquake had already dislodged most of the unstable blocks on those hillslopes.

The internal consistency of the two seismic shaking index maps and the three peaks of Figure 6.31 support the hypothesis of three earthquakes in 6 yr. The ~1842 A.D. Clarence-Elliott earthquake is not included in lists of historical earthquakes (post-1840 A.D.). European colonists had arrived in the coastal settlements of Nelson and Christchurch in 1840, but the first attempt to find a route up the remote Awatere valley was in 1850 (McCaskill, 1969).

The historical Mw magnitude ~7.4 earthquake of 16 October 1848 damaged many cob houses of early settlers. Until recently it was thought to have occurred on the Wairau Valley section of the Alpine fault – the northernmost of the Marlborough district fault zones. Important recent studies (Grapes et al., 1998; Benson et al., 2001) confine the 1848 surface rupture to the eastern half of the Awatere fault. A seismic shaking index map of this event, such as shown in Figure 6.32C, would have provided the same answer. Areas of most intense seismic shaking are restricted to northeast Marlborough and are centered on the Awatere fault.

The largest historic New Zealand earthquake was the Mw magnitude ~8.2 event in the southern part of the North Island on 23 January 1855. Beaches at Turakirae Head were raised as much as 6 m (Hull and McSaveney, 1993) and 12 m of horizontal dis-

placement was noted for a surface rupture on the West Wairarapa fault zone (Grapes and Downes, 1997). The intensity of seismic shaking in northeastern Marlborough (Fig. 6.32D) was about the same as the local 1848 event even though the 1855 earthquake epicenter was ~85 km away. Although maximum values of seismic shaking index are confined to northeastern Marlborough, strong seismic shaking extended far to the southwest. The most distant Figure 6.32D lichenometry site is about 600 km from the earthquake epicenter. More distant sites also have 27.4 mm lichen-size measurements (normalized again to 1991 A.D.), but the total number of lichen-size measurements between 25 and 30 mm at these sites was insufficient to include in Figure 6.32D.

Both the 1848 and 1855 seismic shaking index maps use the percentage of lichens in their respective lichen-size peaks, relative to the total number of rockfalls in the 25–30 mm lichen-size range. This allows construction of a map showing the relative seismic shaking for these two events (Fig. 6.33). Marlborough lichenometry site responses to seismic shaking range greatly. Either earthquake may have caused the larger number of rockfalls. Areas with anomalously low 1855 rockfall abundance might be the result of having most of the unstable blocks being dislodged by the 1848 earthquake. Areas with high 1855 rockfall relative abundance might be the result of having been made unstable by the 1848 earthquake. The southwestern two-thirds of the map clearly shows the progressively greater relative seismic shaking intensity of the 1855 event. Although the 1848 earthquake was large, comparatively it was quite local. The 1855 earthquake was enormous; it maintained hillslope disruptive power as it rumbled farther southwest through the distant Southern Alps.

Perhaps these four recent earthquakes in the eastern part of the transpressional zone between the Pacific and Australian plates are related and occurred as a sequence at 6-year intervals. Their temporal spacing and northward spatial progression support this speculation. Figure 6.33 shows the locations of the primary faults, but the secondary cross faults were involved too. The first was in 1836 on the Conway segment of the Hope fault (Fig. 6.32A). Next was the 1842 earthquake on the Clarence-Elliott faults (Fig. 6.32B). Right-lateral displacement of about 7 m (Nicol and Van Dissen, 2002) indicates a Mw magnitude >7.0 earthquake for this event. The magnitude ~7.4 Marlborough earthquake in 1848 (Fig. 6.32C) had a surface rupture whose southwestern

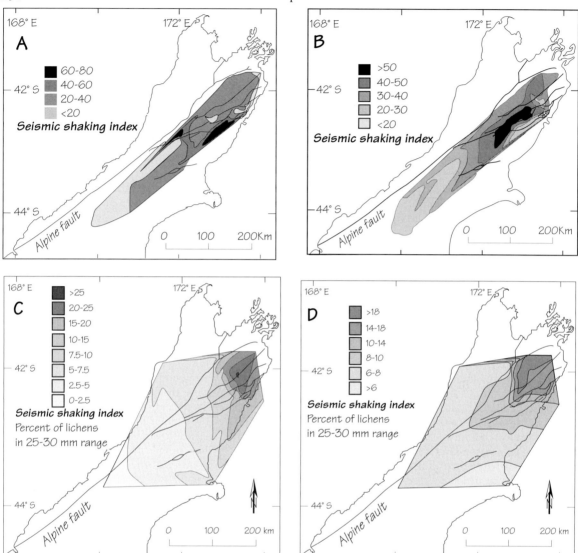

Figure 6.32 Maps of regional variations in seismic shaking based on rockfall abundance at lichenometry sites for specific lichen-size peaks.

A. Map for the 30 mm event indicates two areas of highest seismic shaking near edges of the study region. From Figure 23A of Bull and Brandon (1998).

B. Map for the 29 mm event shows one area of seismic shaking in the center of the study region. From Figure 23B of Bull and Brandon (1998).

C. Map for the 28 mm event shows pattern of seismic shaking associated with the historical Marlborough Mw magnitude ~7.4 earthquake of 16 October 1848 A.D. on the eastern Awatere fault.

D. Map for the 27 mm event shows pattern of seismic shaking associated with the historic Wairarapa Mw magnitude ~8.2 earthquake of 23 January 1855 A.D. The epicenter was across Cook Strait in the North Island about 25 km east of Wellington.

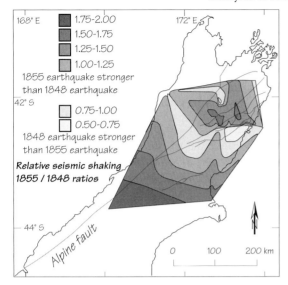

Figure 6.33 Relative intensity of seismic shaking (rockfall block abundance) for the Mw magnitude 8.2 Wairapapa earthquake of 1855 (earthquake epicenter in the North Island), compared to the Mw magnitude 7.4 Marlborough earthquake of 1848 (earthquake epicenter on eastern Awatere fault). Either event can be dominant near the Awatere fault, but seismic shaking during 1855 is progressively more important to the southwest.

end at Barefells pass (Benson et al., 2001) was only 10 km north of the Clarence fault. This 105 km long rupture extends towards the northeast. The M ~8.2 Wairarapa earthquake of 1855 (Fig. 6.32D), described by Van Dissen and Berryman (1996), occurred 6 ½ years after the Marlborough earthquake and an additional 85 km farther northeast. Benson et al., (2001, p. 1,090) note "The dextral strike slip Wairarapa fault can be traced offshore from lower North Island into Cook Strait and to within 20 km of the northeastern end of the Awatere fault (Carter et al., 1988). Although there may be no direct fault connection between the two faults, it might be speculated that the 1848 rupture of the Awatere fault precipitated failure of the Wairarapa fault during the 1855 earthquake." Following their lead, I have added the two earlier earthquakes to the sequence. Unlike the earthquake sequences along the North Anatolian fault of Turkey (Stein et al., 1997), the four events noted here are on fault zones with different slip rates and earthquake recurrence intervals. This sequence – four earth-

quakes at 6-year intervals on different fault zones – is unlikely to be repeated. The well-developed set of cross faults (Fig. 6.4) almost certainly was important in the northeastward propagation of crustal strain between the primary Marlborough faults.

6.2.2 California Earthquakes

Application of the Bull–Brandon (1998) lichenometry approach to earthquake-prone California provided diverse and different landscapes to further develop this new approach to paleoseismology. The long-term goal of a project that began in 1991 is to learn more about prehistorical earthquakes generated by the San Andreas fault.

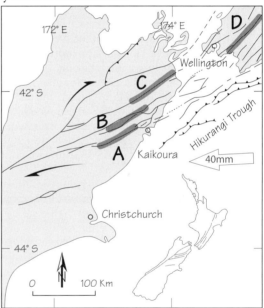

Figure 6.34   Recent large earthquakes in the Marlborough-Wellington, New Zealand, transpressional zone. Locations of primary faults are from Barnes and Audru (1999) and Van Dissen and Yeats (1991). Earthquake sequence includes prehistorical ~1836 A.D. event on the Hope fault (A), prehistorical ~1842 A.D. event on the Clarence-Elliott faults (B), historical 1848 A.D. event on the Awatere fault (C), and the historic 1855 A.D. event on the Wairarapa fault (D).

This plate-bounding structure passes through most of the length of California and is strongly transpressional where it makes a big bend (Fig. 6.35). The large historical earthquakes of 1812 and 1857 occurred adjacent to the Transverse Ranges. Section 4.1 summarizes the tectonic setting of this part of the North America–Pacific Plate boundary. Farther north, late Quaternary compression related to the San Andreas transform boundary created the eastern margin of the Coast Ranges and slip on the fault near San Francisco caused the large historical earthquakes of 1838 and 1906.

We started in the Sierra Nevada microplate, a lofty 650 km long granitic mountain range. Earthquakes emanating from the adjacent Walker Lane–Eastern California shear zone, and from earthquakes farther east in the Basin and Range Province, affect Sierra Nevada geomorphic processes. A Mw magnitude 7.6 earthquake in adjacent Owens Valley in 1872 A.D. was the largest recent local earthquake (Fig. 6.35). San Andreas fault surface-rupture events are quite distant from this stable plutonic block but they too cause rockfalls in the Sierra Nevada.

Work began in the Sierra Nevada instead of the Transverse Ranges and Coast Ranges, which are astride the San Andreas fault, because prior work by lichenometrists and lichenologists (Curry, 1969; Wetmore, 1986: Hale and Cole, 1988) improved our chances of finding suitable lichens.

Diverse climates accompany the 3,000 m altitude range for our lichenometry sites. Orographic controls on precipitation result in contrasting subhumid western slopes and a semiarid rain-shadow along the eastern side of the range. We did not know which lichen(s) would be most suitable for earthquake studies. So we started by measuring the largest lichen on rockfall blocks using four different genera (described by Bull, 2003c, 2004). It seemed like more work than was necessary, especially after needing only *Rhizocarpon* subgenus *Rhizocarpon* in New Zealand. In due course we appreciated that dating earthquake-generated regional rockfall events with several lichen genera provided valuable cross checks. Inclusion of faster-growing genera improved precision of dating. Besides, the diverse climates of the Sierra Nevada restricted lichen choices at many sites to only one or two of the four calibrated genera.

The Sierra Nevada are vastly different than the Southern Alps, so expect dissimilar tectonic, lithologic, and geomorphic settings. This provides an expanded opportunity to explore the potential of lichenometry-based paleoseismology studies in mountain ranges subject to earthquake-generated rockfalls. Earthquakes are generated outside of the stable Sierra Nevada microplate (Fig. 6.35). Massive granitic lithologies with moderate spacing of joints and fractures generate large rockfall blocks that have a favorable FALL unit area for determining largest lichen maximum diameters. Huge blocks were avoided as not being worth the time to search and find their single largest lichen. Joint-face smoothness is not ideal where the plutonic rocks are coarsely crystalline. Porphyritic textures posed sufficient problems that some sites were excluded because of insufficiently smooth substrate surfaces on which to measure otherwise nice lichens.

Two types of earthquake-sensitive landforms were preferred. One was bouldery glacial moraines, especially the recently deposited detritus left by Little Ice Age glacier advances. Such unstable moraines were valuable in our New Zealand studies. Many Sierra Nevada moraines are at altitudes of more than 3,000 m, which has the disadvantage of poor quality thalli in an environment characterized by wind-driven snow. The second type of landform was talus slopes beneath cliffs of U-shaped valleys scoured to great depths by repeated advances of glaciers during the Pleistocene. Typical scenes of grandeur, such as those of Yosemite National Park, give an initial impression of incredible rock mass strength. But sets of exfoliation joints formed parallel to the surfaces of these granitic monoliths after retreat of the ice that carved them (Gilbert, 1904; Bahat et al., 1999). The volume of talus below these cliffs is ample testimony of frequent post-glacial collapse events. Wieczorek and Jäger (1996) conclude that earthquakes did not trigger most of the historical rockfall events, but that earthquakes were responsible for most of the landslide volume that now resides in talus accumulations at the base of cliffs.

The lichenometry earthquake record varies greatly because the Sierra Nevada landscape has the potential to record both local and distant earthquakes. Some sites provide information about local earthquakes such as the 1872 A.D. event. To our considerable surprise other sites were splattered mainly with rockfall blocks generated by distant San Andreas fault earthquakes emanating from the San Francisco Bay region or from even more distant Southern California (Fig. 6.35). But first I outline ways of calibrating lichen growth, once again underscoring the value of tree-ring analyses.

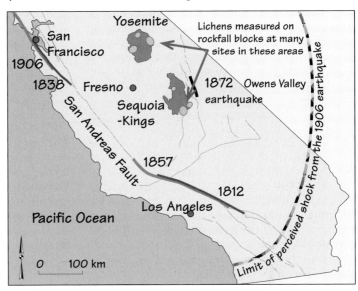

Figure 6.35 California location map showing locations of groups of lichenometry sites in the Sierra Nevada. Extents of historical magnitude Mw ~7.5 San Andreas fault earthquakes are shown by gray lines and the Owens Valley earthquake of 1872 with a black line. Limit of 1906 perceived shock is from Ellsworth (1990). From Bull (Figure 1, 2003a).

6.2.2.1 Calibration of Lichen Growth Rates

Calibration of the growth rate for each lichen genus included tree-ring dated landslide sites. The Figure 6.36 example assumes that a paleoseismologist wants to calibrate the rate of growth for *Rhizocarpon* subgenus *Rhizocarpon* using only sites in Yosemite National Park. Three control points for young sites suggest a possible growth-rate equation, which is confirmed by tree-ring dating of a much older site. Lichen sizes were measured on blocks exposed during the construction of a trail to Nevada Falls in 1940, and for a landslide that fell off a granitic dome named Liberty Cap in 1908. We identified the blocks for a witnessed 1872 coseismic landslide by cross-dating the annual growth rings of trees damaged by the granitic blocks. These three calibration points span only 68 years, which is not sufficient for extrapolation into the distant past. We knew of a large rock avalanche called "The Slide" (Huber et al., 2002) in the headwaters of Piute Creek, a remote part of Yosemite National Park. It had potential as an older calibration site, but only if we could date its time of formation to the year using dendrochronology.

Radiocarbon dating of wood buried by the rock avalanche might seem ideal but has two deficiencies. First, any wood collected for this purpose grew before the landslide event, so can only constrain the approximate time of the cataclysmic slide. Secondly, variations in production of ^{14}C in the atmosphere have varied so much during the past 300 years that one

has to choose between several possible radiocarbon age ranges. Wood sampled from beneath "The Slide" gave Bronson and Watters (1987) calendric radiocarbon age estimates of 1638–1680, 1753–1804, and 1937–1954 A.D. We can dismiss the 1937–1954 possibility, but have no basis for picking between the other two. Even if we guessed correctly, the radiocarbon age estimate would predate the event age. Dendrochronology was ideal for calibrating lichen growth here because damaged, but still living, trees dated the event to the season of a particular year.

"The Slide" stopped abruptly in a forest of hemlock trees. Tree-ring analyses were made by Bill Phillips who cross-dated the patterns of annual growth rings and dated the time of landslide damage to the forest as occurring between the end of the growing season in September 1739 and before the start of renewed growth in June 1740 A.D. This fourth calibration point in Figure 6.36 at 1739 A.D. falls on the same line as the other three points, nicely defining long-term linear lichen growth rate.

Dating "The Slide" was essential for regional calibration of all four lichen genera. Regression of mean lichen size in millimeters, D, and calendric years, t, describes uniform *Rhizocarpon* subgenus *Rhizocarpon* growth as:

$$D = 190 - 0.095\,t \qquad (6.4)$$

n (number of control points) is 5 and r^2 (correlation coefficient) is 0.99. High correlation coefficients for these calibration equations result partly from regressing cumulative time vs cumulative increase in lichen size. Uniform-phase growth is about 9.5 mm per century. Thus lichens growing on substrates exposed in A.D. 1 would have a mean size of 190 mm in 2000 A.D. Similar calibration procedures were used for the other three lichen genera.

The calibration equation for *Acarospora chlorophana* is:
$$D = 225 - 0.114\,t \qquad (6.5)$$
n is 5 and r^2 is 1.0.

The calibration equation for *Lecidea atrobrunnea* is:
$$D = 448 - 0.231\,t \qquad (6.6)$$
n is 6 and r^2 is 0.998.

The calibration equation for *Lecanora sierrae* is:
$$D = 377 - 0.189\,t \qquad (6.7)$$
n is 7 and r^2 is 0.998.

Sizes for earthquake-generated lichen-size peaks of known age (1812, 1857, 1872, 1906) in different parts of the mountain range also fall on the Figure 6.36 calibration regression line. So, each lichen genus used for paleoseismology dating grows

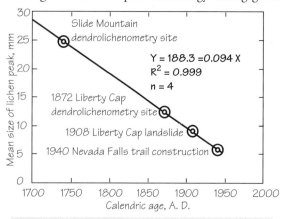

Figure 6.36 Example of calibration procedure to determine lichen growth rates, using lichen-size measurements only from sites in Yosemite National Park whose time of exposure is known to the year. Many lichen sizes were measured and dominant lichen-size peaks were identified for *Rhizocarpon* subgenus *Rhizocarpon*. The 68-year time span of the three historical sites was increased by 132 years when tree-ring dating of a big rock avalanche known as "The Slide" extended the time span for this calibration data set.

at the same rate throughout the 600 m to 3500 m altitude range of the Sierra Nevada study region.

The following discussion about Sierra Nevada paleoseismology has great variety. Lichens on Sierra Nevada rockfall blocks seem to record distant San Andreas fault earthquakes. If so, how do the lichenometry age estimates compare with the dates of historical earthquakes?; or with the most precisely determined radiocarbon dates for prehistorical San Andreas fault earthquakes? We now address these questions in depth. Discussion focuses mainly on two events, the historic earthquake of 1812 A.D. and a prehistorical 1739 A.D. earthquake. This is my first attempt to evaluate rockfall abundance in different topographic and structural settings, in response to seismic waves generated by earthquakes to the east, south, and north. Lichenometry-site locations and important historic surface ruptures are shown in Figure 6.35.

6.2.2.2 Recent Cliff Collapse

Let us start in the South Fork of the Kings River where I was preparing for the 2003 Friends of the Pleistocene field trip. I looked up at the massive north-facing cliffs towering above the Roaring River parking lot. Monotonous gray lichens coat the surface of the jointed, granitic cliff face except for several small whitish patches (Fig. 6.37). Could these be where rock, weakened by expanding joints and fractures, had fallen so recently that the cliff has yet to be re-coated by the ubiquitous gray lichens?

Surely the trip participants would be curious as to why and when parts of the cliff had failed. I'd prefer to be specific rather than make superficial comments such as "the coating of gray lichens shows that these cliffs are fairly stable, but white patches suggest a recent rockfall event; most likely caused by the nearby Mw magnitude 7.6 earthquake of 1872".

The solution was to measure sizes of lichens. Having only 45 minutes before meeting a Park Ranger, I measured only a dozen lichens each for three genera. I hoped that 42 lichen-size measurements might reveal a consistent story about recent cliff failure(s), and that analyses of different lichen genera would confirm a common theme. Would there be a pattern of a few blocks breaking loose from the cliff face each frosty winter? If so, the lichen sizes would be evenly spread out in a histogram. Would all the lichen sizes plot as one peak? This would indicate that the cliff failed only once and was otherwise stable.

The bright yellow crustose lichen, *Acarospora chlorophana* (Fig. 6.39), is one of my favorite lichens (Bull; 2003c, 2004). So I measured largest lichen maximum diameters on rockfall blocks at the start of a short transect across the lower talus slope where snapped limbs and crushed trunks of oak trees at the downslope edge of the talus also declared damage from recent rockfalls. A simple histogram revealed two prominent peaks even with only 14 measurements (Fig. 6.38A). This initial result suggested that rocks fell from the cliff face as discrete events rather than as an annual trickle. Equation 6.5 dates the times of these rockfall events as 1811 and 1739 A.D. ± 10 years. The 2σ–95% confidence level–uncertainties sum the measurement, calibration, and modeling sources of error and are rounded up to the nearest decade.

Our prior work indicates that both of these *Acarospora chlorophana* age estimates match times of regional rockfall events in the Sierra Nevada. A large Southern California earthquake occurred on the San Andreas fault in 1812, and evidence that will be presented later suggests that the Honey Lake fault zone northwest of Reno, Nevada might be the source of the regional seismic shaking during the winter of 1739–1740 A.D. that caused "The Slide".

So, the first analysis indicated that the cliff face failed not once but twice, and apparently from earthquakes whose epicenters were more than 300 km away. Both times were quite a bit older than my simplistic guess made by just looking at the cliffs. Lichen-size measurements for two other genera tested this initial hypothesis.

A few measurements of *Lecidea atrobrunnea* (Fig. 6.40) were made a little further along the transect (Fig. 6.38B). Its equation 6.6 uniform phase rate of growth is 23.1 mm per century as compared to 11.4 mm per century for *Acarospora chlorophana*. This graph reveals a single tall symmetrical lichen-size peak that dates to about 1736 A.D. ± 10 years. Once again this is quite close to the time of the 1739 A.D. regional seismic shaking event.

The third plot is for the slower-growing yellow green *Rhizocarpon* subgenus *Rhizocarpon* (Fig. 6.41) and it is still different. Some of the measurements form a large peak at about 14 mm (Fig. 6.38C), which dates to about 1851 A.D. ± 10 years. This result is clearly different than those of the other two lichen genera, but it too seems to have been earthquake generated. A major earthquake occurred on the southern San Andreas Fault in 1857 A.D. A smaller

Figure 6.37 View of cliffs above the talus slope next to the Roaring River parking lot, South Fork Kings River. Whitish splotches on granitic outcrop are where masses of rock detached along exfoliation joints, in one or several events, and fell to add another increment(s) to the talus accumulating at the edge of the valley floor. The rest of the cliff has been stable for sufficiently long to become coated with gray lichens.

lichen-size peak with an age estimate of 1816 A.D. ±10 years also may record an 1812 A.D. coseismic rockfall. These yellow rhizocarpon measurements were collected at the end of my traverse, and show that I had moved off the 1739 A.D. talus into an area that had been splashed more recently with chunks of rock during the 1857 and 1812 A.D. earthquakes. These age estimates may not be as accurate as those for *Acarospora chlorophana* and *Lecidea atrobrunnea*, because these yellow rhizocarpons have poorer quality thalli and a relatively slower growth rate.

So, even 42 lichen-size measurements were sufficient to revise my erroneous initial impression that the talus slope had been splashed by a single,

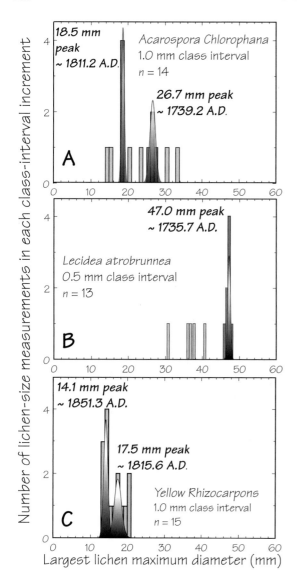

Figure 6.38 Simple histogram graphs for three common crustose lichens reveal clustering of lichen sizes. These lichen-size peaks date the times of the most recent rockfall events that splashed blocks and rock fragments onto a slope at the Roaring River parking lot site, South Fork Kings River.
A. *Acarospora chlorophana.*
B. *Lecidea atrobrunnea.*
C. *Rhizocarpon* subgenus *Rhizocarpon.*

fairly recent, collapse of the cliff face. Three small samples of lichen sizes seemed to record three earthquakes with epicenters far to the north and south of the Kings Canyon.

Where were the anticipated rockfall blocks from the Mw magnitude 7.6 Owens Valley earthquake of 1872 whose epicenter was only about 46 km to the east? Apparently the cliff-parallel orientations of the exfoliation joints and the overall east–west trend of this glaciated valley did not favor many blocks being pried loose by seismic waves coming from the east. The east–west orientation of the joints in these granitic cliffs appears to be more conducive to blocks being pried loose by seismic energy from either the north or south.

The three lichens used in the above analysis are quite different. The bright chartreuse *Acarospora chlorophana* has a smooth surface and no black algal thallus rim. Patches of smooth fungi near the center of the example shown in Figure 6.39A may be newer growth that has repaired damage caused by competing lichens. This appears to be a single lichen because of the overall elliptical shape of the outer margins and the parallel elliptical bands of internal cell structure. We obtain best statistical results by measuring the long axes of the ellipse for each lichen, shown here by the black line. I am confident that the upper left measuring point is intact, but am less sure about the lower right end point. This lowers my quality assessment for this lichen-size measurement.

As a quality-control measure, subjective evaluations are made of the relative quality of each lichen measured in order to answer important questions.
1) Is this really a single thallus, or have several lichens grown together?
2) Is the lichen sufficiently well preserved to distinguish the end points of the longest axis of growth?
3) Are the margins at the two measurement points sharp and well defined?

Quality control numbers range from 1 to 4:
1) A superb lichen that is worthy of a photograph.
2) Close to an ideal lichen-size measurement.
3) Nothing special, but good enough to include in the dataset.
4) We hesitate about including this lichen size in the dataset; its thallus has borderline quality.

The lichen shown in Figure 6.39B is not complete because it is growing at the edge of a block and it has been partially cannibalized by competing lichens. Its nice elliptical shape is diminished by an indentation at C, but this does not bother me because the internal cells – like the Figure 6.39A lichen – have a pattern suggestive of one instead of two lichens. The margin at measurement point A is ideal. Fortunately, the other side of the lichen at B has not been removed. But I do not know how this margin would have appeared if the lower-right portion of the lichen had not been severely degraded. On balance this lichen seems good enough to measure and I give it a quality rating of 3. Your vote might be for a quality 2 or 4 rating because subjective evaluations have quite a bit of operator variance.

A welcome surprise was that the seemingly small density of lichen size measurements was sufficient to consistently identify specific earthquake-generated lichen-size peaks, as described earlier for Figure 6.6. This is in part due to minimal background noise from random-event lichen sizes.

Data density determines choice of class interval, or Gaussian kernel size, when defining lichen-size peaks. It is useful to plot the density of measurements for a site (Fig. 6.42). Measurements made by the 2002 lichenometry short course participants

at the Roaring River site produced lichen-size peaks that fit the regional picture. The 189 Acarospora measurements occur mainly between 20 and 60 mm (Fig. 6.42A), which allows us to surmise that density of <1.0 lichen-size measurements/mm might be risky, so get more data. 1 to 2 measurements/mm is OK if quality of lichen thalli is above average, but >2 lichen-size measurements/mm is preferred. Try using a smaller class interval, or Gaussian kernel size, to separate compound lichen-size peaks without introducing peaks that are mere noise.

Increase of density of measurements improves the reliability of results. Density of lichen-size measurements for a particular analysis can be improved by combining datasets. Two datasets, only 5 km apart, are combined in Figure 6.42B. Total measurements are increased to 409, which increases one's confidence in the analytical results for the 1,000 year time span between 860 and 1868 A.D.

The plots shown in Figure 6.42 are means of 10 mm intervals, so one should expect higher densities in the vicinity of specific lichen-size peaks. The densities of lichen size peaks in Figure 6.38 are

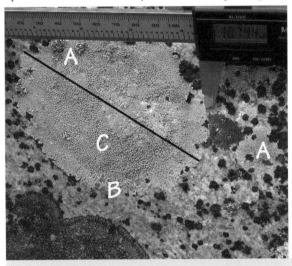
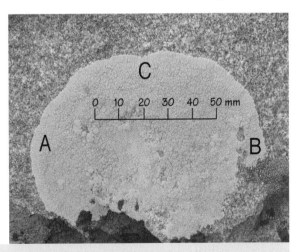

Figure 6.39   *Acarospora chlorophana* on talus in the Kings Canyon of the Sierra Nevada.

Left-side photo. A 1,000 year-old lichen growing at the Roaring River site. (**A**) Lichens are competitive. Two other types of lichens grow on this *Acarospora chlorophana*. This process creates holes in the thallus and embayed margins. (**B**) The strings of fungal cells near the margins have a different structure and color compared to the larger central areoles and apothecia that contain reproductive spore cases at **C**.
Right-side photo. Assessing the quality of this large *Acarospora chlorophana* considers potential problems of the distance between longest axis measurement points **A** and **B**, and evaluating whether indentations like that at **C** suggest merging of several lichens.

good despite quite small numbers of measurements. Densities for the 4 peaks discussed are 4.0, 1.3, 4.0, and 2.3 measurements/mm.

6.2.2.3 Rockfall Processes in Glaciated Valleys

Cliff collapse commonly initiates a sequence of rock-fall events over a time span ranging from days to years, or even decades or centuries (Bull, 2003b). Wieczorek and Snyder (1999) nicely document three such events in 7 months above Curry Village in Yosemite valley (Fig. 6.43). None were earthquake induced. The

first rockfall from the cliffs below Glacier Point was the largest, being an estimated 1576 metric tons. Ice-plugged fractures raised ground-water levels that may have increased seepage forces sufficiently to trigger cliff failure. Large block(s) fell 30 to 45 m down a 75° cliff face to a ledge, broke into smaller chunks, and then fell another 290 m to talus. Block size and velocity were sufficient to remove large trees. Huge prehistorical rockfall blocks partly determined the paths of bouncing blocks that crushed vegetation as they rumbled through the forest.

Figure 6.40 *Lecidea atrobrunnea* with typical large black apothecia and algal thallus rim. Broken areoles reveal white interior color.

Figure 6.41 Small thalli of *Rhizocarpon* sub-genus *Rhizocarpon*. A progression of lichen sizes that approach the favored largest lichen maximum diameter is much better than having only a single lichen to measure on a rockfall block. Measurements include black algal thallus rims.

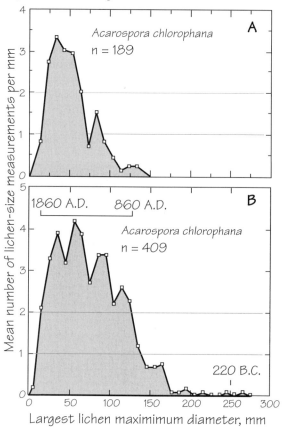

Figure 6.42 Variations in density of measure-ments with increasing lichen size. Midpoints of consecutive 10 mm size increments.
A. With 189 lichen-size measurements, density is >2/mm in the 20 to 65 mm size range.
B. Combined data from two Kings River sites. With 409 lichen-size measurements, density is >2/mm in the 15 to 130 mm size range.

Subsequent rockfalls followed earlier routes. The 25 May 1999 event was much smaller (112 metric tons), but the 13 June 1999 event was of intermediate size (600 metric tons). The rockfall block ballistic splatter pattern was similar to the previous events, and had almost the same extent as the 16 November 1998 event. Some blocks traveled 500 m from the top of the talus, and small fly rocks with ballistic trajectories traveled much further from impact points high on the cliffs. A person measuring lichens a century from now would conclude that this sequence was a single event. This would influence her or his perception of landslide-event size. The measurements used to define the lichen-size peak would come from two sources: rockfall blocks and chips, and from older blocks that had been smashed to create fresh surfaces to be colonized by new lichens. Yes, one rockfall block may record two events.

Landslide damage is impressively different when huge blocks remain coherent until impacting the valley floor. This contrast is underscored by the 1996 rockfall at nearby (Fig. 6.43) Happy Isles (Wieczorek et al., 2000). An arch of exfoliating rock, 150 m long, 10 to 40 m high, and 6 and 9 m thick detached from the cliffs below Glacier Point as two huge blocks. Both blocks accelerated while sliding down a 47° cliff and then fell about 500 m in a ballistic trajectory to a talus slope. Two impacts 13 seconds apart were recorded by seismographs and created an airblast that uprooted and snapped a thousand trees. Then a cloud of pulverized rock descended from the impact site, abrading remnants of trees and depositing gravelly coarse sand.

Rockfalls and other landslides have been studied carefully in Yosemite National Park and a detailed inventory of 519 of them has been compiled (Wieczorek et al., 1992, 1998; Wieczorek and Snyder, 2003). They are a hazard in this glaciated valley (Guzzetti et al., 2003). Three million people visit the park each year and rockfalls have killed 12 and injured 62 of them. Landslides generated by the 1872 earthquake resulted from strong seismic shaking that emanated from Owens Valley adjacent to the eastern flank of the Sierra Nevada (Figs. 6.35, 6.49). John Muir (1901) described truly spectacular coseismic debris slides and rock avalanches in Yosemite Valley.

But do sources of earthquake energy that are more than 200 km away disrupt small parts of this massive granitic landscape that appears incredibly strong? The great San Francisco earthquake of 1906 apparently did not produce rockfalls in Yosemite valley worthy enough to catch the attention of people. Distant seismic shaking events may generate just a few blocks, which can fall at locations out of view of humans. The crash of falling ice during a winter night sounds much like falling rocks, making recognition of rockfall events even more complex.

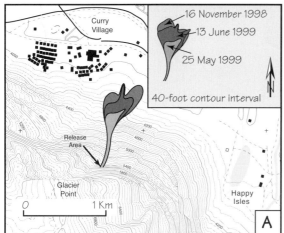
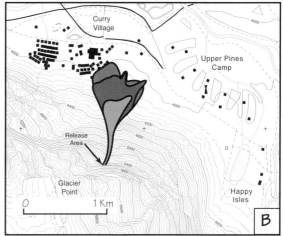

Figure 6.43 Maps of areas of rockfalls (A) and areas splattered with flying rock fragments (B) below the Glacier Point rockfall release area near Camp Curry, Yosemite National Park. Big blocks bounced shorter distances than the 10–20 cm fly-rock fragments. From Figures 2A and 2B of Wieczorek and Snyder, 1999.

Massive granitic cliffs (Huber, 1987) along the sides of glaciated valleys become progressively unstable as exfoliation joints and other fractures form (Fig. 6.44). Exfoliation joints form roughly parallel to a cliff face when melting of glaciers removes lateral support of the valley walls and the surficial portion of massive granitic rock becomes weaker as joints and fractures gradually open. The most recent Pleistocene glaciers of the Tioga glacial advance did not fill valleys with ice to the same level as earlier glacial advances, in part because each episode of glacial erosion lowered the floor of Yosemite valley. So the higher parts of the cliffs have had more time to develop extensive networks of fractures and joints. This is where most late Holocene rockfalls originated.

Some distant earthquakes do indeed cause landslides in the Sierra Nevada. A recent example is the San Simeon Mw magnitude 6.5 earthquake of 21 December 2003, which occurred 270 km southwest of Yosemite valley. This moderate earthquake was felt in Yosemite and even more surprising is that a magnitude 4.1 aftershock on the next day was also felt. Gerald F. Wieczorek (written communication, 26 February 2004) notes that the aftershock coincided with the time of a debris slide from the upper part of Sentinel Creek in Yosemite Valley. A miniscule amount of seismic energy was sufficient to cause part of the landscape to pass through a stability threshold–a crossing that was recorded by a landslide.

This is a nonsteady-state geomorphic process involving two thresholds (Fig. 6.44). The first is the miniscule fracture that signals initial failure of solid rock, and induces a self-enhancing feedback mechanism that progressively decreases rock mass strength. Episodic processes that include frost wedging, seismic shaking, and seepage forces either increase driving forces or decrease resisting forces. These events are superimposed on long term expansion of the network of cliff face fractures – a decrease of rock mass strength that eventually results in collapse of the outcrop. Collapse time for Sierra Nevada cliffs generally is accelerated by one of the episodic processes that move the system across the second threshold – the outcrop block collapse threshold. Paleoseismologists measure the timing of the second threshold by measuring lichens sizes on the resulting landslide.

The role of earthquakes in this process is perhaps more vague than we would like. Two recent

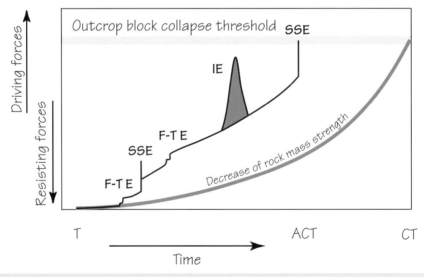

Figure 6.44 Diagram showing interaction of factors leading to collapse of an outcrop block as a block slide, rockfall, or rock avalanche. Block separates from outcrop at threshold crack time, T. Then, the miniscule crack widens at an exponentially increasing rate as propagation of additional fractures by weathering, seismic shaking, and free-thaw episodes (F-T E) decrease the resisting force of rock mass strength. Failure occurs at collapse time, CT, or earlier if accelerated, ACT. The driving forces/resisting forces ratio may be increased temporarily by rainfall- or snowmelt-infiltration events, IE, that increase seepage forces, or by seismic shaking events, SSE.

New Zealand rock avalanches occurred 4 and 6 months after local seismic shaking and it was concluded that they were not the result of earthquakes. The Figure 6.44 model would say otherwise; seismic shaking moved these hillslopes so close to the collapse threshold that the time of failure was greatly hastened. Earthquakes are important in causing the initial miniscule fracture, being part of the processes tending to decrease rock mass strength, and being an obvious triggering mechanism for ultimate outcrop collapse. I suspect that the primary cause of the Sentinel Creek debris slide of 22 December 2003 noted above was a delayed response to the seismic shaking of the previous day's earthquake.

Seismic energy arrives in waves that move cliffs back and forth, opening and closing cracks in the rock. Climbers scaling cliffs during the 1980 Mammoth Lakes and the 1989 Loma Prieta earthquakes saw rocks and rubble drop into fissures that opened and closed with the passage of seismic waves. This input of seismic energy can dislodge parts of cliffs by the "seismic-ratchet" process described in Figure 6.45, causing slabs to fall. Characteristics of individual landslides vary greatly as a function of the height and mass of the landslide source, the steepness of the cliff, and the presence of projecting ledges that can convert big falling blocks into small fragments.

6.2.2.4 San Andreas Fault Earthquakes

We measured the sizes of lichens on rockfall blocks below Middle Brother in Yosemite Valley. This fractured granitic monolith rises 800 to 1000 m above the valley floor and has a well-deserved reputation for being unstable. Wieczorek and Snyder (2003) note that 23 rockfalls have been recorded at Middle Brother, so this would seem to be a good site for testing the hypothesis that distant earthquakes cause landslides in the Sierra Nevada. Decomposition of the probability density plot describes two large rockfall events (Fig. 6.46) estimated to have occurred in 1860 A.D. ± 10 years, and 1812 A.D. ± 10 years. These rockfall events may have been generated by strong ground motions emanating from the distant San Andreas fault earthquakes (Ellsworth, 1990) of 1857 (330 km away) and 1812 (420 km away). An 1857 cliff collapse on the opposite side of the valley is part of the Wieczorek and Snyder (2003) inventory of observed landslides. Two minor Figure 6.46 subpopulations have lichenometry ages of 1914 A.D. ± 10 years and 1833 A.D. ± 10 years,

and may record 1906 and 1838 A.D. San Francisco Bay region earthquakes on the San Andreas fault.

The Middle Brother data suggest that some landslides are coseismic, but we need to find out if seismic shaking really has a truly pervasive influence on the Sierra Nevada rockfall process. The modeling done in Figure 6.47 is similar to that of Figure 6.46 and the large dataset is the combined measurements from 10 lichenometry sites in the central and southern Sierra Nevada. Indeed the lichen-size peaks have times that clearly match the times of historical local or distant earthquakes. The second largest peak that lichenometry dates to about 1837 A.D. was an enigma when this figure was first published in 1996. All I could say about it was "the 1837 A.D. ± 10 years lichen-size peak records a regional rockfall event of unknown cause". Then Toppozada and Borchardt (1998) described a previously unregistered San Andreas fault earthquake that occurred near Hollister and San Francisco in 1838. The epicenter of this earthquake is directly opposite the Sierra Nevada study area (Fig. 6.35), which contributed to the large size of the 1837 lichen-size peak. We now know that all of the lichen-size peaks of Figure 6.47 record regional seismic shaking events.

Regional seismic shaking should decrease with increasing distance from an earthquake epicenter and so should the number of coseismic rockfall blocks. Making maps that show regional variations in seismic shaking index can test this hypothesis. This index is simply the percentage of lichen-size measurements contained within the lichen-size peak relative to the total measurements in a 6 mm wide band of lichen sizes – 3 mm to each side of the peak that we are interested in. The results of two analyses are shown in Figures 6.48A, B. One map is for the historic San Andreas fault earthquake of 1812 A.D. in Southern California and the other map is for a prehistorical earthquake that I presume occurred in 1739 A.D. on the Honey Lake fault zone in northeastern California.

Rockfall abundance for the 1812 event decreases markedly from south to north. This overall pattern is just what one would expect from a large southern California earthquake.

Local details of the 1812-event map are intriguing. The southern part of the area varies from 10 to 20% response to seismic shaking to >50%. I attribute this to the different orientations of the rockfall block source areas at cliffy lichenometry sites in the Kern River gorge. North-facing source areas may

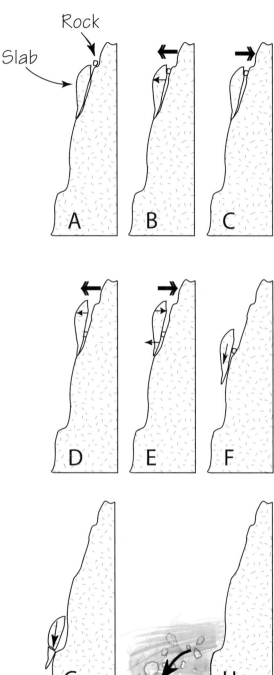

Figure 6.45 "Seismic-ratchet" process of generating landslides in an 800 m high glaciated granodiorite cliff with exfoliation joints. Large horizontal arrows show directions of oscillating seismic forces during an earthquake. Small black arrows show directions of movement for a potential landslide block. Concept courtesy of John Tinsley, U.S. Geological Survey.

A. Cliff-face parallel fractures open gradually over a long time span.

B. Seismic force from the right rotates top of block around a basal pivot point and allows rocks to fall into crack widened by seismic shaking.

C. Oscillation of seismic wave does not close the crack. It is now wedged open by the rock(s).

D. Renewed seismic shaking, perhaps during a subsequent earthquake, further widens the crack and allows the rock(s) to drop further into the wider crack. The rock(s) is now below the center of gravity of the potential landslide block.

E. Reversal of seismic wave rotates the landslide block, reducing its basal support.

F. The block slides down the cliff face, with underlying loose rocks acting as ball bearings, moving away from the cliff face as it strikes projecting outcrops. Block(s) may become ballistic where they slide over steeper parts of the cliff.

G. The accelerating rock mass(es) falls onto a projecting lower part of the cliff, crushing the brittle block into fragments that range in size from huge rockfall blocks to sand grains the size of the minerals composing the granodiorite. Seismic-impact waves propagating back up the slope may trigger additional rockfalls.

H. Landslide movement changes to mainly horizontal when it reaches the valley floor, where it buries trees. Lichens will begin to colonize the fresh rock surfaces after a few years.

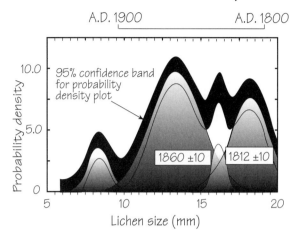

Figure 6.46   Modeling of lichen-size peaks on rockfall blocks at Middle Brother site reveals two large subpopulations close to the times of the 1857 and 1812 earthquakes on southern San Andreas fault. From Figure 3 Bull, (1996a).

well be more sensitive than outcrops facing east or west (see Figures 6.37 and 6.49). Seismic energy from the south would tend to move partially detached blocks away from north-facing cliffs (Fig. 6.45).

The western edge of the seismic shaking index map reveals a slightly higher sensitivity – 30 to 40% as compared to 20 to 30% in the adjacent area to the east. Landform sensitivity to seismic shaking may vary somewhat in the study area. The eastern subarea lichenometry sites include glacial moraines, fractured cliffy mountainsides, and steep debris slopes whose blocks could be set in motion again with seismic shaking. Sites in the western subarea are in glaciated valleys. The massive cliffs give the impression of being very strong, but they have pervasive exfoliation joints that parallel cliff faces. The Figure 6.48A map suggests that such joints are responsible for increased sensitivity to seismic shaking when compared to sites along the crest and east side of the mountain range. Middle Brother in Yosemite Valley (Fig. 6.46) is an example of a very unstable cliff face that has both exfoliation joints parallel to the cliff face and numerous fractures oriented in other directions.

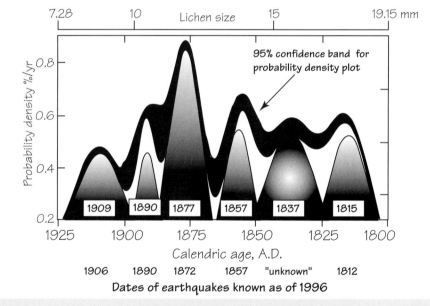

Figure 6.47 Modeled times of rockfalls for combined dataset of Rhizocarpon subgenus Rhizocarpon from 10 Sierra Nevada sites are clustered, which requires regional causes. 1872 and 1890 are local earthquakes and 1812, 1857, and 1906 are San Andreas fault earthquakes. Five dates have an accuracy of 2.2 ± 3.5 yr. The A.D. 1837 ± 10 yr lichen-size peak recorded a regional rockfall event of "unknown cause" in the opinion of Bull (1996), but it turned out to be caused by the San Andreas earthquake of 1838 that was discovered later by Toppozada and Borchardt (1998). From Figure 4, Bull (1996a).

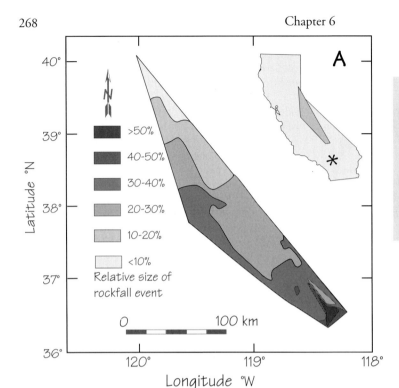

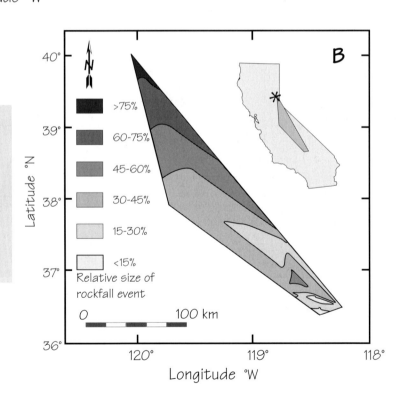

Figure 6.48 Variation in seismic shaking index for two regional rockfall events. The * symbol in the inset map approximates the earthquake epicenter location.

A. The 1812 A.D. earthquake on the Mojave segment of the San Andreas fault in southern California.

Figure 6.48 Variation in seismic shaking index for two regional rockfall events. The * symbol in the inset map approximates the earthquake epicenter location.

B. The 1739 A.D. earthquake in the northern Sierra Nevada on the Honey Lake fault zone of the Walker Lane–Eastern California shear zone.

Recognition of the large 1739 A.D. earthquake is the result of diverse studies by many people. An early speculation was that "The Slide" rock avalanche was a consequence of Little Ice Age variations in snowpack and ground-water table. As such it would be a very large, but local, landslide event. Then, site-by-site, we gradually realized that lichensize peaks dating to this time (Figs. 6.38A, B) were common throughout the Sierra Nevada.

Then a lucky breakthrough occurred. I pitched my tent near a volcanic neck in the Fort Sage Mountains of the Basin and Range Province just east of the northern Sierra Nevada. Jim Brune and I measured the sizes of *Acarospora chlorophana* that consisted mainly of a dominant lichen-size peak dating to about 1737 A.D. (Bull, 2003a, Figures 8–10). The volcanic neck is only 8 km from the Honey Lake fault zone, a major right-lateral strike-slip fault of the Walker Lane tectonic belt. It has ruptured stream deposits at least four times during the Holocene and has an estimated slip rate of 2 m/1,000 years (Wills and Borchardt, 1993). I knew little about the Honey Lake fault zone. Was it really the source of large earthquakes during the late Holocene?

The mainstay of paleoseismology – radiocarbon-dated faulted stratigraphy – provides essential information to answer this question. Pleistocene Lake Lahontan receded from this inter-montane basin about 12,000 years ago. The uppermost lake-bed deposits are used by paleoseismologists as a widespread stratigraphic marker of known recent age. Then a stream alternated between depositing sandy layers on top of the lake clays, and cutting deep channels like the one that now exposes the fluvial strata, soils, and faults shown in Figure 6.50. These intervals of nondeposition allowed weathering processes to create incipient soil profiles; each defines a former land surface. General dating control is good. A volcanic ash spewed by Crater Lake volcano in Oregon (the former Mount Mazama stratovolcano) occurs as a layer below the streambank exposure of Long Valley Creek fluvial deposits, so all strata and faults shown in Figure 6.50 are younger than 6,800 years. A single radiocarbon sample higher in the stratigraphic section dates to about 5,700 years ago.

The nice work by Wills and Borchardt (1993) shows that faulting repeatedly broke through to surface, 1 through 4, oldest to youngest (Fig. 6.50).

Figure 6.49 Digital image of the south-central Sierra Nevada. RR is the mouth of the Roaring River. The different orientations of the cliffs flanking the gorges of the Kern and Kings Rivers may influence rockfall responses to seismic shaking being transmitted north–south as compared to east–west.

Seismic shaking by earthquake 1 liquefied saturated sand and jetted it through a fissure to the surface where a fountain spread wet sand out as a low circular mound. This formerly level surface has been tilted by surface-rupture events 2, 3, and 4, which ruptured mainly in a right-lateral sense. Deposition of the surficial sediments occurred just before modern stream-channel entrenchment and so recently that a soil profile has yet to form. Wills and Borchardt believe that event 4 occurred "within the past few hundred years". This strikes me as a good candidate for my postulated 1739 A.D. surface-rupture event. There may have been more than four events since the mid-Holocene on the Honey Lake fault zone if earthquakes occurred during times of nondeposition. A surface rupture tomorrow would rupture through to the same land surface as event 4, and both would appear in the future stratigraphic record as being the result of a single earthquake event.

Was the epicenter of the 1739 A.D. regional seismic shaking event really this far north? A seismic shaking index map is one way to test this hypothesis. The map shown in Figure 6.48B is the first map depicting areal variations in seismic shaking for a prehistorical earthquake in California. The intensity of seismic shaking decreases progressively towards the south, and makes a nice contrast with the pattern for the 1812 earthquake, which decreases progressively towards the north (Fig. 6.48A). All this diverse evidence supports the opinion that an earthquake on the Honey Lake fault zone in 1739 A.D. caused "The Slide" in Yosemite National Park.

6.2.2.5 Lichenometry and Precise Radiocarbon Dating Methods

We conclude by testing the lichenometry method of dating earthquakes. The traditional approach in paleoseismology is to dig a trench across a scarp created by an active fault, describe the ruptured layers, and constrain the times of surface ruptures by radiocarbon dating of organic matter that grew either before or after each surface-rupture event. Determining how much "before or after" requires careful work and considerable intuitive skill. The importance of this approach to paleoseismology was initiated by the work of Kerry Sieh at Pallett Creek on the Mojave segment of the San Andreas fault in southern California (Sieh et al., 1989). His studies have become the hallmark for paleoseismology investigations, because of his skills in the field and the precision (lengthy and low counting background) of his radiocarbon age estimates. The site is unusual in that many surface-rupture events are recorded during the past 1,500 years and only one earthquake on this part of the San Andreas fault may have occurred during a time of nondeposition of marshy sediment.

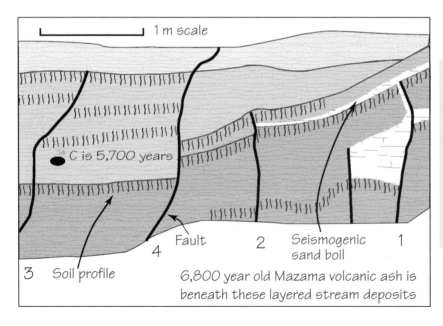

Figure 6.50 Cross section of a streambank of Long Valley Creek showing displacement of former land surfaces (soil profiles) by four recent earthquakes on the Honey Lake fault zone in northeastern California. From Figure 4 and text of Wills and Borchardt (1993).

Pallett Creek earthquake dates provide an opportunity to assess lichenometry's potential for future studies of prehistorical San Andreas fault earthquakes. Historical San Andreas fault earthquakes emanating from both central and southern California caused Sierra Nevada rockfalls that can be dated with lichenometry. If appropriate, future lichenometry investigations of San Andreas fault earthquakes would include the Transverse Ranges as well as the Sierra Nevada.

Can we really record and date prehistorical San Andreas fault earthquakes throughout the central and southern Sierra Nevada? Are the lichenometry age estimates sufficiently accurate for us to trust them? Which lichen genera should we use? Conversely, a good way to appraise the quality of traditional stratigraphic paleoseismology work done at Pallett Creek is to check it against the much different geomorphic approach to paleoseismology. Table 6.5 shows a splendid match between the times of San Andreas

Calendric 14C ages for earthquakes	Kern River Acarospora chlorophana	Kern River Rhizocarpon subgenus Rhizocarpon	Kings River Acarospora chlorophana	Rock Creek Lecanora sierrae	Rock Creek Acarospora chlorophana	Yosemite Rhizocarpon subgenus Rhizocarpon	Tioga Pass Lecidea atrobrunnea	Mean lichenometry calendric age
[1906.30]		1905 ± 3		1907 ± 3		1906 ± 3	1901 ± 3	1904.8 ± 1.5
[1857.02]	1864 ± 3	1860 ± 3		1859 ± 2	1858 ± 3	1860 ± 3	1849 ± 2	1858.3 ± 1.1
[1838.47]		1839 ± 3	1837 ± 3	1841 ± 3	1836 ± 3	1838 ± 3	1841 ± 3	1838.7 ± 1.2
[1812.95]	1816 ± 5		1811 ± 4	1816 ± 3	1811 ± 5	1812 ± 3	1810 ± 3	1812.6 ± 1.7
1688 ±13 (1686 ±8)	1690 ± 6	1686 ± 6	1697 ± 8	1699 ± 4	1692 ± 6	1697 ± 6	1689 ± 7	1693 ± 2
1480 ±15 (1470±20)	1488 ± 9	1477 ± 9	1482 ± 10	1485 ± 7	1490 ± 9		1489 ± 10	1485 ± 4
1346 ±17	1344 ± 10	1340 ± 11	1335 ± 12	1351 ± 8	1341 ± 10			1342 ± 5
1100 ±65	1094 ± 12		1086 ± 12		1091 ± 12		1102 ± 20	1093 ± 8
1048 ±33	1054 ± 12	1041 ± 16	1037 ± 14		1052 ± 12			1046 ± 8
997 ±16	993 ± 13		997 ± 14				1000 ± 21	997 ± 12
797 ±22	797 ± 16		781 ± 18					789 ± 14
734 ±13	745 ± 17		741 ± 20		745 ± 17		736 ± 26	742 ± 11
671 ±13			654 ± 22					654 ± 22
Distance 0 km	230 km	250 km	290 km	340 km	340 km	400 km	410 km	

Table 6.5 Comparisons of lichenometry ages for Sierra Nevada regional rockfall events with dates of historical [ ] earthquakes, and with precise radiocarbon ages for surface-rupture events at Pallett Creek (Sieh et al., 1989) and Wrightwood ( ) (Fumal et al., 1993) sites on the Mojave segment of the San Andreas fault. Lichenometry sites are 230 to 410 km from the Mojave segment. All uncertainties are 2σ (95%). Lichenometry uncertainties include errors for lichen-size measurement, decomposition of probability density plots, smoothing function, and spread of regression 95% lines based on slope of regression. Average uncertainty for mean lichenometry age estimate is directly proportional to $n^{0.5}$, where n is the number of lichenometry age estimates.

fault earthquakes and lichenometry age estimates for rockfall events at seven sites. Cross-checks include using four genera of lichens.

The four historical earthquakes of Table 6.5 can be used to ascertain the accuracy of lichenometric age estimates of Sierra Nevada coseismic rockfall events. Mean lichenometry age estimates are within 1.1 to 1.7 years of the known ages, an accuracy that should please paleoseismologists. Every Table 6.5 Pallett Creek event is matched by a lichenometry-determined time of regional seismic shaking in the Sierra Nevada. Lichenometry indeed has the capability of dating times of exposure of rock surfaces > 1,000 years, but only where numerous rockfall blocks have large lichens.–

All four lichen genera can provide accurate age estimates. *Lecanora sierrae* is the most fragile and its thalli margins may not be consistently sharp, so it is unlikely to date events older than 700 years. *Rhizocarpon* subgenus *Rhizocarpon* is the slowest growing lichen and can date old events. Lichenometrists

use it more than any other lichen on a world-wide basis. It is unfortunate that the Sierra Nevada is a bit too dry, hot, or windy at the high altitudes for it to have the superb thalli found in places like Sweden. The result is that thallus quality is usually not particularly good where it is the dominant lichen: high altitude Little Ice Age glacial moraines and partly shaded low-altitude, wetter microclimate hillslope sites. The combination of a very slow growth rate and mainly quality 3 and 4 thalli may reduce the precision of age estimates made with the yellow Rhizocarpons. *Lecidea atrobrunnea* not only has great thallus quality but also is the fastest growing lichen; both characteristics favor precise dating of events older than 1,200 years. Although it occurs at mid- to high-altitude sites it generally is present in modest abundance, and one has to be careful not to measure several look-alike lichens, such as the brown Rhizocarpons. *Acarospora chlorophana* appears to have only one visual species in the field and its bright chartreuse color makes it stand out during a reconnaissance. It has consistently good

Radiocarbon ages + Kerry Sieh's field intuition of Pallett Creek ages – 1	Glenn Biasi Bayesian modeling of Pallett Creek ages – 2	Biasi's ages minus Sieh's ages in years	Bull's Sierra Nevada lichenometry rockfall calendric ages – 3	Bull's ages minus Sieh's ages in years
1688 A.D. ±13*	1693	+5	1693	+5
1480 A.D. ±15	1546	+66	1485	+5
1346 A.D. ±17	1360	+14	1342	–4
1100 A.D. ±65	1087	–13	1093	–7
1048 A.D. ±33	1062	+14	1046	–2
997 A.D. ±16	954	–43	997	0
797 A.D. ±22	840	+43	797	0
734 A.D. ±13	764	+30	742	+8

* One of several possible radiocarbon ages. Kerry Sieh did not have convincing stratigraphic information to indicate that this was a Pallett Creek earthquake date. It is present at the Wrightwood site.

Table 6.6 Comparison of three ways to estimate calendric ages of surface-rupture events of prehistorical earthquakes at the Pallett Creek paleoseismology site on the Mojave segment of the San Andreas fault, southern California.
1) Precise calibrated radiocarbon ages applied to stratigraphic interpretations (Sieh et al., 1989).
2) Mathematical modeling of calibrated ^{14}C age ranges for layers that are stratigraphically above or below the position of the land surface at the time of a prehistorical earthquake (Biasi and Weldon, 1994; Biasi et al., 2002).
3) Sierra Nevada regional rockfall events dated by lichenometry (Table 6.5).

thallus margins and grows slowly, but does not occur at wet or low-altitude sites. Superb huge thalli occur on enormous rockfall blocks in the Kings Canyon so it may be the oldest lichen in the Sierra Nevada.

In hindsight, which lichen would I now use to start a project? I still see an advantage of choosing between the several crustose lichens whose growth rates have been calibrated because a genus that may be perfect at one site might not even grow at the next site. Top ranking for paleoseismology studies goes to *Acarospora chlorophana* and *Lecidea atrobrunnea* (Fig. 6.38A,B for example). *Rhizocarpon* subgenus *Rhizocarpon* is reliable for surface exposure dating over long time spans. I would use *Lecanora sierrae* only where the other lichens are largely absent.

Mathematical Bayesian modeling of calibrated radiocarbon age estimates of San Andreas fault earthquakes (Biasi and Weldon, 1994; Biasi and Weldon, 1998; Biasi et al., 2002) has been assumed to be an improvement on Kerry Sieh's stratigraphic interpretations of precise calibrated radiocarbon ages. Comparisons of the age estimates for Mojave segment San Andreas fault earthquakes by Sieh, Biasi, and Bull (Table 6.6) tests the idea that Bayesian modeling improves the accuracy of radiocarbon dating. The lichenometry and Kerry Sieh age estimates depart from one another by an average of only 3.9 years, which implies nice precision and possibly reasonable accuracy by both dating methods. Sieh's dates typically coincide with times of regional rockfall events in the Sierra Nevada caused by seismic shaking emanating from the south. Figure 6.48A is an example for the 1812 A.D. earthquake. The average departure for the Bayesian modeling dates from Kerry Sieh's set of age estimates is much larger, being 28.5 years. Many of Biasi's age estimates occur between times of regional seismic shaking, or happen to coincide with regional rockfall events whose seismic shaking did not emanate from the south. I infer that such Bayesian modeling of radiocarbon age estimates should not be used unless it can be shown that both Bill Bull's geomorphic approach to paleoseismology and Kerry Sieh's stratigraphic analyses are invalid.

6.3 Summary

The stratigraphic, geomorphic, and dendrochronologic approaches used in this chapter to paleoseismology are robust. Lichenometry applications to paleoseismology are not restricted to working at sites along the active fault traces, but coseismic mass move-

ment rock surfaces with usable lichens are required. Tree-ring analyses are the best way to cross-check lichenometry age estimates. Lichenometry has now reached the stage where a day's work may provide precise and accurate dating of the times of 20 coseismic rockfalls and landslides at a single site. Calibrated lichen growth rates and large regional datasets in New Zealand (Bull and Brandon, 1998), California (Bull, 1996a), and Sweden (Bull et al., 1995) allow paleoseismologists to visit new sites in earthquake-prone mountain ranges and efficiently assess surface-exposure ages.

Spatial variations in coseismic rockfall abundance can be used to describe patterns of prehistorical seismic shaking, and locate the most likely fault zone for a particular earthquake, and assess the likely direction of surface-rupture propagation. Lichenometry dating of earthquakes in any seismically active region can be tested against the local historical earthquake record. Application of the Bull–Brandon lichenometry method to other earthquake-prone areas complements paleoseismic inferences from stratigraphic investigations. Several diverse project areas should be considered.

A project in Southern California, USA would start by cross-checking the calibrations for four genera of Sierra Nevada lichens. These lichens grow in the San Bernardino and San Gabriel Mountains. Lichenometry sites would straddle the plate-bounding San Andreas fault. These data, combined with the Sierra Nevada dataset, would be used to make maps depicting seismic shaking for both historical and prehistorical earthquakes. This would answer the important question as to whether or not there is a characteristic style and magnitude of rupture for the Mojave segment of the San Andreas fault.

Lichenometry may be the best tool to address a longstanding question in the Pacific Northwest, USA. Was the great subduction thrust earthquake of 1700 A.D. a single surface rupture that extended all away from British Columbia, Canada to Northern California? Or was this event the result of two or more major earthquakes? A successful lichenometry project would show the pattern(s) of seismic shaking, but could only separate the times of major earthquakes if they occurred more than 4 years apart. Collection of basic data would be restricted to the higher mountains and to the drier east side of the Cascade range, because all of the Olympic Peninsula as well as the lower altitudes of the Coast and Cascade ranges have too wet a climate for numerous, good quality, crustose lichens. Cross-checks provided by dating with more than one genus of lichens would be advisable.

The Andes of South America are well known for truly large earthquakes. Largely pristine, coseismic mass movements at a variety of altitudes should be ideal calibration sites for lichen growth of the past 400 years since arrival of colonists. Steep, unstable high mountains should be an advantage when making maps depicting strength of seismic shaking for some of the world's largest recorded earthquakes. Previous tree-ring analyses of earthquakes in Chile would provide useful cross-checks. Dendroseismology should be part of this project.

Lichenometrists working in the earthquake-prone Mediterranean Sea region may prefer to use slow growing lichens on carbonate-rock substrates. Calibration using historical coseismic landslides may encompass more than 1,000 years in Turkey and Greece. Regions of identical lichen-growth rate may be relatively small because of the considerable spatial variation in both temperature and precipitation. Furthermore one would have to be cautious about the imprint of human activities everywhere.

The different approaches and prospects noted briefly in the above sample of future lichenometry investigations also apply to much of Asia, including Russia, Japan, and the Himalayas.

References Cited

Abrahams, A.D., editor, 1994, Steepland geomorphology: Geomorphology, v. 9, p. 169–260.

Adams, J., 1980, Paleoseismicity of the Alpine fault seismic gap, New Zealand: Geology v. 8, p. 72–76.

Adams J., 1981, Earthquake-dammed lakes in New Zealand: Geology, v. 9, p. 215–219.

Adams, K.D., and Wesnousky, S.G., 1999, The Lake Lahontan highstand; Age, surficial characteristics, soil development, and regional shoreline correlation: Geomorphology, v. 30, p. 357–392.

Adams, K.D., Wesnousky, S. G., and Bills, B.G., 1999, Isostatic rebound, active faulting, and potential geomorphic effects in the Lake Lahontan basin, Nevada and California: Geological Society of America Bulletin, v. 111, p. 1739–1756.

Ahnert, F., 1970, Functional relationships between denudation, relief, and uplift in large mid-latitude drainage basins: American Journal of Science, v. 268, p. 243–263.

Aki, K., 1966, Generation and propagation of G waves from the Niigata earthquake of June 16, 1964. Part 2, Estimation of earthquake moment, released energy, and stress-strain drop from the G wave spectrum: University of Tokyo, Earthquake Research Institute Bulletin, v. 44, p. 73–88.

Aki, K., 1969, Analysis of the seismic coda of local earthquakes as scattered waves: Journal of Geophysical Research, v. 74, p. 615–631.

Aki, K., 1984, Asperities, barriers, and characteristic earthquakes: Journal of Geophysical Research, v. 89, p. 5867–5872.

Albertson, P.E., and Patrick, D.M., 1996, Lower Mississippi River tributaries; contributions to the collective science concerning the "Father of Waters": in Geology in the Lower Mississippi Valley; implications for engineering, the half century since Fisk, 1944, Saucier, R.T., Smith, L.M., and Autin, W.J, editors, Engineering Geology, v. 45, p. 383–413.

Allen, C.R., 1968, The tectonic environments of seismically active and inactive areas along the San Andreas Fault system: in Proceedings of the Conference on Geological Problems of the San Andreas System, W.R. Dickinson and A. Granz, editors, Stanford University Publications in Geological Sciences, v. 11, p. 70–82.

Allen, C.R., 1975, Geological criteria for evaluating seismicity: Geological Society of America Bulletin, v. 86, p. 1041–1057.

Allen, C.R., Engen, G.R., Hanks, T.C., Nordquist, J.M., and Thatcher, W.R., 1971, Main shock and larger aftershocks of the San Fernando earthquake, February 9 through March 1, 1971: in The San Fernando, California, earthquake of February 9, 1971, U.S. Geological Survey Professional Paper 733, p. 17–20.

Allen, J.R.L., 1974, Reaction, relaxation and lag in natural systems; general principles, examples and lessons: Earth Science Reviews, v. 10, p. 263–342.

Allen, P.A., and Densmore, A. L., 2000, Sediment flux from an uplifting fault block: Basin Research, v. 12, p. 367–380.

Alley, R.B., 2000, The Younger Dryas cold interval as viewed from central Greenland: Quaternary Science Reviews, v. 19, p. 213–226.

Amit, R., Harrison, J.B.J., and Enzel, Y., 1995, Use of soils and colluvial deposits in analyzing tectonic events, The southern Arava rift, Israel: Geomorphology, v. 12, p. 91–107.

Amit, R., Harrison, J.B.J., Enzel, Y., and Porat, N., 1996, Soils as a tool for estimating ages of Quaternary fault scarps in a hyperarid environment, The southern Arava Valley, the Dead Sea rift, Israel: Catena, v. 28, p. 21–45.

Amit, R., Zilberman, E., Enzel, Y., and Porat, N., 2002, Paleoseismic evidence for time dependency of seismic response on a fault system in the southern Arava Valley, Dead Sea rift, Israel: Geological Society of America Bulletin, v. 114, p. 192–206.

Amoroso, L., Pearthree, P.A., and Arrowsmith, R., 2004, Paleoseismology and Neotectonics of the Shivwits Section of the Hurricane Fault, Northwestern Arizona: Bulletin of the Seismological Society of America, v. 94, p. 1919–1942.

Anderson, H., and Webb, T., 1994, New Zealand seismicity; patterns revealed by the upgraded National Seismograph Network: New Zealand Journal of Geology and Geophysics, v. 37, p. 477–493.

Anderson, M.G., and Brooks, S.M, editors, 1996, Advances in hillslope processes: British Geomorphological Research Group Symposia Series. 12; John Wiley and Sons, Chichester, United Kingdom, 1306 p.

Anderson, R.S., and Densmore, A.L., 1997, Tectonic geomorphology of the Ash Hill fault, Panamint Valley, California: Basin Research, v. 9, p. 53–63.

Andrews D.J., and Bucknam, R.C., 1987, Fitting degradation of shoreline scarps by a nonlinear diffusion model: Journal of Geophysical Research, v. 92, p. 12,857–12,867.

Andrews, D.J., and Hanks, T.C., 1985, Scarp degraded by linear diffusion: inverse solution for age: Journal of Geophysical Research, v. 90, p. 10,193–10,208.

Armstrong, R.L., 1972, Low-Angle (Denudation) Faults, Hinterland of the Sevier Orogenic Belt, Eastern Nevada and Western Utah: Geological Society of America Bulletin, v. 83, p. 1729–1754.

Arrowsmith, J.R., and Rhodes D.D., 1994, Original forms and initial modifications of the Galway Lake Road scarp formed along the Emerson fault during the 28 June 1992 Landers, California, earthquake: Bulletin of the Seismological Society of America, v. 84, p. 511–527.

Arrowsmith, J.R., Rhodes, D.D., and Pollard, D.D., 1998, Morphologic dating of scarps formed by repeated slip events along the San Andreas Fault, Carrizo Plain, California: Journal of Geophysical Research, v. 103, p. 10,141–10,160.

Atwater, B.F., and Yamaguchi, D.K., 1991, Sudden, probably coseismic submergence of Holocene trees and grass in coastal Washington state: Geology, v. 19, p. 706–09.

Atwater, B.F., and 15 others, 1995, Summary of coastal geologic evidence for past great earthquakes at the Cascadia Subduction Zone: Earthquake Spectra, v.11, p. 1–18.

Atwater, B.F., Stuvier, M., and Yamaguchi, D.K., 1991, Radiocarbon test of earthquake magnitude at the Cascadia Subduction Zone: Nature, v. 353, p. 156–158.

Atwater, B.F., Trumm, D.A., Tinsley, J.C.III, and Stein, R.S., 1990, Alluvial plains and earthquake recurrence at the Coalinga anticline: in Rymer, M.J., Ellsworth, W.L., The Coalinga, California earthquake of May 2, 1983, U.S. Geological Survey Professional Paper 1487, p. 273–299.

Atwater, T., 1970, Implications of plate tectonics for the Cenozoic tectonic evolution of western North America: Geological Society of America Bulletin, v. 81, p. 3,513–3,536.

Atwater, T., and Stock, J., 1998, Pacific–North America plate tectonics of the Neogene southwestern United States, An update: International Geology Review, v. 40, p. 375–402.

Augustinus, P.C., 1992, The influence of rock mass strength on glacial valley cross-profile morphometry, A case study from the Southern Alps, New Zealand: Earth Surface Processes and Landforms, v. 17, p. 39–51

Augustinus, P.C., 1995, Glacial valley cross-profile development: the influence of in situ rock stress and rock mass strength, with examples from Southern Alps, New Zealand: Geomorphology, v. 11, p. 87–97.

Avouac, J.P., 1993, Analysis of scarp profiles; evaluation of errors in morphologic dating: Journal of Geophysical Research, v. 98, p. 6745–6754.

Bachman, G.O., and Machette, M.N., 1977, Calcic soils and calcretes in the southwestern United States: U.S. Geological Survey Open-File Report 77-794, 162 p.

Bagnold, R.A., 1977, Bed-load transport by natural rivers: Water Resources Research, v. 13, p. 303–312.

Bahat, D., Grossenbacher, K., and Karasaki, K., 1999, Mechanism of exfoliation joint formation in granitic rocks at Yosemite National Park: Journal of Structural Geology, v. 21, p. 85–96.

Baillie, M.G.L., 1995, A slice through time, dendrochronology, and precision dating, Batsford Ltd., London, 176 p.

Baker, V.R., and Costa, J.E., 1987, Flood power: in Mayer, Larry and Nash, David editors, Catastrophic Flooding, London, Allen and Unwin: Annals of the Association of American Geographers, v. 78, p. 746–748.

Baker, V.R., and Kale, V.S., 1998, The role of extreme floods in shaping bedrock channels: in Tinkler, K., and Wohl, E., editors, Rivers over rock: Fluvial processes in bedrock channels: Geophysical Monograph Series, v. 107, American Geophysical Union, Washington, D.C., p. 153–165.

Baker, V.R., and Kochel, R.C., 1988, Flood sedimentation in bedrock fluvial systems, in Baker, V.R., Kochel, R.C., and Patton, P.C., editors: Flood geomorphology, John Wiley and Sons, New York, p. 123–137.

Ballantyne, C.K., 2002a, A general model of paraglacial landscape response: The Holocene, v. 12, p. 371–376.

Ballantyne, C.K., 2002b, Paraglacial geomorphology: Quaternary Science Reviews, v. 21, p. 1935–2017.

Bard, E., Hamelin, B. Fairbanks, R.G., and Zindler, A. 1990, Calibration of the ^{14}C timescale over the past 30,000 years using mass spectrometric U-Th ages from Barbados corals: Nature, v. 345, p. 3513–3536.

Barnard, P., L.A. Owen, and Finkel, R. C., 2006, Quaternary fans and terraces in the Khumbu Himal south of Mount Everest: their characteristics, age and formation: Journal of the Geological Society of London, v. 163, p. 383–399.

Barnes, P.M., and Audru, J.C., 1999, Recognition of active strike-slip faulting from high-resolution marine seismic reflection profiles, Eastern Marlborough fault system, New Zealand: Geological Society of America Bulletin, v. 111, p. 538–559.

Barnes, P.M., Sutherland R., and Delteil J., 2005, Strike-slip structure and sedimentary basins of the southern Alpine Fault, Fiordland, New Zealand: Geological Society of America Bulletin, v. 117, p. 411–435.

Barrell, D.J.A., Alloway, B.V., Shulmeister, J., and Newnham, R.M., editors 2005, Towards a climate event stratigraphy for New Zealand over the past 30,000 years: Institute of Geological and Nuclear Sciences, Science Report 2005/07, 12 p.

Barrell, J., 1917, Rhythms and measurement of geologic time: Geological Society of America Bulletin, v. 28, p. 745–904.

Barrows, A.G., Kahle, J.E., Weber, F.H., and Saul, R.B., 1973, Map of Surface Breaks Resulting From the San Fernando, California, Earthquake of February 9, 1971: in: San Fernando, California, earthquake of February 9, 1971; v. 3, National Oceanic and Atmospheric Administration, Washington, D.C., p. 127–135.

Bayarsayhan, C., Bayasgalan, A., Enhtuvshin, B., Hudnut, Kenneth W., Kurushin, R. A., Molnar, P., and Ölziybat, M., 1996, Gobi-Altay, Mongolia, earthquake as a prototype for southern California's most devastating earthquake: Geology v. 24, p. 579–582.

Bayasgalan A., Jackson J., Ritz J-F., and Carretier S., 1999, Forebergs, Flower structures, and the development of large intracontinental strike-slip faults: the Gurvan Bogd fault system in Mongolia: Journal of Structural Geology, v. 21, p. 1285–1302.

Beaty, C.B., 1961, Topographic effects of faulting, Death Valley, California: Association of American Geographers Annals, v. 51, p. 234–240.

Begin, Z.B., 1988, Application of a diffusion-erosion model to alluvial channels which degrade due to base-level lowering: Earth Surface Processes and Landforms, v. 13, p. 487–500.

Begin, Z.B., 1992, Application of quantitative morphologic dating to paleoseismicity of the northwestern Negev, Israel: Israel Journal of Earth Sciences, v. 41, p. 123–137.

Benedict, J.B., 1967, Recent glacial history of an alpine area in the Colorado Front Range, U.S.A., Establishing a lichen growth curve: Journal of Glaciology, v. 6, p. 817–832.

Bennett, R.A., Davis, J.L., and Wernicke, B.P., 1999: Present-day pattern of Cordilleran deformation in the western United States: Geology, v. 27, p. 371–374.

Benson, A.M., Little, T.A., Van Dissen, R.J., Hill, N., and Townsend, D.B. , 2001, Late Quaternary paleoseismic history and surface rupture characteristics of the eastern Awatere strike-slip fault, New Zealand: Geological Society of America Bulletin, v. 113, p. 1079–1091.

Benson L.V., and Thompson, R.S., 1987a, Lake-level variation in the Lahontan basin for the past 50,000 years: Quaternary Research, v. 2, p. 69–85.

Benson, L.V., and Thompson, R.S., 1987b, The physical record of lakes in the Great Basin, in North America and adjacent oceans during the last deglaciation: Ruddiman, W.F., and Wright, H.E., Jr., editors, The Geology of North America, v. K-3, The Geological Society of America, Boulder Colorado, p. 241–260.

Berggren, W. A., Hilgen, F. J., Langereis, C. G., Kent, D. V., Obradovich, J. D., Raffi, I., Raymo, M. E., and Shackleton, N. J., 1995, Late Neogene chronology, New perspectives in high-resolution stratigraphy: Geological Society of America Bulletin, v. 107, p. 1272–1287.

Berger, A., 1988, Milankovitch theory and climate: Reviews of Geophysics, v. 26, p. 624–657.

Berryman, K., Cooper, A.F., Norris, R.J., Sutherland, R, and Villamor, P., 1998, Paleoseismic investigation of the Alpine Fault at Haast and Okuru, Geological Society of New Zealand Miscellaneous Publication 101A, p. 44.

Berryman, K.R., Beanland, S., Cooper, A.F., Cutten, H.N., Norris, R.J., and Wood, P.R., 1992, The Alpine Fault, New Zealand, variation in Quaternary structural style and geomorphic expression: Annals Tectonica, v. 6, p. 126–163.

Biasi, G.P., and Weldon, R.J., II, 1994, Quantitative refinement of calibrated ^{14}C distributions: Quaternary Research, v. 41, p. 1–18.

Biasi, G.P., and Weldon, R.J., 1998, Paleoseismic date refinement and implications for seismic hazard estimation: in Dating and Earthquakes: Review of Quaternary Geochronology and its Application to Paleoseismology, J.M. Sowers, J.S. Noller and W.R. Lettis, editors, 4-39-4-48, Nuclear Regulatory Commission Report NUREG/CR-5562, p. 3-61–3-66..

Biasi, G.P., Weldon, R.J, Fumal, T.E., and Seitz, G.C., 2002, Paleoseismic event dating and the conditional probability of large earthquakes on the southern San Andreas fault, California: Bulletin of the Seismological Society of America, v. 92, p. 2761–2781.

Bielecki, A.E., and Mueller, K.J., 2002, Origin of terraced hillslopes on active folds in the southern San Joaquin Valley, California: Geomorphology, v. 42, p. 131–152.

Bierman, P., and Caffee, M.W., 2002, Cosmogenic exposure and erosion history of Australian bedrock landforms: Geological Society of America Bulletin, v. 114, p. 787–803.

Bierman, P., and Turner, J., 1995, ^{10}Be and ^{26}Al Evidence for exceptionally low rates of Australian bedrock erosion and the likely existence of pre-Pleistocene landscapes: Quaternary Research, v. 44, p. 378–382.

Bierman, P., Zen, E., Pavich, M., and Reusser, L.J., 2004, The incision history of a passive margin river, the Potomac near Great Falls, *in* Southworth, S., and Buron, W., editors, Geology of the National Capital Region, Field Trip Guidebook for NE/SE Geological Society of America Meeting, p. 191–122.

Bird, P., 1991, Lateral extrusion of lower crust from under high topography: *in* The isostatic limit, Journal of Geophysical Research, v. 96, p. 10,275–10,286.

Birkeland, P.W., 1981, Soil data and the shape of the lichen growth-rate curve for the Mount Cook area: New Zealand Journal of Geology and Geophysics, v. 23, p. 443–445.

Birkeland, P.W., 1999, Soils and geomorphology, Third edition: New York, Oxford University Press, 448 p.

Blackwelder, E., 1934, Talus slopes in the Basin and Range province: Proceedings of Geological Society of America for 1934, p. 317.

Bloom, A.L., Broecker, W.S., Chappell, J., Matthews, R.K., and Mesolella, K.J., 1974, Quaternary sea level fluctuations on a tectonic coast: new $^{230}Th/^{234}U$ dates from Huon Peninsula, New Guinea: Quaternary Research, v. 4, p.185–205.

Blum, M.D., and Valastro, S., 1994, Late Quaternary sedimentation, lower Colorado River, Gulf Coastal Plain of Texas: Geological Society of America Bulletin, v. 106, p. 1002–1016.

Bradshaw, G.A., and Zoback, M.D., 1988, Listric normal faulting, stress refraction, and the state of stress in the Gulf Coastal Basin: Geology v. 16, p. 271–274.

Bronson, B.R., and Watters, R.J., 1987, The effects of long term slope deformations on the stability of granitic rocks of the Sierra Nevada, California: Proceedings of the 23rd Engineering Geology and Soils Engineering Symposium, Logan, Utah State University, p. 203–217.

Brook, M.S., Kirkbride, M.P., and Brock, B.W., 2004, Rock strength and development of glacial valley morphology in the Scottish Highlands and Northwest Iceland: Geografiska Annaler, v. 86, p. 225–237.

Brown, W.M., and Taylor, B.D., 1982, Inland control structures: *in* Sediment management for southern California mountains, coastal plains and shoreline: CalTech Environmental Quality Laboratory Report 17-D, Pasadena, California, 26 p.

Brown R.W., Summerfield M.A., and Gleadow A.J.W., 2002, Denudational history along a transect across the Drakensberg Escarpment of southern Africa derived from apatite fission track thermochronology: Journal of Geophysical Research, v. 107: 2350. doi:10.1029/2001JB000745.

Brune, J.N., 1968, Seismic moment, seismicity, and rate of slip along major fault zones: Journal of Geophysical Research, v. 73, p. 777–784.

Brunsden, D., 1980, Applicable models of long-term landform evolution: Zeitschrift für Geomorphologie, v. 36, p. 16–26.

Brunsden, D., and Thornes, J.B., 1979, Landscape sensitivity and change: Transactions of the Institute of British Geographers, v. 4, p. 463–484.

Bucknam, R.C., and Anderson, R.E., 1979, Estimation of fault scarp ages from a scarp-height-slope-angle relationship: Geology, v. 7, p. 11–24.

Bucknam, R.C., Hemphill-Haley, E., and Leopold, E.B., 1992, Abrupt uplift within the past 1700 years at southern Puget Sound, Washington: Science, v. 258, p. 1611–1614.

Bull, W.B., 1962, Relations of alluvial-fan size and slope to drainage basin size and lithology in western Fresno County, California: U.S. Geological Survey Professional Paper 450B, p. 51–53.

Bull, W.B., 1964a, Geomorphology of segmented alluvial fans in western Fresno County, California: U.S. Geological Survey Professional Paper 352E, p. 89–129.

Bull, W.B., 1964b, History and causes of channel trenching in western Fresno County, California: American Journal of Science, v. 262, p. 249–258.

Bull, W.B., 1975a, Allometric change of landforms: Geological Society of America Bulletin, v. 86, p. 1489–1498.

Bull, W.B., 1975b, Land subsidence due to ground-water withdrawal in the Los Banos-Kettleman City area –Part 2– Subsidence and compaction of deposits: U.S. Geological Survey Professional Paper 437F, 90 p.

Bull, W.B., 1977a, Landforms that do not tend toward a steady state: *in* Theories of landform development, W.N. Melhorn and R. Flemal, editors, Proceedings 6th Annual Geomorphology Symposium, State University New York at Binghamton, p. 111–128.

Bull, W.B., 1977b, Tectonic geomorphology of the Mojave Desert, California: U.S. Geological Survey Contract Report 14-08-001-G-394; Office of Earthquakes, Volcanoes, and Engineering, Menlo Park, California, 188 p.

Bull, W.B., 1978, Geomorphic tectonic activity classes of the south front of the San Gabriel Mountains, California: U.S. Geological Survey Contract Report 14-08-001-G-394, Office of Earthquakes, Volcanoes, and Engineering, Menlo Park, California, 100 p.

Bull, W.B., 1984, Tectonic geomorphology: Journal of Geological Education, v. 32, p. 310–324.

Bull, W.B., 1985, Correlation of flights of global marine terraces: *in* Tectonic geomorphology, M. Morisawa and J. Hack, editors, Proceedings 15th Annual Geomorphology Symposium, State University of New York at Binghamton, George Allen and Unwin, Hemelhempstead, England, p. 129-152.

Bull, W.B.,1988, Relative rates of long-term uplift of mountain fronts: *in* Directions in paleoseismology (A. J. Crone and E M.. Omdahl, editors), Proceedings of Conference XXXIX, U. S. Geological Survey National Earthquakes-Hazards Reduction Program, U. S. Geological Survey Open-File Report 87-673, p. 192–202.

Bull, W.B., 1990, Stream-terrace genesis-implications for soil development: Geomorphology, v. 3, p. 351–368

Bull, W.B., 1991, Geomorphic responses to climatic change: New York, Oxford University Press, 326 p.

Bull, W.B., 1996a, Dating San Andreas fault earthquakes with lichenometry: Geology, v. 24, p. 111–114.

Bull, W.B., 1996b, Global climate change and active tectonics: effective tools for teaching and research: Geomorphology (Special Issue) v. 16, p. 217–232.

Bull, W.B., 1996c, Prehistorical earthquakes on the Alpine fault, New Zealand: Journal of Geophysical Research, Solid Earth, Special Section "Paleoseismology", v. 101, p. 6037–6050.

Bull, W.B., 1997, Discontinuous ephemeral streams: Geomorphology, v. 19, p. 227–276.

Bull, W.B., 2000, Correlation of fluvial aggradation events to times of global climate change: *in* Noller, J. S, Sowers, J. M., and Lettis, W. R., editors, Quaternary Geochronology: Methods and applications; American Geophysical Union Reference Shelf Series, v. 4, p. 456–464.

Bull, W.B., 2003a, Guide to Sierra Nevada lichenometry: *in* Tectonics, climate change, and landscape evolution in the southern Sierra Nevada, California: Greg Stock, editor; 2003 Pacific Cell Friends of the Pleistocene field trip, Sequoia and Kings Canyon, Appendix 8, p. 100-121. This article can be downloaded from (http://activetectonics.com/downloads/BullFOPlichenometry.pdf)

Bull, W.B., 2003b, Lichenometry dating of coseismic changes to a New Zealand landslide complex: Annals of Geophysics: v. 46, p. 1155–1167.

Bull, W.B., 2003c, Choices and calibration of lichens for dating geomorphic processes in the Sierra Nevada of California: 7 p. This unpublished article can be downloaded from (http://activetectonics.com/downloads/MeasuringSNlichens.pdf)

Bull, W.B., 2004, Sierra Nevada earthquake history from lichens on rockfall blocks: Sierra Nature Notes, v. 4, 20 p. can be downloaded from (http://www.yosemite.org/naturenotes/LichenIntro.htm)

Bull, W.B., 2005, Recent Alpine fault earthquakes: Geo-

logical Society of New Zealand Annual Meeting Abstracts, p. 11.

Bull, W.B., and Brandon, M. T., 1998, Lichen dating of earthquake-generated regional rockfall Events, Southern Alps, New Zealand: Geological Society of America Bulletin, v. 110, p. 60–84.

Bull, W.B., and Cooper, A.F., 1986, Uplifted marine terraces along the Alpine fault, New Zealand: Science, v. 234, p. 1225–1228.

Bull, W.B., and Cooper, A.F., 1988, Uplifted marine terraces; uplift rates: Response to Technical Comment by C.M. Ward, Science, v. 236, 803–805.

Bull, W.B., and Knuepfer, P.L.K., 1987, Adjustments by the Charwell River, New Zealand, to uplift and climatic changes: Geomorphology, v. 1, p. 15–32.

Bull, W.B., and McFadden, L.D., 1977, Tectonic geomorphology north and south of the Garlock fault, California: *in* Geomorphology in arid regions, D.O. Doehring, editor, Proceedings 8th Annual Geomorphology Symposium, State University New York at Binghamton, p. 115–137.

Bull, W.B., and Pearthree P.A., 1988, Frequency and size of Quaternary surface ruptures of the Pitaycachi fault, northeastern Sonora, Mexico: Bulletin of the Seismological Society of America, v. 78, p. 956–978.

Bull, W.B., Schlyter, P., and Brogaard, S., 1995, Lichenometric analysis of the Kaerkerieppe slush-avalanche fan, Kaerkevagge, Sweden: Geografiska Annaler, v. 77A, p. 231–240.

Bullard, T.F., and Lettis, W.R., 1993, Quaternary fold deformation associated with blind thrust faulting, Los Angeles basin, California: Journal of Geophysical Research, v. 98, p. 8349–8369.

Burbank, D.W., and Anderson, R.S., 2001, Tectonic geomorphology: Blackwell Science, Oxford, 274 p.

Burbank, D.W., and Beck, R. A., 1991, Rapid, long-term rates of denudation: Geology, v. 19, p. 1169–1172.

Burchfiel, B.C., Molnar, P., Zhang, P., Deng, Q., Zhang, W., and Wang, Y., 1995, Example of a supradetachment basin within a pull-apart tectonic setting: Mormon Point, Death Valley, California: Basin Research, v. 7, p. 199–214.

Burchfiel, B.C., and Stewart, J.H., 1996, "Pull-apart" origins of the central segment of Death Valley, California: Geological Society of America Bulletin, v. 77, p. 439–442.

Cadena, A.M., Rubin, C.M., Rockwell, T.K., Walls, C., Lindvall, S., Madden, C., Khatib, F., and Owen, L., 2004, Late Quaternary activity of the Pinto Mountains fault at the Oasis of Mara: Implications for the Eastern California Shear Zone: Geological Society of America

Abstracts with Programs, v. 36, p. 137.

Camelbeeck, T., Martin, H., Vanneste, K., Verbeeck, K., and Meghraoui, M., 2001, Morphometric analysis of active normal faulting *in* slow-deformation areas: examples in the Lower Rhine Embayment: Geologie En Mijnbouw-Netherlands, v. 80, p. 95–107.

Carretier S., Ritz J.F., Jackson J., and Bayasgalan A., 2002, Morphologic dating of cumulative reverse fault scarp, examples from the Gurvan Bogd fault system, Mongolia: Geophysical Journal International, v. 148, p. 256–277.

Carson, M.A. 1971, The mechanics of erosion: Pion, London, 174 p.

Carson, M.A., and Kirkby, M.J., 1972, Hillslope form and processes: Cambridge University Press, Cambridge, 475 p.

Carter, L., Lewis, K.B., and Davey, F., 1988, Faults in Cook Strait and their bearing on the structure of central New Zealand: New Zealand Journal of Geology and Geophysics, v. 31, p. 431–446.

Castle, R.O., Church, J.P., Elliott, M.R., and Morrison, N.L., 1975, Vertical crustal movements preceding and accompanying the San Fernando earthquake of February 9, 1971; a summary: Tectonophysics, v. 29; Recent crustal movements, p. 127–140.

Cerling, T.E., 1990. Dating geomorphic surfaces using cosmogenic He: Quaternary Research, v. 33, p. 148–156.

Cerling, T.E., Webb, R.H., Poreda, R.J., Rigby, A.D., and Melis, T.S., 1999. Cosmogenic He ages and frequency of late Holocene debris flows from Prospect Canyon, Grand Canyon, USA: Geomorphology, v. 27, p. 93–111.

Chadwick, O.A., Hecker, S., and Fonseca, J., 1984, A soils chronosequence at Terrace Creek: U.S. Geological Survey Open-File Report 84-90, 29 p.

Chappell, J., 1983, A revised sea-level curve for the last 300,000 years from Papua New Guinea: Search, v. 14, p. 99–101.

Chappell, J., 2001. Sea level change through the last glacial cycle: Science, v. 292, p. 679–686.

Chappell J., Ota Y., and Berryman, K., 1996, Late Quaternary coseismic uplift history of Huon Peninsula, Papua New Guinea: Quaternary Science Reviews, v. 15, p. 7–22.

Chappell, J., and Shackleton, N.J., 1986, Oxygen isotopes and sea level: Nature, v. 324, p. 137–140.

Chinn, T.J.H., McSaveney, M.J., and McSaveney, E.R., 1992, The Mount Cook rock avalanche of 14 December, 1991: Institute of Geological and Nuclear Sciences Ltd., Lower Hutt, New Zealand, Information

Series 20.

Chorley, R.J., Schumm, S.A., and Sugden, D.E., 1984, Geomorphology: Methuen, London, 605 p.

Christensen, M.N., 1966, Late crustal movements in the Sierra Nevada of California: Geological Society of America Bulletin, v. 77, p. 163–182.

Cichanski, M., 2000, Low-angle, range-flank faults in the Panamint, Inyo, and Slate ranges, California; Implications for recent tectonics of the Death Valley region: Geological Society of America Bulletin, v. 112, p. 871–883.

Cifuentes, I.L., 1989, The 1960 Chilean earthquakes: Journal of Geophysical Research, v. 94, p. 665–680.

Clark, J.A., 1998, Morphologic dating of fault scarps along the Socorro Canyon Fault, Socorro, NM: *in* J.B.J. Harrison, editor, Soil, water, and earthquakes around Socorro, New Mexico, Field Trip Guidebook for Rocky Mountain Cell, Friends of the Pleistocene, September 10–13, Socorro, New Mexico.

Colman, S.M., 1987, Limits and constraints of the diffusion equation in modeling geological processes of scarp degradation: *in* Crone, A.J., Omdahl, E.M, editors, Proceedings of Conference XXXIX, Directions in Paleoseismology: U.S. Geological Survey Open-File Report 87-0673, p. 311–316.

Colman, S.M., and Watson, K., 1983, Age estimated from a diffusion equation model for scarp degradation: Science, v. 221, p. 263–265.

Coney, P.J., and Harms, T.A., 1984, Cordilleran metamorphic core complexes, Cenozoic extensional relics of Mesozoic compression: Geology, v. 12, p. 550–554.

Cook, E. R., and Kairiukstis, L.A., editors, 1990, Methods of dendrochronology: Boston, Massachusetts, Kluwer Academic Publishers, 394 p.

Cooke, A.V., and Warren, A., 1973, Geomorphology *in* Deserts: University of California Press, Berkeley, 394 p.

Cooke R.U., and Doornkamp, J.C. 1990. Geomorphology in environmental management, 2nd edition. Oxford: Clarendon Press, 410 p.

Cooke, R.U., Warren, A., and Goudie, A.S, 1993, Desert Geomorphology London: University College London Press, 512 p.

Cooper, A.F., and Bishop, D.G., 1979, Uplift rates and high level marine platforms associated with the Alpine fault at Okuru River, south Westland, *in* Walcott, R.I., and Cresswell, M.M., editors, The origin of the Southern Alps: Royal Society of New Zealand Bulletin 18, p. 35–43.

Cowan, H.A., 1991, The North Canterbury Earthquake of September 1, 1888, Journal of the Royal Society of

New Zealand, v. 21, p. 1–12.

Cowan, H.A., Nicol, A., and Tonkin, P., 1996, A comparison of historical and paleoseismicity in a newly formed fault zone and a mature fault zone, North Canterbury, New Zealand: Journal of Geophysical Research, v. 101, p. 6021–6036.

Crone, A.J., and Haller, K.M., 1989, Segmentation of Basin and Range normal faults: examples from east-central Idaho and southwestern Montana, in Schwartz, D.P., and Sibson, R. H., editors, Proceedings of Workshop XLV, Fault segmentation and controls of rupture initiation and termination: U.S. Geological Survey Open-File Report 89-315, p. 110–130.

Crone, A.J., Machette, M.N., Bonilla, M.G., Lienkaemper, J.J., Pierce, K.L., Scott, W.E., and Bucknam, R.C., 1987, Surface faulting accompanying the Borah Peak earthquake and segmentation of the Lost River fault, central Idaho: Seismological Society of America Bulletin, v. 77, p. 739–770.

Crook, R., Jr., Allen, C.R., Kamb, B., Payne, C.M., and Proctor, R.J., 1987, Quaternary geology and seismic hazard of the Sierra Madre and associated faults, western San Gabriel Mountains: U.S. Geological Survey Professional Paper 1339, p. 27–63.

Crosby, B.T., and Whipple, K.X., 2006, Knickpoint initiation and distribution within fluvial networks; 236 waterfalls in the Waipaoa River, North Island New Zealand: Geomorphology, v. 82, p. 16–38.

Culling, W.E.H., 1960, Analytical theory of erosion: Journal of Geology, v. 68, p. 336–344.

Culling, W.E.H., 1963, Soil creep and the development of hillside slopes: Journal of Geology, v. 71, p. 127–161.

Culling, W.E.H., 1965, Theory of erosion on soil-covered slopes: Journal of Geology, v. 73, p. 230–254.

Curry, R.R., 1969, Holocene climatic and glacial history of the central Sierra Nevada, California: in S.A. Schumm and W.C. Bradley, editors: Geological Society of America Special Paper 123, p. 1–47.

Dadson, S.J., and 11 others, 2003, Links between erosion, runoff variability and seismicity in the Taiwan orogen: Nature, v. 426, p. 648–651.

Dadson, S.J., and 10 others, 2004, Earthquake-triggered increase in sediment delivery from an active mountain belt: Geology, v. 32, p. 733–736.

Davies, J.L., and Williams, M.A.J., editors, 1978, Landform evolution in Australasia: Australia National University Press, Canberra, Australia. 376 p.

Davis, G.H., 1984, Structural Geology of Rocks and Regions: New York, John Wiley and Sons, 492 p.

Davis, G. H., and Coney, P. J., 1979, Geologic development of the Cordilleran metamorphic core complexes: Geology, v. 7, p. 120–124.

Davis, T. L., and Namson, J. S., 1994, A balanced cross-section of the 1994 Northridge earthquake, southern California: Nature, v. 372 p. 167–169.

Davis, T.L., Namson, J., and Yerkes, R.F., 1989, A cross section of the Los Angeles area: Seismically active fold and thrust belt, the 1987 Whittier Narrows earthquake, and seismic hazard: Journal of Geophysical Research, v. 94, p. 9644–9664.

Davis, W.M., 1890. The rivers of northern New Jersey, with notes on the classification of rivers in general: National Geographic Magazine, v. 2, p. 81–110.

Davis, W.M., 1902, Base-level, grade, and peneplain: Journal of Geology, v. 10, p. 77–111.

Davis, W.M., 1903, The mountain ranges of the Great Basin: Harvard University Museum of Comparative Zoology Bulletin, v. 42, p. 129–177.

DeMets, C., Gordon, R.G., Argus, D.F., and Stein, S., 1990, Current plate motions: Geophysics Journal International, v. 101, p. 425–478.

DeMets, C., Gordon, R.G., Argus, D.F., and Stein, S., 1994, Effect of recent revisions to the geomagnetic reversal timescale on estimates of current plate motions, Geophysical Research Letters, v. 21, p. 2191–2194.

Demsey, K.A., 1987, Holocene faulting and tectonic geomorphology along the Wassuk Range, west-central Nevada: University of Arizona, Geosciences Department, Tucson, Arizona, M.S. prepublication manuscript, 64 p.

de Polo, C.M., 1998, A reconnaissance technique for estimating the slip rates of normal-slip faults in the Great Basin, and application to faults in Nevada, U.S.A. Ph.D. dissertation, Reno, University of Nevada, 381 p.

de Polo, C.M., and Anderson, J. G., 2000, Estimating the slip rates of normal faults in the Great Basin, USA: Basin Research, v. 12, p. 227–240.

de Polo, C.M., Clark, D.G., Slemmons, D.B., and Ramelli, A. R., 1991, Historical surface faulting in the Basin and Range Province, western North America–implications for fault segmentation: Journal of Structural Geology, v. 13, p. 123–136.

Dickey, D.D., Carr, W.J., and Bull, W.B., 1980, Geologic map of the Parker NW, Parker, and parts of the Whipple Mountains SW and Whipple Wash Quadrangles, California and Arizona: U.S. Geological Survey Miscellaneous Investigations Series Map I-1124.

Dickinson, W.R., 1996, Kinematics of transrotational tectonism in the California Transverse Ranges and its contribution to cumulative slip along the San Andreas transform fault system: Geological Society of America

Special Paper 305, 46 p.

Dickinson, W.R., and Wernicke, B.P., 1997: Reconciliation of San Andreas slip discrepancy by a combination of interior Basin and Range extension and transrotation near the coast: Geology, v. 25, p. 663–665.

Dodge, R.L., and Grose, L.T., 1980, Tectonic and geomorphic evolution of the Black Rock fault, northwestern Nevada, *in* Earthquake hazards along the Wasatch Sierra-Nevada frontal fault zones: U.S. Geological Survey Open-File Report 80-801, P.C. Andrise, editor, p. 494–508.

Dohrenwend, J.C., 1978, Systematic valley asymmetry in the central California Coast Ranges: Geological Society of America Bulletin, v. 89, p. 891–900.

Dokka, R.K., and Ross, T.M., 1995, Collapse of southwestern North America and the evolution of early Miocene detachment faults, metamorphic core complexes, the Sierra Nevada orocline, and the San Andreas fault system: Geology, v. 23, p. 574–578.

Dokka, R.K., and Travis, C.J., 1990a, Late Cenozoic strike slip faulting in the Mojave Desert, California: Tectonics, v. 9, p. 311–340.

Dokka, R.K., and Travis, C.J., 1990b, Role of the eastern California shear zone in accommodating Pacific-North American plate motion: Geophysical Research Letters, v. 17, p. 1323–1326.

Dolan, J. F., Sieh K., Rockwell, T. K., Yeats, R. S., Shaw, J., Suppe, J., Huftile, G., and Gath, E., 1995, Prospects for larger or more frequent earthquakes in greater metropolitan Los Angeles, California: Science, v. 267, p. 199–205.

Dolan, R., Howard, A., and Trimble, D., 1978, Structural control of the rapids and pools of the Colorado River in the Grand Canyon: Science, v. 202, p. 629–631.

Downs, G.L., 1995, Atlas of isoseismal maps of New Zealand earthquakes: New Zealand Institute of Geological and Nuclear Sciences Monograph 11, 304 p.

Ducea, M.N., and Saleeby, J.B., 1996, Buoyancy sources for a large, unrooted mountain range, the Sierra Nevada, California: evidence from xenolith thermobarometry: Journal of Geophysical Research, v. 101, p. 8229–8244.

Ducea, M.N., and Saleeby, J.B., 1998, A case for delamination of the deep batholithic crust beneath the Sierra Nevada: International Geology Review, v. 40, p. 78–93.

Duncan, R.P. 1989, An evaluation of errors in tree age estimates based on increment cores in kahikatea (*Dacrycarpus dacrydioides*): New Zealand Natural Sciences, v. 16, p. 31–37.

Dunne, T., and Black, R.D. 1970, Partial area contributions to storm runoff in a small New England watershed: Water Resources Research, v. 6, p. 1296–1311.

Dunwiddie, P.W., 1979, Dendrochronological studies of indigenous New Zealand trees: New Zealand Journal of Botany, v. 17, p. 251–266.

Duvall, A., Kirby, E., and Burbank, D., 2004, Tectonic and lithologic controls on bedrock channel profiles and processes in coastal California: Journal of Geophysical Research, v. 109, F03002, doi:10.1029/2003JF000086.

Eaton, B.C., and Church, M., 2004, A graded stream response relation for bed load–dominated streams: Journal of Geophysical Research, v. 109, F03011, doi:10.1029/2003JF000062.

Edwards, R.L., Chen, J.H., and Wasserburg, G.J., 1987a, ^{238}U/^{234}U-^{230}Th/^{232}Th systematics and the precise measurement of time over the past 500,000 years: Earth and Planetary Science Letters, v. 81, 175–192.

Edwards, R.L., Chen, J.H., Ku, T.L., and Wasserburg, G.J., 1987b, Precise timing of the last interglacial period from mass spectrometric determination of thorium-230 in corals: Science, v. 236, p. 1547–1552.

Ekström, G., Stein, R. S., Eaton J. P., and Eberhart-Phillips, D., 1992, Seismicity and geometry of a 110-km long blind thrust fault, 1, The 1985 Kettleman Hills, California, earthquake: Journal of Geophysical Research, v. 97, p. 4843–4864.

Ellis, M.A., Densmore, A.L., and Anderson, R.S., 1999, Development of mountainous topography in the Basin Ranges, USA: Basin Research, v. 11, p. 21–41.

Ellsworth, W.L., 1990, Earthquake history, 1769-1989 *in* R.E. Wallace, editor, The San Andreas Fault System: U.S. Geological Survey Professional Paper 1515, p. 153–187.

England, P., and Molnar, P., 1990, Surface uplift, uplift of rocks, and exhumation of rock Geology, v. 18, p. 1173–1177.

Enzel, Y., Amit, R., Porat, N., Zilberman, E., and Harrison, J.B.J., 1996, Estimating the ages of fault scarps in the Arava, Israel: Tectonophysics, v. 253, p. 305–317.

Eusden, J.D., Jr., Koons, P.O., Pettinga, J.R., and Upton, P., 2005b, Structural geology and geodynamic modeling of linking faults in the Marlborough fault zone, South Island, New Zealand: New Zealand Geological Society Annual Meeting abstracts, p. 26.

Eusden, J.D. Jr., Pettinga, J.R., and Campbell, J.K., 2000. Structural evolution and landscape development of a collapsed transpressive duplex on the Hope Fault, North Canterbury, New Zealand: New Zealand Journal of Geology and Geophysics, v. 43, p. 391–404.

Eusden, J.D. Jr., Pettinga, J.R., and Campbell, J.K.,

2005a, Structural collapse of a transpressive hanging-wall fault wedge, Charwell region of the Hope Fault, South Island, New Zealand: New Zealand Journal of Geology and Geophysics, v. 48, p. 295–309.

Farmer, G. L., Glazner, A. F., and Manley, C. R., 2002, Did lithospheric delamination trigger late Cenozoic potassic volcanism in the southern Sierra Nevada, California?: Geological Society of America Bulletin, v. 114, p. 754–768.

Fenton, C.R., Webb, R.H., Cerling, T.E., Poreda, R.J., and Nash, B.P., 2002, Cosmogenic ^{3}He ages and geochemical discrimination of lava-dam out-burst-flood deposits in western Grand Canyon, Arizona, in House, K. et al., editors, Paleoflood Hydrology, American Geophysical Union, p. 191–215.

Ferguson, R.I., 2005, Estimating critical stream power for bedload transport calculations in gravel-bed rivers: Geomorphology, v. 70, p. 33–41.

Flenley, J., and Bush, M., 2006, Tropical Rainforest Responses to Climatic Change, Springer Praxis Books, 400 p.

Fisk, H.N., 1944, Geological investigation of the alluvial valley of the lower Mississippi River: Mississippi River Commission, U.S. Army Corps of Engineers, Vicksburg, Mississippi.

Fisk, H.N., 1947, Fine-grained alluvial deposits and their effects on Mississippi River activity: Mississippi River Commission, Vicksburg, Mississippi, 82 p.

Florensov, N. A., and Solonenko, V. P., editors, 1963, The Gobi-Altai earthquake: Moscow, Akademiya Nauk USSR (in Russian; English translation, 1965, U.S. Department of Commerce, Washington, D.C.), 424 p.

Fonseca, J., 1988, The Sou Hills -- A barrier to faulting in the central Nevada seismic belt: Journal of Geophysical Research, v. 93, p. 475–489.

Forman, S.L., Nelson, A.R., and McCalpin, J.P., 1991, Thermoluminescence dating of fault-scarp-derived colluvium-Deciphering the timing of paleoearthquakes on the Weber segment of the Wasatch fault zone, north-central Utah: Journal of Geophysical Research, v. 96, no. B1, p. 595–605.

Frank, E.C., and Lee, R. 1966, Potential beam irradiation on slopes, Tables for 30° to 50° latitude: Fort Collins, Colorado, U.S. Forest Service Research Paper RM-18, 116 p.

Frankel, K.L., and Dolan, J.F., 2007, Characterizing arid-region alluvial fan surface roughness with airborne laser swath mapping digital topographic data: Journal of Geophysical Research – Earth Surface, doi:10.1029/ 2006JF000644

Frankel, K.L., and Pazzaglia, F.J., 2005, Tectonic geomorphology, drainage basin metrics, and active mountain fronts: Geografia Fisica e Dinamica Quaternaria, v. 28, p. 7–21.

Frankel, K.L., and Pazzaglia, F.J., 2006, Mountain fronts, base-level fall, and landscape evolution; insights from the southern Rocky Mountains: in Tectonics, climate, and landscape evolution, S. Willett, N. Hovius, M. Brandon, and D. Fisher, editors: Geological Society of America Special Paper 398.

Frankel, K.L., and 10 other authors, (in press for 2007), Cosmogenic ^{10}Be and ^{36}Cl geochronology of offset alluvial fans along the northern Death Valley fault zone: Implications for transient strain in the eastern California shear zone: Journal of Geophysical Research–Solid Earth. doi:10.1029/ 2006JB004350.

Fraser, G.D., Witkind, I.J., and Nelson, W.H., 1964, A geological interpretation of the Epicentral area – the dual-basin concept, the Hebgen Lake, Montana, earthquake of August 17, 1959: U.S. Geological Survey Professional Paper 435-J, p. 99–106.

Freed, A.M., and Lin J., 2002, Accelerated stress buildup on the southern San Andreas fault and surrounding regions caused by Mojave Desert earthquakes: Geology, v. 30 p. 571–574.

Fritts, H.C., 1976, Tree rings and climate, Academic Press, New York, 567 p.

Fumal, T.E., Pezzopane, S.K., Weldon, R.J., II, and Schwartz, D.P., 1993, A 100-year average recurrence interval for the San Andreas fault at Wrightwood, California: Science, v. 259, p. 199–203.

Gallup, C.D., Edwards, R.L., and Johnson, R.G., 1994, The timing of high sea levels over the past 200,000 years: Science, v. 263, p. 796–800.

Gammond, J.F., 1994, normal faulting and tectonic inversion driven by gravity in a thrusting regime: Journal of Structural Geology, v. 16, p. 1–9.

Garfunkel, Z., 1974, Model for the late Cenozoic tectonic history of the Mojave Desert, California, and its relation to adjacent regions: Geological Society of America Bulletin, v. 85, p. 1931–1944.

Gellis, A.C., Emmett, W.W., and Leopold, L.B., 2005, Channel and hillslope processes revisited in the Arroyo de los Frijoles watershed near Santa Fe, New Mexico: U.S. Geological Survey Professional Paper 1704, 53 p.

Giaccio, B,. Galadini, F., Sposato A., Messina, P., Moro, M., Zreda, M., Cittadini, A., Salvi, S., and Todero, A., 2003, Image processing and roughness analysis of exposed bedrock fault planes as a tool for paleoseismological analysis: results from the Campo Felice fault (central Apennines, Italy): Geomorphology, v. 49, p. 281–301.

Gilbert, G.K., 1877, Geology of the Henry Mountains (Utah): U.S. Geographical and Geological Survey of the Rocky Mountain Region, U.S. Government Printing Office, Washington, D.C., 170 p.

Gilbert, G.K., 1890, Lake Bonneville: U.S. Geological Survey Monograph 1, 438 p.

Gilbert, G.K., 1904, Domes and dome structures of the high Sierra: Geological Society of America Bulletin, v. 15, p. 29–36.

Gilbert, G.K., 1928, Studies of basin-range structures: U.S. Geological Survey Professional Paper 153, p. 1–92.

Gilchrist, A.R., Kooi, H., and Beaumont, C., 1994, Post-Gondwana geomorphic evolution of southwestern Africa: implications for the controls on landscape development from observations and numerical experiments: Journal of Geophysical Research, v. 99, p. 211–228.

Gilchrist, A.R., and Summerfield, M.A., 1991, Erosion, isostasy and landscape evolution: Earth Surface Process and Landforms, v. 16, p. 555–562.

Gilchrist, A.R., Summerfield, M.A., and Cockburn, H.A.P., 1994, Landscape dissection, isostatic uplift, and the morphologic development of orogens: Geology, v. 22, p. 963–966.

Gile, L.H., and Grossman, R.R., 1979, The Desert Project Soil Monograph, U.S. Soil Conservation Service, 964 p.

Gile, L.H., Hawley, J.W., and Grossman, R.R., 1981, Soils and geomorphology in the Basin and Range area of southern New Mexico-guidebook to the Desert Project: New Mexico Bureau of Mines and Mineral Resources Memoir 39, 218 p.

Glenn, N.F., Streutker D.R. Chadwick D.J., Thackray, G.D., and Dorsch, S.J., 2006, Analysis of LiDAR-derived topographic information for characterizing and differentiating landslide morphology and activity: Geomorphology, v. 73, p. 131–148.

Goff, J.R., and McFadgen B.G. 2002. Seismic driving of nationwide changes in geomorphology and prehistoric settlement – a 15th Century New Zealand example: Quaternary Science Reviews, v. 21, 2313–2320.

Goff, J.R. Wells, A. Chague-Goff, C. Nichol, S.L. and Devoy, R.J., 2004, The elusive AD 1826 tsunami, south Westland, New Zealand: New Zealand Geographer, v. 60, p. 14–25.

Goldrick, G., and Bishop, P., 1995, Differentiating the roles of lithology and uplift in the steepening of bedrock river long profiles; an example from southeastern Australia: Journal of Geology, v. 103, p. 227–231.

Gosse, J.C., and Phillips, F. M., 2001, Terrestrial in situ cosmogenic nuclides: theory and application: Quaternary Science Reviews, v. 20, p. 1475–1560.

Goudie, A.S. 1978, Dust storms and their geomorphological implications: Journal of Arid Environments, v. 1, p. 291–310.

Graf, W.L., 1980, The effect of dam closure on downstream rapids: Water Resources Research, v. 16, p. 129–136.

Graf, W.L., 1982, Spatial variations of fluvial processes in semiarid lands, in C.E. Thorn (editor), Space and time in geomorphology: Proceedings of the 12th Annual Binghamton Symposium, Allen and Unwin, Boston, p. 193–217.

Graf, W.L., 1983, Downstream changes in stream power, Henry Mountains: Association of American Geographers Annals, v. 73, p. 373–387.

Grant, L.B., and Lettis, W.R, 2002, Introduction to the special issue on paleoseismology of the San Andreas Fault system: Seismological Society of America Bulletin, v. 92, p. 2551–2554.

Grant, L.B., and Sieh, K.E., 1993, Stratigraphic evidence for 7 meters of dextral slip on the San Andreas fault during the 1857 earthquake in the Carrizo Plain: Seismological Society of America Bulletin, v. 83, 619–635.

Grantz, A., editor, 1971, The San Fernando, California, earthquake of February 9, The San Fernando, California, earthquake of February 9, 1971: U.S. Geological Survey Professional Paper 733, 254 p.

Grapes, R., and Downes, G. 1997, The 1855 Wairarapa, New Zealand, earthquake-analysis of historical data: Royal Society of New Zealand Bulletin, v. 30, No.4, p. 271–368.

Grapes, R., Little, T.A., and Downes G., 1998, Rupturing of the Awatere Fault during the 1848 October 16 Marlborough earthquake, New Zealand; historical and present day evidence: New Zealand Journal of Geology and Geophysics, v. 41, p. 387–399.

Gregory, K.M., and Chase, C.G., 1994, Tectonic and climatic significance of a late Eocene low relief, high-level geomorphic surface, Colorado: Journal of Geophysical Research, v. 99, p. 20,141–20,160.

Griffiths, G.A., 1981, Some suspended sediment yields from South Island catchments, New Zealand: Water Resources Bulletin, v. 17, p. 662–671.

Griffiths, G.A., and Glasby, G.P., 1985, Input of river-derived sediments to the New Zealand continental shelf; I. Mass. Estuarine: Coastal and Shelf Science, v. 21, p. 773–787.

Griffiths, P.G., Webb, R.H., and Melis, T.S., 1996, Initiation and frequency of debris flows in Grand Canyon, Arizona: U.S. Geological Survey Open-File Report 96-491, 35 p.

Griffiths, P.G., Webb, R.H., and Melis, T.S., 2004, Frequency and initiation of debris flows in Grand Canyon, Arizona: Journal of Geophysical Research, v. 109, F04002, doi:10.1029/2003JF000077.

Guzzetti, F, Reichenbach, P., and Wieczorek, G.F., 2003, Rockfall hazard and risk in Yosemite Valley, California, USA: Natural Hazards and Earth System Sciences, v. 3, p. 491–503.

Hack, J.T., 1957, Studies of longitudinal stream profiles in Virginia and Maryland: U.S. Geological Survey Professional Paper 294-B, p. 45–97.

Hack, J.T., 1960, Interpretation of erosional topography in humid temperate regions: American Journal of Science (Bradley Volume), v. 258-A, p. 80–97.

Hack, J.T., 1965, Geomorphology of the Shenandoah Valley, Virginia and West Virginia, an origin of the residual ore deposits: U.S. Geological Survey Professional Paper 484, 84 p.

Hack, J.T., 1973, Stream-profile analysis and stream-gradient index: U.S. Geological Survey Journal of Research, v. 1, p. 421–429.

Hack, J.T., 1982, Physiographic divisions and differential uplift in the piedmont and Blue Ridge: U.S. Geological Survey Professional Paper 1265, 49 p.

Hale, M.E., and Cole, M., 1988, Lichens of California: University of California Press, Berkeley, 254 p.

Hamblin, W. K., 1963, Transition between the Colorado Plateau and the Basin and Range in southern Utah and northern Arizona: Geological Society of America Special Paper, 85 p.

Hamblin, W. K., 1965, Origin of "reverse drag" on the downthrown side of normal faults: Geological Society of America Bulletin, v. 74, p. 1145–1164.

Hamblin, W.K., 1976, Patterns of displacement along the Wasatch fault: Geology, v.4, p. 619–622.

Hamblin, W.K., and Rigby, J.K., 1968, Guidebook to the Colorado River, Part 1 – Lees Ferry to Phantom Ranch in Grand Canyon National Park, Provo, Utah: Brigham Young University Geology Studies, v. 15, part 5, 84 p.

Hancock, G.S., and Anderson, R.S., 2002: Numerical modeling of fluvial strath-terrace formation in response to oscillating climate: Geological Society of America Bulletin, v. 114, p. 1131–1142.

Hancock, G.S., Anderson, R.S., and Whipple, K.X., 1998, Beyond power: Bedrock river incision process and form, in Tinkler, K., and Wohl, E., editors, Rivers over rock: Fluvial processes in bedrock channels: Geophysical Monograph Series, v. 107, American Geophysical Union, Washington, D.C., p. 35–60.

Hancock, P.L., Yeats, R.S., and Sanderson, D.J, editors, 1991, Characteristics of active faults: Journal of Structural Geology, v. 13, 248 p.

Hancox, G.T., McSaveney, M.J., Manville, V.R., and Davies, T.R., 2005, The October 1999 Mt. Adams rock avalanche and subsequent landslide dam-break flood and effects in Poerua River, Westland, New Zealand: New Zealand Journal of Geology and Geophysics, v. 48, p. 683–706.

Handwerger, D.A., Cerling, T.E., and Bruhn, R.L., 1999, Cosmogenic 14C in carbonate rocks: Geomorphology v. 27, p. 13–24.

Hanks, T.C., 2000, The age of scarplike landforms from diffusion-equation analysis, in: J.S. Noller, J.M. Sowers, W.R. Lettis, editors, Quaternary Geochronology: Methods and Applications, AGU Reference Shelf 4, American Geophysical Union, Washington, D.C., p. 313–338.

Hanks, T.C., and Andrews, D.J., 1989, Effect of far-field slope on morphological dating of scarplike landforms: Journal of Geophysical Research, v. 94, p. 565–573.

Hanks, T.C., Bucknam, R.C., LaJoie, K.R., and Wallace, R.E., 1984, Modification of wave-cut and faulting-controlled landforms: Journal of Geophysical Research, v. 89, p. 5771–5790.

Hanks, T.C., and Kanamori, H., 1979, A moment magnitude scale: Journal of Geophysical Research, v. 84, p. 2348–2350.

Hanks, T.C., and Wallace, R.E., 1985, Morphological analysis of the Lake Lahontan shoreline and beachfront fault scarps, Pershing County, Nevada: Seismological Society of America Bulletin, v. 75, p. 835–846.

Harden, J.W., 1982, A quantitative index of soil development from field descriptions: examples from a chronosequence in central California: Geoderma, v. 28, p. 1–28.

Harden, J.W., 1987, Soil development in granitic alluvium near Merced, California, U.S. Geological Survey Bulletin 1590-A, 65 p.

Harden, J.W., Taylor, E.M., Hill, C.L., Mark, R.K., McFadden, L.D., Reheis, M.C., Sowers, J.M., and Wells, S.G., 1991, Rates of soil development from four soil chronosequences in the southern Great Basin: Quaternary Research, v. 35, p. 383–399.

Harding, D.J., and Berghoff, G.S., 2000, Fault scarp detection beneath dense vegetation cover: Airborne LiDAR mapping of the Seattle fault zone, Bainbridge Island, Washington State: Proceedings of the American Society of Photogrammetry and Remote Sensing Annual Conference, Washington, D.C., May, 2000, 9 p.

Harding, T. P., 1974, Petroleum traps associated with wrench faults: American Association of Petroleum Geologists Bulletin, v. 58, p. 1290–1304.

Harrington, C.D., Whitney, J.W., Jull, J.T., and Phillips, W., 1999, Cosmogenic Dating and Analysis of Scarps Along the Solitario Canyon and Windy Wash Faults, Yucca Mountain, Nevada: Chapter G *in* Geologic and Geophysical Characterization Studies of Yucca Mountain, Nevada, A Potential High-Level Radioactive-Waste Repository, J. W. Whitney and W. R. Keefer, Editors, U.S. Geological Survey Digital Data Series 58, Version 1.0.

Harrison, R.W., Palmer, J.R., Hoffman, Vaughn, D., J.D., Forman, S.L., McGeehin, J., and Frederiksen, N., 1997, Profiles and documentation of fault-exploration trenches in the English Hill area, Scott City 7.5-minute quadrangle, Missouri: U.S. Geological Survey Open-File Report, 96 p.

Hart, E.W., and Bryant, W.A., 1997, Fault-Rupture Hazard Zones in California; Alquist-Priolo Earthquake Fault Zoning Act with Index to Earthquake Fault Zones Maps: California Division of Mines and Geology (now the California Geological Survey) Special Publication 42 (with supplements added in 1999 – refer to http://www.consrv.ca.gov/cgs/rghm/ap/index.htm

Hart, E.W., Bryant, W.A., and Treiman, J.A., 1993, Surface faulting associated with the June 1992 Landers earthquake, California: California Geology, v. 46, p. 10–16.

Hart, E.W., Bryant, W.A., Wills, C.J., Treiman, J.A., and Kahle, J.E.,1989, Summary Report: Fault Evaluation Program, 1987-1988, Southwestern Basin and Range Region and Supplemental Areas. Depart of Conservation, Division of Mines and Geology Open-File Report 89–16.

Hasbargen, L.E., and Paola C., 2000, Landscape instability in an experimental drainage basin: Geology, v. 28, p. 1067–1070.

Haugerud, R.A., Harding, D.J., Johnson, S.Y., Harless, J.L., Weaver, C.S., and Sherrod, B.L., 2003, High-Resolution LiDAR Topography of the Puget Lowland, Washington – A Bonanza for Earth Science: GSA Today, v. 13, p. 4–10.

Hauksson, E, Jones, L.M., Davis, T.L., Hutton, L.K., Brady, A.G., Reasenberg, P.A., Michael, A.J., and Yerkes, R.F., 1988, The 1987 Whittier Narrows earthquake in the Los Angeles metropolitan area, California: Science, v. 239, p.1409–1412.

Hayakawa, S.I., 1949, Language in thought and action. Harcourt, Brace and Company: New York.

Haynes, C.V., Jr., 1987, Curry Draw, Cochise County, Arizona; A Late Quaternary Stratigraphic Record of Pleistocene Extinction and Paleo-Indian Activities: in

Cordilleran Section, Geological Society of America Centennial Field Guide, v. 1 p. 23–28.

Heaton, T., 1982, The 1971 San Fernando earthquake: A double event?: Seismological Society of America Bulletin, v. 72, p. 2037–2062.

Heaton, T.H., and Helmberger, D.V., 1979, Generalized ray models of the San Fernando earthquake: Bulletin of the Seismological Society of America, v. 69, p. 1311–1341.

Heaton, T.H., and Kanamori, H., 1984, Seismic potential associated with subduction in the northwestern United States: Bulletin of Seismological Society of America, v. 74, p. 933–942.

Hecker, S., 1993, Quaternary tectonics of Utah with emphasis on earthquake-hazard characterization: Utah Geological Survey Bulletin 127, 157 p.

Hecker, S., 1985, Timing of Holocene faulting in part of the seismic belt, west-central Nevada: University of Arizona, Geosciences Department, Tucson, AZ, M.S. prepublication manuscript, 42 p.

Hecker, S., 1993, Quaternary tectonics of Utah with emphasis on earthquake-hazard characterization: Utah Geological Survey Bulletin 127, 2 plates, scale 1:500,000, 257 p.

Heirtzler, J.R., and Frawley, J.J., 1994, New gravity model for earth science studies: GSA Today, v. 4, p. 269–270.

Hill, M.L., 1984, Earthquakes and folding, Coalinga, California: Geology, v. 12, p. 711–712.

Hill, R.L., and Beeby, D.J., 1977, Surface faulting associated with the 5.2 magnitude Galway Lake earthquake of May 31, 1975: Mojave Desert, San Bernardino County, California: Geological Society of America Bulletin, v. 88, p. 1378–1384.

Hilley, G., Arrowsmith, J., and Amoroso, L., 2001, Interaction between normal faults and fractures and fault scarp morphology: Geophysical Research Letters, v. 28, doi:10.1029/2001GL012876.

Hirano, M., 1968, A mathematical model of slope development-an approach to the analytical theory of erosional topography: Journal of Geosciences Osaka City University, v. 11, p. 13–52.

Hirano, M., 1975, Simulation of development process of interfluvial slopes with reference to graded form: Journal of Geology, v. 83, p. 113–123.

Hogg, A.G., McCormac, F.G., Higham, T.F.G., Reimer, P.J., Baillie, M.G.L., and Palmer, J.G., 2002, High-precision radiocarbon measurements of contemporaneous tree-ring dated wood from the British Isles and New Zealand, AD 1850–950: Radiocarbon, v. 44, p. 633–640.

Holloway, J.T., 1957, Charles Douglas – Observer extraordinary: New Zealand Journal of Forestry, v. 4, p. 35–40.

Holm, D.K., Norris, R.J., and Craw, D., 1989, Brittle and ductile deformation in a zone of rapid uplift; Central Southern Alps, New Zealand: Tectonics, v. 8(2), 153–168.

Hooke, R.LeB., 1967, Processes on arid-region alluvial fans: Journal of Geology, v. 75, p. 438–460.

Hooke, R.LeB., 1972, Geomorphic evidence for late-Wisconsin and Holocene tectonic deformation, Death Valley, California: Geological Society of America Bulletin, v. 83, p. 2073–2098.

Horton, R.E., 1945, Erosional development of streams and their drainage basins: hydrophysical approach to quantitative morphology: Geological Society of America Bulletin, v. 56, p. 275–370.

Hough, S.E., 1995, Earthquakes in the Los Angeles Metropolitan region: a fractal distribution of rupture sizes?: Science, v. 267, p. 211–213.

Hough, S.E., 1996, The case against huge earthquakes: Seismological Research Letters, v. 67, p. 3–4.

Hovius, N., Stark, C.P., Chu, H.T., and Lin, J.C., 2000, Supply and removal of sediment in a landslide-dominated mountain belt: Central Range, Taiwan: Journal of Geology, v. 108, p. 73–89.

Howard, A.D., 1959, Numerical systems of terrace nomenclature; a critique: Journal of Geology, v. 67, p. 239–243.

Howard, A.D., 1994, A detachment limited model of drainage basin evolution: Water Resources Research, v. 30, p. 2261–2285.

Howard, A., and Dolan, R., 1981, Geomorphology of the Colorado River in Grand Canyon: Journal of Geology, v. 89, p. 269–297.

Howard, A.D., and Kerby G., 1983, Channel changes in badlands: Geological Society of America Bulletin, v. 94, p. 739–752.

Huber, N.K., 1987, The geologic story of Yosemite National Park: U.S. Geological Survey Bulletin 1595, 64 p.

Huber, N.K., Phillips, W.M., and Bull, W.B., 2002, The Slide, Yosemite National Park, CA, Yosemite Association: Yosemite, v. 64, no. 3, p. 2–4.

Hudnut, K.W., and 10 others, 1996, Coseismic displacements of the 1994 Northridge, California, earthquake: Bulletin of the Seismological Society of America, v. 86, p. 19–36 in Northridge earthquake special issue.

Huftile, G.J., and Yeats, R.S., 1996, Deformation rates across the Placerita (Northridge Mw = 6.7 aftershock zone) and Hopper Canyon segments of the western Transverse Ranges deformation belt: Seismological Society of America Bulletin, v. 86, p. S3–S18.

Hull A.G., and McSaveney M.J., 1993, A 7000-year record of great earthquakes at Turakirae Head, Wellington, New Zealand: New Zealand Earthquake Commission Research Paper 011, Project 93/139.

Ikeda, Y., 1983, Thrust front migration and its mechanisms – evolution of intraplate thrust fault systems: Bulletin of the Geography Department, University of Tokyo, v. 15, p. 125–159.

Ikeda, Y., and Yonekura, N., 1979, Dislocation fault models of the San Fernando, California, earthquake of 1971; bending of fault plane and its tectonic significance: Seismological Society of Japan, v. 32, p. 477–488.

Innes, J.L., 1985, Lichenometry: Progress in Physical Geography, v. 9, p. 187–254.

Israelson, C., and Wohlfarth, B., 1999. Timing of the Last Interglacial high sea level on the Seychelles Islands, Indian Ocean: Quaternary Research, v. 51, p. 306–316.

Jackson, J., Norris R., and Youngson, J.H., 1996, The structural evolution of active fault and fold systems in central Otago, New Zealand: Evidence revealed by drainage patterns: Journal of Structural Geology, v. 18, p. 217–234.

Jackson, J., 2002: Strength of the continental lithosphere: Time to abandon the jelly sandwich?: GSA Today, v. 12, p. 4–10.

Jacoby, G.C., 1997, Application of tree ring analysis to paleoseismology: Reviews of Geophysics, v. 35 , p. 109–124.

Jacoby, G.C., Bunker, D.E., and Bensen, B.E., 1997, Tree-ring evidence for an A.D. 1700 Cascadia earthquake in Washington and northern Oregon: Geology, v. 25, p. 999–1002.

Jacoby, G., Carver, G., and Wagner, W., 1995, Trees and herbs killed by an earthquake ~300 yr. ago at Humboldt Bay, California: Geology, v. 23, p. 77–80.

Jacoby, G.C., Sheppard, P.R., and Sieh, K.E., 1988, Irregular recurrence of large earthquakes along the San Andreas fault: Evidence from trees: Science, v. 241, no. 4862, p. 196–199.

Janda, R.J., 1965, Quaternary alluvium near Friant, California, in Northern Great Basin and California, 7th International Association of Quaternary Research Congress, USA, Guidebook for Field Conference 1, p. 128–133.

Jennings, C.W., 1994, Fault Activity Map of California and Adjacent Areas: California Department of Conservation, Division of Mines and Geology, Geologic Data

288 References Cited

Map No. 6, Scale 1:750,000.

Johnson, R.G., 1982, Brunhes-Matuyama magnetic reversal dated at 790,000 years B.P. by marine-astronomical correlations: Quaternary Research, v. 17, p. 135–147.

Jones, C.H., Farmer, G.L., and Unruh, J.R., 2004, Tectonics of Pliocene removal of lithosphere of the Sierra Nevada, California: Geological Society of America Bulletin, v. 116, p. 1408–1422.

Jones, C.H., Hanimori, K., and Roeker, S.W., 1994, Missing roots and mantle "drips": Regional Pn and teleseismic arrival times in the southern Sierra Nevada and vicinity, California: Journal of Geophysical Research, v. 99, p. 4567–4601.

Jones, C.H., Unruh, J., and Sonder, L.J., 1996, The role of gravitational potential energy in active deformation in the southwestern United States: Nature, v. 381, p. 37–41.

Jones, C.H., Wernicke, B.P., Farmer, G.L., Walker, J.D., Coleman, D.S., McKenna, L.W., and Perry, F.V., 1992, Variations across and along a major continental rift; An interdisciplinary study of the Basin and Range Province, western USA: Tectonophysics, v. 213, p. 57–96.

Kamb, B., Silver, L.T., Abrams, M.J. Carter, B.A., Jordan, T.H., and Minster, J.B., 1971, Pattern of faulting and nature of fault movement in the San Fernando earthquake: in The San Fernando, California, earthquake of February 9, 1971, U. S. Geological Survey Professional Paper, p. 41–54.

Janda, R.J., 1965, Quaternary alluvium near Friant, California, in Northern Great Basin and California, 7th International Association of Quaternary Research Congress, USA, Guidebook for Field Conference 1, p. 128–133.

Keefer, D.K., 1984, Landslides caused by earthquakes: Geological Society of America Bulletin, v. 95, p. 406–421.

Keefer, D.K., 1994, The importance of earthquake-induced landslides to longterm slope erosion and slope-failure hazards in seismically active regions: Geomorphology, v. 10, p. 265–284.

Keller, E.A., and Pinter, N., 2002, Active Tectonics – Earthquakes, Uplift, and Landscape (2nd edition): Prentice Hall, London, 362 p.

Keller, E.A., and Rockwell, T.K., 1984, Tectonic geomorphology, Quaternary chronology and paleoseismology: in J.E. Costa and P.J. Fleisher, editors, Developments and Applications of Geomorphology, Springer-Verlag, Berlin, p. 203–239.

Keller, E.A., Seaver, D.B., Laduzinsky, D.L., Johnson, DL., and Ku, T.L., 2000, Tectonic geomorphology of active folding over buried reverse faults: San Emigdio Mountain front, southern San Joaquin Valley, California: Geological Society of America Bulletin v. 112, p. 86–97.

Keller, E.A., Zapeda, R.L., Rockwell, T.K., Ku, and T.L., Dinklage, W.S., 1998, Active tectonics at Wheeler Ridge southern San Joaquin Valley, California: Geological Society of America Bulletin, v. 110, p. 298–310.

Kieffer, S.W., 1985, The 1983 hydraulic jump in Crystal Rapid – Implications for river running and geomorphic evolution in the Grand Canyon: Journal of Geology, v. 93, p. 385–406.

King, G., and Stein, R.S., 1983, Surface folding, river terrace deformation rate and earthquake repeat time in a reverse faulting environment; The Coalinga California earthquake of May 1983: California Division of Mines and Geology, v. 66, p. 165–176.

King, L.C. 1942, South African scenery; a textbook of geomorphology: Oliver and Boyd, Edinburgh, 64 figs., 144 plates.

King, L.C. 1968, Scarps and tablelands: Zeitschrift fur Geomorphologie, v. 12; p. 114–115.

Kirby, E., Whipple, K.X., Tang, W., and Chen, Z., 2003, Distribution of active rock uplift along the eastern margin of the Tibetan Plateau, Inferences from bedrock channel longitudinal profiles: Journal of Geophysical Research, v. 108, (B4), p.. 2217, doi:10.1029/2001JB000861.

Kirk, R.M. 1977, Rates and forms of erosion on intertidal platforms at Kaikoura Peninsula, South Island, New Zealand: New Zealand Journal of Geology and Geophysics, v, 20, p. 571–613.

Kirkby, M.J., 1971, Hillslope process-response models based on the continuity equation: Transactions of the Institute of British Geographers Special Publication No. 3, p. 15–30.

Kirkby, M. 1992, An erosion-limited hillslope evolution model: Catena Supplement, v. 23, p. 157–187.

Klinger, R.E., 1999, Tectonic geomorphology along the Death Valley fault system – Evidence for recurrent late Quaternary activity in Death Valley National Park, in Slate, J.L., editor, Proceedings of Conference on Status of Geologic Research and Mapping, Death Valley National Park: U.S. Geological Survey Open-File Report 99- 153, p. 132–140.

Klinger, R.E., 2001, Evidence for large dextral offset near Red Wall Canyon, in Machette, M.N., Johnson, M.L., and Slate, J.L., editors, Quaternary and Late Pliocene geology of the Death Valley region: Recent Observations on Tectonics, Stratigraphy, and Lake Cycles (Guidebook for the 2001 Pacific Cell – Friends of the

Pleistocene fieldtrip): U.S. Geological Survey Open-File Report 01-51, p. A32–A37.

Knuepfer, P.L.K., 1984, Tectonic geomorphology and present-day tectonics of the Alpine shear system, South Island, New Zealand: University of Arizona Ph.D. thesis, Tucson, Arizona, 489 p.

Knuepfer, P.L.K., 1988, Estimating ages of later Quaternary stream terraces from analysis of weathering rinds and soils: Geological Society of America Bulletin, v. 100, p. 1224–1236.

Knuepfer, P.L.K., 1989. Implications of the characteristics of endpoints of historical surface fault ruptures for the nature of fault segmentation: U.S. Geological Survey Open-File Report 89-315, p.193–228.

Knuepfer, P.L.K., 1992, Temporal variations in latest quaternary slip across the Australian-Pacific Plate boundary, northeastern South Island, New Zealand: Tectonics, v. 11, p. 449–464.

Kooi, H., and Beaumont, C., 1994, Escarpment evolution on high-elevation rifted margins; insights derived from a surface processes model that combines diffusion, advection, and reaction: Journal of Geophysical Research, v. 99, p. 12,191–12,210.

Koons, P.O., 1989, The topographic evolution of collisional mountain belts; A numerical look at the Southern Alps, New Zealand: American Journal of Science, v. 289, p. 1041–1069.

Korup, O., 2005, Large landslides and their effect on sediment flux in South Westland, New Zealand: Earth Surface Processes and Landforms, v. 30, p. 305–323.

Kramer, P.J., and Kozlowski, T.T., 1979, Physiology of woody plants: Orlando, Academic Press, Florida, 811 p.

Kurushin R.A., Bayasgalan A., Ölziybat M., Enhtuvshin B., Molnar P., Bayarsayhan C., Hudnut W. K., Lin, J., 1998, The Surface Rupture of the 1957 Gobi-Altay, Mongolia, Earthquake: Geological Society of America Special Paper 320, 144 p.

Kurz, M.D., Colodner, D., Trull, T.W., Moore, R.B., and O'Brien, K., 1990, Cosmic ray exposure dating with in situ produced cosmogenic ^{3}He; results from young Hawaiian lava flows: Earth and Planetary Science Letters, v. 97, p. 177–189.

Lague, D., Hovius, N., and Davy, P., 2005, Discharge, discharge variability, and the bedrock channel profile: Journal of Geophysical Research, v. 110, F04006, doi:10.1029/2004JF000259.

Lajoie, K.R., 1986, Coastal tectonics, in Active Tectonics: National Academy Press, Washington, D.C., p. 95–124.

Lal, D., 1991, Cosmic ray labelling of erosion surfaces, in situ production rates and erosion models: Earth and Planetary Science Letters, v. 104, p. 424–439.

Lambeck, K., and Chappell, J., 2001, Sea level change through the last glacial cycle, Science: v. 292, p. 679–686.

Lavé, J., and Burbank, D., 2004, Denudation processes and rates in the Transverse Ranges, southern California, Erosional response of a transitional landscape to external and anthropogenic forcing: Journal of Geophysical Research, v. 109, F01006, doi:10.1029/2003JF000023.

Le, K., Lee, L., Owen, L.A., Finkel, R., 2007, Late Quaternary slip rates along the Sierra Nevada frontal fault zone, California; Slip partitioning across the western margin of the Eastern California Shear Zone-Basin and Range Province: Geological Society of America Bulletin, v. 119, p. 240–256. doi:10.1130/B25960.1

Lee, H. K., and Schwarcz, H.P., 1996, Electron spin resonance plateau dating of periodicity of activity on the San Gabriel fault zone, southern California: Geological Society of America Bulletin, v. 108, p. 735–746.

Lee, J., and Lister, G.S., 1992, Late Miocene ductile extension and detachment faulting, Mykonos, Greece: Geology, v. 20, p.121–124.

Lee, J., Spencer, J.Q., and Owen, L.A., 2001, Holocene slip rates along the Owens Valley fault, California: implications for the recent evolution of the Eastern California Shear Zone: Geology, v. 29, p. 819–822.

Leonard, E.M., 2002, Geomorphic and thermal forcing of late Cenozoic warping of the Colorado Piedmont: Geology, v. 34, p. 595–598.

Leopold, L.B., 1969, The rapids and pools – Grand Canyon, in The Colorado River Region and John Wesley Powell: U.S. Geological Survey Professional Paper 669, p. 131–145.

Leopold, L.B., 1992, Base level rise; gradient of deposition: Israel Journal of Earth Sciences, v. 41, p. 57–64.

Leopold, L.B., and Bull, W.B., 1979, Base level, aggradation, and grade: Proceedings of American Philosophical Society, v. 123, p. 168–202.

Leopold, L.B., and Langbein, W.B., 1962, The concept of entropy in landscape evolution: U.S. Geological Survey Professional Paper 500-A, 20 p.

Leopold, L.B., and Maddock, T., Jr., 1953, The hydraulic geometry of stream channels and some physiographic implications: U.S. Geological Survey Professional Paper 252, 56 p.

Leopold, L.B., and Miller, J.P., 1954, A post-glacial chronology for some alluvial valleys in Wyoming: U.S. Geological Survey Water-Supply Paper 1261, 90 p.

Leopold, L.B., Wolman, M.G., and Miller, J.P., 1964,

Fluvial processes in geomorphology: W.H. Freeman, San Francisco, 522 p.

Lettis, W.R., 1982, Late Cenozoic stratigraphy and structure of the western margin of the central San Joaquin Valley, California: U.S. Geological Survey Open-File Report 82-526, 203 p.

Lettis, W.R., 1985, Late Cenozoic stratigraphy and structure of the west margin of the central San Joaquin Valley, California, *in* Soils and Quaternary geology of the southwestern United States: Geological Society of America Special Paper 203, D.L. Weide, editor, p. 97–114.

Lettis, W. R., and Hanson, K. L., 1991, Crustal strain partitioning; implications for seismic-hazard assessment in western California: Geology, v. 19, p. 559–562.

Lifton, N.A., Jull, A.J.T., and Quade, J., 2001, A new extraction technique and production rate estimate for in situ cosmogenic ^{14}C in quartz: Geochimica et Cosmochimica Acta, v. 65, p. 1953–1969.

Lin, J., and Stein, R.S., 2006, Seismic Constraints and Coulomb Stress Changes of a Blind Thrust Fault System, 1, Coalinga and Kettleman Hills, California: U.S. Geological Survey Open-File Report 2006-1149, 17 p.

Lindvall, S.C., and Rockwell, T.K., 1995, Holocene activity of the Rose Canyon fault in San Diego, California: Journal of Geophysical Research, v. 100, p. 24,121–24,132.

Lindvall, S.C., Rockwell, T.K., and Rubin, C.M., 2000, Collaborative paleoseismic studies along the 1999 Hector Mine earthquake surface rupture and adjacent faults: Southern California Earthquake Center Annual Report, 6 p.

Lister, G. S., Etheridge, N. A., and Symonds, P. A., 1986, Detachment faulting and the evolution of passive continental margins: Geology, v. 14, p. 246–250.

Litchfield, N.J., and Berryman, K.R., 2005. Correlation of fluvial terraces within the Hikurangi Margin, New Zealand: implications for climate and base-level controls: Geomorphology, v. 68, p. 291–313.

Locke, W.W., III, Andrews, J.T., and Webber, P.J., 1979, A manual for lichenometry: British Geomorphological Research Group Technical Bulletin 26, 47 p.

Louderback, G.D., 1904, Basin Range structure of the Humboldt region, Nevada: Geological Society of America Bulletin, p. 289–346.

Lund, W.R., Pearthree, P.A., Amoroso, L., Hozik, M.J., and Hatfield S.C., 2001, Paleoseismic investigation of earthquake hazard and longterm movement history of the Hurricane Fault, southwestern Utah and northwestern Arizona, Final Technical Report, U.S. Geolog-

ical Survey National Earthquake Hazards Reduction Program, 71.

Mabbutt, J.A., 1966, Mantle-controlled planation of pediments: American Journal of Science, v. 264, p. 78–91.

Machette, M.N., 1978, Dating Quaternary faults in the southwestern United States by using buried calcic paleosols: U.S. Geological Survey Journal of Research, v. 6, p. 369–381.

Machette, M.N., 1982, Quaternary and Pliocene faults in the La Jencia and southern part of the Albuquerque-Belen basins, New Mexico: Evidence of fault history from fault morphology and Quaternary geology: New Mexico Geological Society 33rd Annual Field Conference Guidebook, p. 161–169.

Machette, M.N., 1986, History of Quaternary offset and paleoseismicity along the La Jencia fault, central Rio Grande rift, New Mexico: Bulletin of Seismological Society of America, v. 76, p. 259–272.

Machette, M.N., and Brown, W.M., 1995, Utah braces for the future: U. S. Geological Survey Fact Sheet, 2 p.

Machette, M.N. Haller, K.M., Ruleman, C.A., Mahan, S., and Okumura, K., 2005 Evidence for late Quaternary movement on the Clan Alpine fault, west-central Nevada Trench logs, location maps, and sample and soil descriptions: U.S. Geological Survey Scientific Investigations Map 2189–http://pubs.usgs.gov/sim/2005/2891/

Machette, M., Johnson, L., and Slate, J. L., editors, 2001, Quaternary and Late Pliocene Geology of the Death Valley Region; Recent Observations on Tectonics, Stratigraphy, and Lake Cycles: *in* Guidebook for the 2001 Pacific Cell – Friends of the Pleistocene Fieldtrip: U.S. Geological Survey Open-File Report 01-51, 254 p.

Machette, M.N., Personius, S.F., Nelson, A.R., 1992, Paleoseismology of the Wasatch fault zone; a summary of recent investigations, interpretations, and conclusions: in Assessment of regional earthquake hazards and risk along the Wasatch Front, Utah, Gori-Paula L., and Hays W.W., editors, U.S. Geological Survey Professional Paper 1500A, p. A1–A71.

Machette, M.N., Personius, S.F., Nelson, A.R., Schwartz, D.P., and Lund, W.R., 1989, Segmentation models and Holocene movement history of the Wasatch fault zone, Utah: U.S. Geological Survey Open-File Report 89-0315, p. 229–245.

Machette, M.N., Personius, S.F., Nelson, A.R., Schwartz, D.P., and Lund, W.R., 1991, The Wasatch fault zone, Utah—Segmentation and history of Holocene movement: Journal of Structural Geology, v. 13, no. 2, p. 137–149.

Mackin, J.H., 1948, Concept of the graded river: Geological Society of America Bulletin, v. 59, p. 463–512.

Manley, C.R., Glazner, A.F., and Farmer, G.L., 2000, Timing of volcanism in the Sierra Nevada of California: evidence for Pliocene delamination of the batholithic root?: Geology, v. 28, p. 811–814.

Martin, Y., and Church, M., 1997, Diffusion in landscape development models: On the nature of basic transport relations: Earth Surface Processes and Landforms, v. 22, p. 273–279.

Mason, D., Little, T., Van Dissen, R., 2006, Rates of active faulting during late Quaternary fluvial terrace formation at Saxton River, Awatere fault, New Zealand: Geological Society of America Bulletin. doi:10.1130/B25961.

Massong, T.M., and Montgomery, D.R., 2000, Influence of sediment supply, lithology, and wood debris on the distribution of bedrock and alluvial channels: Geological Society of America Bulletin, v. 112, p. 591–599.

Matmon, A., Bierman, P.R., Larsen, J., Southworth, S., Pavich, M, and Caffee, M., 2003, Temporally and spatially uniform rates of erosion in the southern Appalachian Great Smoky Mountains: Geology, v. 31, p. 155–158.

Matsuda, T., 1978, Estimation of future destructive earthquakes from active faults on land in Japan: in Earthquake Precursors; Proceedings of the US-Japan seminar on theoretical and experimental investigations of earthquake precursors, Kisslinger, C., and Suzuki, Z., editors, Center for Academic Publications of Japan, Tokyo, p. 251–260.

Matthews, J.A. 1974, Families of lichenometric dating curves from the Storbreen gletschervorfeld, Jutunheimen, Norway: Norsk Geografiska Tidsskrift, v. 28, p. 215–235.

Matti, J.C., Morton, D.M., and Cox, B.F., 1985, Distribution and geologic relations of fault systems in the vicinity of the central Transverse Ranges, southern California: U.S. Geological Survey Open File Report 85-365, 23 p.

Mattson A, and Bruhn, R.L., 2001, Fault slip rates and initiation age based on diffusion equation modeling: Wasatch Fault Zone and eastern Great Basin: Journal of Geophysical Research, v. 106, p. 13,739–13,750.

Maxon, J.H., 1950, Physiographic features of the Panamint Range, California: Geological Society of America Bulletin, v. 61, p. 99–114.

Mayer, L., 1979, The evolution of the Mogollon Rim in central Arizona, in Plateau Uplift Mode and Mechanism, T.R. McKetchin and R.B. Merill, Kisslinger, C., and Suzuki, Z., editors, Tectonophysics, v. 61, p. 49–62.

Mayer, L, 1984, Dating Quaternary fault scarps formed in alluvium using morphologic parameters: Quaternary Research, v. 22, p. 300–313.

Mayer, L., 1986, Tectonic geomorphology of escarpments and mountain fronts: in Active Tectonics: National Academy Press, Washington, D.C., p. 125–135.

Mayer, L., 1987, Sources of error in morphologic dating of fault scarps: in Crone, A.J., Omdahl, E.M. Kisslinger, C., and Suzuki, Z., editors, Proceedings of Conference XXXIX, Directions in Paleoseismology: U.S. Geological Survey Open-File Report 8787-0673, p. 302–310.

McCalpin, J.P., 1996, Field techniques in paleoseismology, Chapter 2 in McCalpin, J.P., Kisslinger, C., and Suzuki, Z., editors, Paleoseismology: Academic Press, New York, p. 33–84.

McCalpin, J.P., and Forman, S.L., 1991, Late Quaternary faulting and thermoluminescence dating of the East Cache fault zone, north-central Utah: Bulletin of the Seismological Society of America, v. 81, p. 139–161.

McCarroll, D., and Nesje, A., 1996, Rock surface roughness as an indicator of degree of rock surface weathering: Earth Surface Processes and Landforms, v. 21, p. 963–977.

McCarthy, J., and Parsons, T., 1994, Insights into the kinematic Cenozoic evolution of the Basin and Range-Colorado Plateau transition from coincident seismic refraction and reflection data: Geological Society of America Bulletin, v.106, p. 747–759.

McCaskill, L.W., 1969, Molesworth, Wellington, A.H. Reed, 292 p.

McFadden, L.D., Tinsley, J.C., and Bull, W.B., 1982, Late Quaternary pedogenesis and alluvial chronologies of the Los Angeles basin and San Gabriel Mountains areas, southern California, in J.C. Tinsley, J.C. Matti, and L.D. McFadden editors, Late Quaternary Pedogenesis and Alluvial Chronologies of the Los Angeles and San Gabriel Mountains areas, southern California, and Holocene Faulting and Alluvial Stratigraphy Within the Cucamonga Fault Zone; Field Trip 12, Cordilleran Section of the Geological Society of America, p. 1–13.

McGill, S.F., and Rubin, C.M. 1999, Surficial slip distribution on the central Emerson fault during the 28 June 1992 Landers earthquake, California: Journal of Geophysical Research, v. 104, p. 4811–4833.

McGill, S.F., and Sieh, K.E., 1991, Surficial offsets on the central and eastern Garlock fault associated with prehistoric earthquakes: Journal of Geophysical Research, v. 96, p. 21,597–21,621.

McGill, S., and Sieh, K., 1993, Holocene slip rate of the central Garlock Fault in southeastern Searles Valley,

California: Journal of Geophysical Research, v. 98, p. 14,217–14,231.

McMillan, M.E., C.L. Angevine, and P.L. Heller, 2002, Postdepositional tilt of the Miocene-Pliocene Ogallala Group on the western Great Plains: Evidence of late Cenozoic uplift of the Rocky Mountains: Geology, v. 30, p. 63–66.

McNutt, S.R., and Sydnor, R.H., editors, 1990, The Loma Prieta (Santa Cruz Mountains), Santa Clara and Santa Cruz Counties, California, Earthquake of 17 October 1989: California Division of Mines and Geology Special Publication 104, 150 p.

McSaveney, M.J. 1992, A manual of weathering-rind dating of grey sandstone of the Torlesse Supergroup, New Zealand: Institute of Geological and Nuclear Sciences Report 92/4, Institute of Geological and Nuclear Sciences Ltd., Lower Hutt, New Zealand, 52 p.

McSaveney, M.J., 2002 Recent rockfalls and rock avalanches in Mount Cook National Park, New Zealand, p. 35–70, in Evans, S.G. DeGraff, J.V. Kisslinger, C., and Suzuki, Z., editors, Catastrophic landslides: effects, occurrence, and mechanisms, Geological Society of America, Boulder, Colorado, Reviews in Engineering Geology 15.

Melis, T.S., and Webb, R.H., 1993, Debris flows in Grand Canyon National Park, Arizona–Magnitude, frequency and effects on the Colorado River, in Shen, H.W., Su, S.T., and Wen, F., Kisslinger, C., and Suzuki, Z., editors, Hydraulic Engineering '93: American Society of Civil Engineers, New York, Proceedings of the ASCE Conference, San Francisco, California, p. 1290–1295.

Melton, M.A., 1960, Origin of the drainage of southeastern Arizona: Arizona Geological Society Digest, v. 3, p. 113–122.

Melton, M.A., 1965, Debris-covered hillslopes of the southern Arizona desert-consideration of their stability and sediment contribution: Journal of Geology, v. 73, p. 715–729.

Menges, C.M., 1987, Temporal and spatial segmentation of the Pliocene-Quaternary fault rupture along the western Sangre de Cristo mountain front, northern New Mexico: U.S. Geological Survey Open-File Report 87-673, p. 203–222.

Menges, C.M, 1990a, Late Quaternary fault scarps, mountain front landforms, and Pliocene-Quaternary segmentation on the range-bounding fault zone, Sangre de Cristo Mountains, New Mexico: Reviews in Engineering Geology, v. 8, p. 131–156.

Menges, C.M., 1990b, Late Cenozoic rift tectonics and mountain-front landforms of the Sangre de Cristo Mountains near Taos, northern New Mexico: in New

Mexico Geological Society Guidebook, 41st Field Conference, p. 113–122.

Merritts, D.J., and Vincent, K.R., 1989, Geomorphic response of coastal streams to low, intermediate, and high rates of uplift Mendocino triple junction region, northern California: Geological Society of America Bulletin, v. 101, p. 1373–1388.

Merritts, D.J., Vincent, K.R., and Wohl, E.E., 1994, Long river profiles, tectonism, and eustasy; A guide to interpreting fluvial terraces: Journal of Geophysical Research, v. 99, p. 14,031–14,050.

Miller, M.M., Webb, F.H., Townsend, D., Golombek, M.P., and Dokka, R.K., 1993, Regional coseismic deformation from the June 28, 1992, Landers, California, earthquake; results from the Mojave GPS network: Geology, v. 21, p. 868–872.

Mitchell, S.G., Matmon, A., Bierman, P.R., Enzel, Y., Caffee, M., and Rizzo, D., 2001, Displacement history of a limestone normal fault scarp, northern Israel, from cosmogenic 36Cl: Journal of Geophysical Research, v. 106, p. 4247–4264.

Molnar, P., and England, P., 1990, Late Cenozoic uplift of mountain ranges and global climatic change: Chicken or egg: Nature, v. 346, p. 29–34.

Molnar, P., and Lyon-Caen, H., 1988, Some simple physical aspects of the support, structure and evolution of mountain belts: Geological Society of America Special Paper 218, p. 179–207.

Monastero, F.C., Katzenstein, A.M., Miller, J.S., Unruh, J.R., Adams, M.C., and Richards-Dinger, K., 2005, The Coso geothermal field; A nascent metamorphic core complex: Geological Society of America Bulletin, v. 117, p. 1534–1553.

Montgomery, D.R., 1994, Valley incision and the uplift of mountain peaks: Journal of Geophysical Research, v. 99, p. 13,913–13,921.

Montgomery, D.R., 2002, Valley formation by fluvial and glacial erosion: Geology, v. 30, p. 1047–1050.

Montgomery, D.R., Abbe, T.B., Buffington, J.M., Peterson, N.P., Schmidt, K.M., and Stock, J.D., 1996, Distribution of bedrock and alluvial channels in forested mountain drainage areas: Nature, v. 381, p. 587–589.

Montgomery, D.R., and Dietrich, W.E., 1992, Channel initiation and the problem of landscape scale: Science, v. 255, p. 826–830.

Montgomery, D.R., and Greenberg, H.M., 2000, Local relief and the height of Mount Olympus: Earth Surface Processes and Landforms, v. 25, p. 385–396.

Moon, B.P., 1984, Refinement of a technique for determining rock mass strength for geomorphological purposes: Earth Surface Processes and Landforms, v. 9, p.

189–193.

Mori, J., Wald, D.J., and Wesson, R.L., 1995, Overlapping fault zones of the 1971 San Fernando and 1994 Northridge, California earthquakes: Geophysical Research Letters, v. 22, p. 1033–1036.

Morisawa, M.E., 1968, Streams, their dynamics and morphology: McGraw-Hill, New York, 175 p.

Morton, D.M., Miller, F.K., and Smith, C.C., 1980, Photo reconnaissance maps showing young-looking fault features in the southern Mojave Desert, California: U.S. Geological Survey Miscellaneous Field Studies Map MF-1051, 7 sheets, scale 1:24,000, 1:62,500.

Morton, D.M., and Matti, J.C., 1987, The Cucamonga fault zone; geologic settings and Quaternary history: U.S. Geological Survey Professional Paper 1339, p. 179–203.

Muir, J., 1901, Our national parks, Chapter VIII, The Fountains and Streams of the Yosemite National Park, Houghton-Mifflin, New York, 370 p.

Mueller, J.E., 1972, Re-evaluation of the relationship of master streams and drainage basins: Geological Society of America Bulletin, v. 83, p. 3471–3474.

Munsell Color Company, Inc., 1992, Munsell soil color charts: Baltimore.

Namson, J.S., and Davis, T.L., 1988, Seismically active fold and thrust belt in the San Joaquin Valley, central California: Geological Society of America of Bulletin, v. 100, p. 257–273.

Namson, J.S., Davis, T.L., and Lagoe, M.B., 1990, Tectonic history and thrust-fold deformation style of seismically active structures near Coalinga, chap. 6 of Rymer, M.J., and Ellsworth, W.L., editors, The Coalinga, California, Earthquake of May 2, 1983: U.S. Geological Survey Professional Paper 1487, p. 79–96.

Nash, D.B., 1980, Morphological dating of degraded normal fault scarps: Journal of Geology, v. 88, p. 353–360.

Nash, D.B., 1984, Morphologic dating of fluvial terrace scarps and fault scarps near West Yellowstone, Montana: Geological Society of America Bulletin, v. 95, p. 1413–1424.

Nash, D.B., 1986, Morphologic dating and modeling degradation of fault scarps, in Wallace, R.E., editor, Active Tectonics: National Academy Press, Washington, D.C., p. 181–194.

Nash, D.B., 1987, Reevaluation of the linear-diffusion model for morphologic dating of scarps: in Crone, A.J., and Omdahl, E.M, editors, Proceedings of Conference XXXIX XXXIX, Directions in Paleoseismology: U.S. Geological Survey Open-File Report 87, 87-0673, p. 325–338.

Nash, D.B., and Beaujon, J.S., 2006, Modeling degradation of terrace scarps in Grand Teton National Park, USA: Geomorphology, v. 75, p. 400–407.

Nelson, A.R., Johnson, S.Y., Wells, R.E., Pezzopane, S.K., Kelsey, H.M., Sherrod, B.L., Bradley, L., Koehler, R.D., III, Bucknam, R.C., Haugerud, R., and Laprade, W.T., 2002, Field and laboratory data from an earthquake history study of the Toe Jam Hill fault, Bainbridge Island, Washington: U.S. Geological Survey Open-File Report 02-0060.

Nicol, A., and Campbell, J.K. 2001, The impact of episodic fault-related folding on late Holocene degradation terraces along Waipara River, New Zealand: New Zealand Journal of Geology and Geophysics, v. 44, p. 145–156.

Nicol, A., and Van Dissen, R.V., 2002, Up-dip partitioning of displacement components on the oblique-slip Clarence fault, New Zealand: Journal of Structural Geology, v. 24, p. 1521–1535.

Nicol, A., and Wise, D.U., 1992, Paleostress adjacent to the Alpine fault of New Zealand: Journal of Geophysical Research, v. 97, p. 17,685–17,692.

Niviere, B., and Marquis, G., 2000. Evolution of terrace risers along the Upper Rhine Graben inferred from morphologic dating methods; evidence of climatic and tectonic forcing: Geophysical Journal International, v. 141, p. 577–594.

Noble, L.F., 1926, The San Andreas rift and some other active faults in the desert region of southeastern California: Carnegie Institution of Washington Yearbook No. 25, Washington, D.C., p. 415–428.

Noller, J.S, Sowers, J.M., and Lettis, W.R., editors, 2000, Quaternary Geochronology: Methods and Applications; American Geophysical Union Reference Shelf Series, v. 4, 582 p.

Norris, R.J., and Cooper, A.F., 1995, Origin of small-scale segmentation and transpressional thrusting along the Alpine fault, New Zealand: Geological Society of America, Bulletin v. 107, p. 231–240.

Norris, R.J., and Cooper, A.F., 1997, Erosional control on the structural evolution of a transpressional thrust complex on the Alpine fault, New Zealand: Journal of Structural Geology, v. 19, p. 1323–1342.

Norris, R.J., and Cooper, A.F., 2001, Late Quaternary slip rates and slip partitioning on the Alpine fault, New Zealand: Journal of Structural Geology, v. 23, p. 507–520.

Norris, R.J., Cooper, A.F., Wright, T., and Berryman, K. 2001. Dating of past Alpine fault ruptures in south Westland: Earthquake Commission Report 99/341.

Norris, R., Cooper, A., Wright, C., Wright, T., Berryman, K., Sutherland, R., and Vilamor, P., 2001, Late Quaternary slip rates and prehistoric earthquakes on the Alpine Fault; *in* N.J. Litchfield, compiler, Ten years of paleoseismology in the ILP: progress and prospects. 17-21 December 2001, Kaikoura, New Zealand, Programme and Abstracts, Institute of Geological and Nuclear Sciences Information Series 50, 133 p.

Norton, D., 1983a, A dendroclimatic analysis of three indigenous tree species, South Island, New Zealand, University of Canterbury Ph.D. dissertation, 439 p.

Norton, D., 1983b, Population dynamics of subalpine *Libocedrus bidwillii* forests in the Cropp River Valley, Westland, New Zealand: New Zealand Journal of Botany, v. 21, p. 127–134.

Nur, A., Ron, H., and Beroza G., 1993a, The nature of the Landers-Mojave earthquake line: Science, v. 261, p. 201–203.

Nur, A., Ron, H., and Beroza, G., 1993b, Landers-Mojave earthquake line: A new fault system?: GSA Today, v. 3, p. 254–258.

Nur, A., Ron, H., and Scotti, O., 1986, Fault mechanics and the kinematics of block rotation: Geology, v. 14, p. 746–749.

O'Neal, M.A., and Schoenenberger, K.B., 2003, A *Rhizocarpon geographicum* growth curve for the Cascade Range of Washington and northern Oregon, USA: Quaternary Research, v. 60, p. 233–241.

Oakeshott, G.B., 1958, Geology and mineral resources of San Fernando quadrangle, Los Angeles County, California: California Division of Mines Bulletin 172, 147 p.

Oguchi, T., 1996, Relaxation time of geomorphic responses to Pleistocene-Holocene climatic change climatic change: Transactions of the Japanese Geomorphological Union, v. 17, no. 4, p. 309–321.

Okaya, D., Stern, T., Holbrook, S., and Van Avendonk, H.J., Davey, H. F., and Henrys, S., 2003, Imaging a plate boundary using double-sided onshore-offshore seismic profiling: The Leading Edge, v. 22, no. 3, p. 256–262.

Ollier, C.D., 1982, The Great Escarpment of eastern Australia; tectonic and geomorphic significance: Journal of the Geological Society of Australia, v. 29 p. 13–23.

Oskin, M., Mukhopadhyay, S., and Iriondo, A., (in press for 2007), Slip rate of the Calico fault; Implications for geologic versus geodetic rate discrepancy in the Eastern California shear zone: Journal of Geophysical Research – Solid Earth

Ota, Y., Pillans, B., Berryman, K., Fujimori, T., Miyauchi, T., Berger, G., Beu., A.G., and Climo, F.M., 1996, Pleistocene coastal terraces of Kaikoura Peninsula and the Marlborough coast, South Island, New Zealand: New Zealand Journal of Geology and Geophysics, v. 39, p. 51–73.

Ouchi, S., 2005, Development of offset channels across the San Andreas fault: Geomorphology, v. 70, p. 112–128.

Owen, L.A., Windley, B.F., Cunningham, W.D., Badamgarav, J., and Dornjnamjaa, D., 1997, Quaternary alluvial fans in the Gobi of southern Mongolia, evidence for neotectonics and climate change: Journal of Quaternary Science, v. 12, p. 239–252.

Pain, C.F., 1985, Cordilleran metamorphic core complexes in Arizona; a contribution from geomorphology: Geology, v. 13, p. 871–874.

Pain, C.F., and Bowler, J.M., 1973, Denudation following the November 1970 earthquake at Madang, Papua New Guinea: Zeitschrift fur Geomorphologie, v. 18, supplement, p. 91–104.

Pantosti D., D'Addezio, G., and Cinti, F.R., 1993b, Paleoseismological evidence of repeated large earthquakes along the 1980 Irpinia earthquake fault: Annali di Geofisica, v. 36, p. 321–330.

Pantosti, D., Schwartz. D.P., and Valensise, G.,1993a, Paleoseismology along the 1980 Irpinia earthquake fault and implications for earthquake recurrence in the southern Apennines: Journal of Geophysical Research, v. 98, p. 6561–6577.

Paola, C., and Mohrig D., 1996, Palaeohydraulics revisited; Palaeoslope estimation in coarse-grained braided rivers: Basin Research, v.8, p. 243–254.

Parker, R.S., 1977, Experimental study of basin evolution and its hydrologic implications, Ph.D. thesis: Fort Collins, Colorado State University, 331 p.

Parsons, A.J., and Abrahams, A.D., 1984, Mountain mass denudation and piedmont formation in the Mojave and Sonoran deserts: American Journal of Science, v. 284, p. 255–271.

Pazzaglia, F.J., 2004, Landscape evolution models: Developments in Quaternary Science, Elsevier, v. 1, p. 247–273. doi:10.1016/S1571-08866(03)01012-1.

Pazzaglia, F.J., and Brandon, M. T., 2001, A fluvial record of long-term steady-state uplift and erosion across the Cascadia forearc high, western Washington State: American Journal of Science, v. 301, p. 385–431.

Pazzaglia, F.J., and Gardner, T.W., 1993, Fluvial Terraces of the Susquehanna River: Geomorphology, v. 8, p. 83–113.

Pazzaglia, F.J., and Gardner, T.W., 1994, Late Cenozoic flexural deformation of the middle US Atlantic Margin: Journal of Geophysical Research, v. 99, p. 12,143–12,157.

Pazzaglia, F.J., Gardner, T.W., and Merritts, D.J., 1998, Bedrock fluvial incision and longitudinal profile development over geologic time scales determined by fluvial terraces: *in* Rivers Over Rock, edited by Keith J. Tinkler and Ellen E. Wohl, American Geophysical Union, Washington DC, v. 107, p. 207–235.

Pearce, A.J., and Watson, A.J., 1986, Effects of earthquake-induced landslides on sediment budget and transport over 50 years: Geology, v. 14, p. 52–55.

Pearthree, P.A., 1990, Geomorphic analysis of young faulting and fault behavior in central Nevada: Tucson, University of Arizona, Ph.D Dissertation, 212 p.

Pearthree, P.A., and Calvo, S.S., 1987, The Santa Rita fault zone; evidence for large magnitude earthquakes with very long recurrence intervals, Basin and Range province of southeastern Arizona: Bulletin of the Seismology Society of America, v. 77, p. 97–116.

Pearthree, P.A., Lund, W.R., Stenner, H.D., and Everitt, B.L., 1998, Paleoseismic investigation of the Hurricane fault in southwestern Utah and northwestern Arizona: Arizona Geological Survey and Utah Geological Survey Final Technical Report to the U.S. Geological Survey National Earthquake Hazard Reduction Program, Award No. 1434-HQ-97-GR-03047, 131 p.

Pelletier, J.D., DeLong, S.B., Al-Suwaidi, A.H., Cline, M., Lewis, Y., Psillas, J.L., and Yanites B., 2006, Evolution of the Bonneville shoreline scarp in west-central Utah: Comparison of scarp-analysis methods and implications for the diffusion model of hillslope evolution: Geomorphology, v. 74, p. 257–270.

Peteet, D.M., 2000, Sensitivity and rapidity of vegetational response to abrupt climate change: in Proceedings of the National Academy of Sciences of the United States of America, v. 97, p. 1359–1361.

Pettinga, J.R., and Wise D.U., 1994, Paleostress adjacent to the Alpine fault: Broader implications from fault analysis near Nelson, South Island, New Zealand: Journal of Geophysical Research, v. 99, p. 2727–2736.

Pettinga, J.R., Yetton, M.D., Van Dissen, R.J., and Downes, G., 2001, Earthquake source identification and characterisation for the Canterbury region, South Island, New Zealand: Bulletin of the New Zealand Society for Earthquake Engineering, v. 34, p. 282–317.

Phillips, F.M., Ayarbe, J.P., Harrison, J.B.J., and Elmore, D., 2003, Dating rupture events on alluvial fault scarps using cosmogenic nuclides and scarp morphology: Earth and Planetary Science Letters, v. 215, p. 203–218.

Phillips, F.M., Stone, W.D., and Fabryka-Martin, J.T., 2001, An improved approach to calculating low-energy cosmic-ray neutron fluxes near the land-atmosphere interface, Chemical Geology: v. 175, p. 689–701.

Phillips, W.M., McDonald, E.V., Reneau, S.L., and Poths, J., 1998. Dating soils and alluvium with cosmogenic ^{21}Ne depth profiles: case studies from the Pajarito Plateau, New Mexico. USA: Earth and Planetary Science Letters, v. 160, p. 209–223.

Pierce, K.L., and Colman, S.M., 1986, Effect of height and orientation (microclimate) on geomorphic degradation rates and processes, late-glacial terrace scarps in central Idaho: Geological Society of America Bulletin, v. 97, p. 869–885.

Pierce, K.L., and Scott, W.E., 1982, Pleistocene episodes of alluvial-gravel deposition, southeastern Idaho: *in* Bonnichsen, B., and Breckenridge, R.M., editor, Cenozoic Geology of Idaho: Idaho Bureau of Mines and Geology Bulletin 26, p. 685–702.

Pigati, J., Lifton, N., Jull, A.J.T., and Quade, J., 2005, Extracting in situ cosmogenic ^{14}C from olivine, significance for the CRONUS-Earth Project: EOS Transactions, v. 86, Abstract U23B-06.

Porter, S.C., 1981. Lichenometric studies in the Cascade Range of Washington: establishment of *Rhizocarpon geographicum* growth curves at Mount Rainier: Arctic and Alpine Research, v. 13, p. 11–23.

Powell, J.W., 1875, Exploration of the Colorado River and its canyons: Dover Publications, New York, 400 p.

Prentice, C.S., Kendrick, K.J., Berryman, K., Bayasgalan, A., Ritz, J.F., and Spencer, J.Q., 2002, Prehistoric ruptures of the Gurvan Bulag fault, Gobi Altay, Mongolia: Journal of Geophysical Research, v. 107, p. 2321, doi:10.1029/2001JB000803.

Proctor, R.J., Crook, R. Jr., McKeown, M.H., and Moresco, R.L. 1972, Relation of known faults to surface ruptures, 1971 San Fernando earthquake, Southern California: Geological Society of America Bulletin, v. 83, p. 1601–1618.

Putkonen J., and O'Neal, M. 2006, Degradation of unconsolidated Quaternary landforms in the western North America: Geomorphology, v. 75, p. 408–419.

Ramasamy, S.M., 1989, Morphotectonic evolution of east and west coast of peninsular India: Geological Survey of India Special Publication 24, p. 336–1161.

Rapp, A., 1981, Alpine debris flows in north Scandinavia, morphology and dating by lichenometry: Geografiska Annaler, v. 63A, p. 183–196.

Raymo, M.E., and Ruddiman, W.F., 1992, Tectonic forcing of late Cenozoic climate: Nature, v. 359, p. 117–122.

Reheis, M.C., and Kihl, R., 1995, Dust deposition in southern Nevada and California, 1984–1989, relations to climate, source area, and lithology: Journal of Geophysical Research, v. 100D5, p. 8893–8918.

Reheis, M.C., Slate, J.L., and Sawyer, T.L., 1995, Geologic map of late Cenozoic deposits and faults in parts of the Mt. Barcroft, Piper Peak, and Soldier Pass 15' quadrangles, Esmeralda County, Nevada, and Mono County, California: U.S. Geological Survey Miscellaneous Investigations Map I-2464, 2 sheets.

Reusser L.J., Bierman P.R., Pavich M.J., Zen E., Larsen J., and Finkel R., 2004, Rapid late Pleistocene incision of Atlantic passive-margin river gorges: Science, v. 305, p. 499–502.

Rigon, R., Rodriguez-Iturbe, I., Maritan, A., Giacometti, A., Tarboton, D.G., and Rinaldo, A. 1996, On Hack's law: Water Resources Research, v. 32, p. 3367–3374.

Rockwell, T.K., 2000, Use of soil geomorphology in fault studies: in Quaternary Geochronology: Methods and Applications, J.S. Noller, J.M. Sowers, and W.R. Lettis, editors, AGU Reference Shelf 4, American Geophysical Union, Washington D.C., p. 273–292.

Rockwell, T.K., Keller, E. A., Clark, M. N., and Johnson D. L., 1984. Chronology and rates of faulting of Ventura River terraces, California: Geological Society of America Bulletin, v. 95, p. 1466–1474.

Rockwell, T.K., Keller, E.A., and Dembroff, G.R., 1988, Quaternary rate of folding of the Ventura Avenue anticline, western Transverse Ranges, southern California: Geological Society of America Bulletin, v. 100, p. 850–858.

Rockwell, T.K., and Pinault, C.T., 1986, Holocene slip events on the southern Elsinore fault, Coyote Mountains, southern California: in Guidebook and Volume on Neotectonics and Faulting in Southern California (P. Ehlig, editor), Cordilleran Section, Geological Society of America, p. 193–196.

Roering, J.J., Almond, P., Tonkin, P., and McKean, J., 2002, Soil transport driven by biological processes over millennial time scales: Geology, v. 30, p. 1115–1118.

Roering, J.J., Almond, P., Tonkin, P., and McKean, J., 2004, Constraining climatic controls on hillslope dynamics using a coupled model for the transport of soil and tracers: Application to loess-mantled hillslopes, South Island: New Zealand Journal of Geology and Geophysics, v. 109, F01010, doi:10.1029/2003JF000034.

Rotstein, Y., Combs, J., and Biehler, S., 1976, Gravity investigation in the southeastern Mojave Desert, California: Geological Society of America Bulletin, v. 87, p. 981–993.

Royse, C.F., and Barsch, D. 1971, Terraces and pediment terraces in the southwest, an interpretation: Geological Society of America Bulletin, v. 82, p. 3,177–3,182.

Rubin, C.M., Lindvall, S.C., and Rockwell, T.K., 1998, Evidence for large earthquakes in metropolitan Los Angeles: Science, v. 281, p. 398–402.

Rubin, C.M., and Sieh, K.E., 1997, Long dormancy, low slip rate and similar slip per event for the Emerson fault, Eastern California shear zone: Journal of Geophysical Research, v. 102, p. 15,319–15,333.

Ruppert, S., Fliedner, M.M., and Zandt, G., 1998, Thin crust and active upper mantle beneath the southern Sierra Nevada in the western United States: Tectonophysics, v. 286, p. 237–252.

Russell, R.J. 1909, Geomorphic of a climatic boundary: Science, v. 74, p. 484–485.

Rymer, M.J., and Ellsworth, W.L., 1990, The Coalinga, California earthquake of May 2, 1983: U.S. Geological Survey Professional Paper 1487, 417 p.

Saleeby, J.M. Ducea M.N., and Clemens-Knott, D., 2003, Production and loss of high-density batholithic root – southern Sierra Nevada, California: Tectonics, v. 22, p. 1064, doi:10.1029/2002TC001374.

Saleeby, J., and Foster, Z., 2004, Topographic response to mantle lithosphere removal in the southern Sierra Nevada region, California: Geology, v. 32, p. 245–248.

Sauber, J., Thatcher, W., and Solomon, S.C., 1986, Geodetic measurement of deformation in the central Mojave Desert, California: Journal of Geophysical Research, v. 91, p. 12683–12693.

Saucier, R.T., 1994, Geomorphology and Quaternary geologic history of the Lower Mississippi Valley: Vicksburg, Mississippi, U.S. Army Engineer Waterways Experiment Station, v. I, 398 p.

Saucier, R.T., 1996, A contemporary appraisal of some key Fiskian concepts with emphasis on Holocene meander belt formation and morphology: in Geology in the Lower Mississippi Valley; Implications for Engineering, the Half Century Since Fisk, 1944; Saucier, R.T., Smith, L.M., and Autin, W.J., editors, Engineering Geology, v. 45, p. 67–86.

Savage, J.C., Lisowski, M., and Prescott, W.H., 1990, An apparent shear zone trending north-northwest across the Mojave Desert into Owens Valley, eastern California: Geophysical Research Letters, v. 17, p. 2113–2116.

Savage, J.C., Prescott, W.H., and Gu, Guohua, 1986, Strain accumulation in southern California 1973-1984: Journal of Geophysical Research, v. 91, p. 7455–7473.

Schreurs, G., 1994, Experiments on strike-slip faulting and block rotation: Geology, v. 22, p. 567–570.

Schumm, S.A., 1956, The role of creep and rainwash on the retreat of badland slopes: American Journal of Science, v. 254, p. 693–706.

Schumm, S.A., 1973, Geomorphic thresholds and complex response of drainage systems, *in* M. Morisawa, editor, Fluvial Geomorphology: Binghamton, State University of New York Publications in Geomorphology, 4th Annual Meeting, p. 299–310.

Schumm, S.A., and Chorley, R.J., 1964, The fall of Threatening Rock: American Journal of Science, v. 262, p. 1041–1054.

Schumm, S.A., and Lichty, R.W., 1965, Time, space and causality in geomorphology: American Journal of Science, v. 263, p. 110–119.

Schumm, S.A, and Mosley, M.P., and Weaver, W.E., 1987, Experimental fluvial geomorphology: Wiley, New York, 413 p.

Schwartz, D.P., and Coppersmith, K.J., 1984, Fault behavior and characteristic earthquakes: examples from the Wasatch and San Andreas fault zones: Journal of Geophysical Research, v. 89, p. 5681–5698.

Schwartz, D.P., and Coppersmith, K.J., 1986, Seismic hazards; New trends in analysis using geologic data: Active Tectonics, National Academy Press, Washington, D. C., p. 215–230.

Schwartz, D.P., and Crone, A.J., 1985, The 1983 Borah Peak earthquake: A calibration event for quantifying earthquake recurrence and fault behavior on Great Basin normal faults, *in* Proceedings of Workshop XXVIII on the Borah Peak, Idaho, Earthquake: U.S. Geological Survey Open-File Report 85-290, p. 153–160.

Schweig, E.S., 1989, Basin-range tectonics in the Darwin Plateau, southwestern Great Basin, California: Geological Society of America Bulletin, v. 101, p. 652–662.

Scientists of the U.S. Geological Survey and the Southern California Earthquake Center, 1994, The magnitude 6.7 Northridge, California, earthquake of 17 January 1994: Science, v. 266, p. 389–397.

Scientists from the U.S. Geological Survey, Southern, California Earthquake Center, and California Division of Mines and Geology, 2000, Preliminary report on the 16 October 1999 M7.1 Hector Mine, California, earthquake: Seismological Research Letters, v. 71, no. 1, p. 11–23.

Scott, K.M., and Williams, R.P., 1978, Erosion and sediment yields in the Transverse Ranges, southern California: U.S. Geological Survey Professional Paper 1030, 38 p.

Scott, W.E., McCoy, W.D., Shroba, R.R., and Rukbin, M., 1983, Reinterpretation of the exposed record of the last two cycles of Lake Bonneville, Western United States: Quaternary Research, v. 20, p. 261–285.

Scott, W.E., Pierce, K.L., and Hait, M.H., Jr., 1985,

Quaternary tectonic setting of the 1983 Borah Peak earthquake, central Idaho: Bulletin of the Seismological Society of America, v. 75, p. 1053–1066.

Seidl, M.A., and Dietrich, W.E., 1992, The problem of channel erosion into bedrock, *in* Schmidt, K.H., and de Ploey, J., editors, Functional geomorphology: Catena, v. 23, p. 101–124.

Selby, M.J., 1982a, Controls on the stability and inclinations of hillslopes on hard rock: Earth Surface Processes and Landforms, v. 7, p. 449–467.

Selby, M.J., 1982b, Hillslope materials and processes, Oxford University Press, New York, 264 p.

Shackelford, T.J., 1980, Tertiary tectonic denudation of a Mesozoic-early Tertiary(?) gneiss complex, Rawhide Mountains, western Arizona: Geology, v. 8, p. 190–194.

Shackleton, N.J., 1987, Oxygen isotopes, ice volumes and sealevel: Quaternary Science Reviews, v. 6, p. 183–190.

Shafiqullah, M., Damon, P.E., Lynch, D.J., Reynolds, S.J., Rehrig, W.A., and Raymond, R.H., 1980, K-Ar geochronology and geologic history of southwestern Arizona and adjacent areas: Arizona Geological Society Digest, v. 12, p. 201–260.

Sharp, R.V., 1975, Displacement on tectonic ruptures: *in* San Fernando, California, earthquake of 9 February 1971: California Division of Mines and Geology Bulletin 196, p. 187–194.

Sharp, R.V., 1982, Comparison of 1979 faulting with earlier displacements in the Imperial Valley, California earthquake of October 15, 1979: U.S. Geological Survey Professional Paper 1254, p. 213–221.

Sheppard P.R., and Jacoby G.C., 1989, Application of tree-ring analysis to paleoseismology, two case studies: Geology, v. 17, p. 226–229.

Shrestha, R.L., Carter, W.E., Lee, M., Finer, P., and Sartori, M., 1999, Airborne laser swath mapping: Accuracy assessment for surveying and mapping: Journal of the American Congress on Surveying and Mapping, v. 59, p. 83–94.

Shroder, J.F. 1980. Dendrogeomorphology, review and new techniques of tree-ring dating: Progress in Physical Geography, v. 4, p.161–188.

Sieh, K.E., 1978a, Central California foreshocks of the great 1857 earthquake: Seismological Society of America Bulletin, v. 68, p. 1731–1749.

Sieh, K., 1978b, Slip on the San Andreas fault associated with the great 1857 earthquakes: Seismological Society of America Bulletin, v. 67, p. 1421–1428.

Sieh, K.E., 1981, A review of geological evidence for recurrence times for large earthquakes: *in* Earthquake

Prediction – an International Review, American Geophysical Union Maurice Ewing Series 4, p. 181–207.

Sieh, K.E., 1984, Lateral offsets and revised dates of large prehistoric earthquakes at Pallett Creek, southern California: Journal of Geophysical Research, v. 89, p. 7641–7670.

Sieh, K.E., and 19 other authors, 1993, Near-field investigations of the Landers earthquake sequence, April to July 1992: Science, v. 260, p. 171–176.

Sieh, K.E., and Jahns R.H., 1984, Holocene activity of the San Andreas Fault at Wallace Creek, California: Geological Society of America Bulletin, v. 95, p. 883–896.

Sieh, K.E., Stuiver, M., and Brillinger, D., 1989, A more precise chronology of earthquakes produced by the San Andreas fault in southern California: Journal of Geophysical Research, v. 94, p. 603–623.

Sieh, K.E., and Wallace, R.E., 1987, The San Andreas fault at Wallace Creek, San Luis Obispo County, California: in Geological Society of America Centennial Field Guide, Cordilleran Section, p. 233–238.

Sieh, K.E., and Williams, P.L., 1990, Behavior of the southernmost San Andreas fault during the past 300 years: Journal of Geophysical Research, v. 95, p. 6629–6645.

Sieh, K.E., and Jahns, R.H., 1984, Holocene activity of the San Andreas fault at Wallace Creek, California: Geological Society of America Bulletin, v. 95, p. 883–896.

Sietz, G., Weldon, R., II, and Biasi, G.P., 1997, The Pitman Canyon paleoseismic record, A re-evaluation of the southern San Andreas fault segmentation: Journal of Geodynamics, v. 24, p. 129–138.

Silverman, B.W., 1986, Density estimation for statistics and data analysis: Chapman and Hall, London, 175p.

Sklar, L.S., and Dietrich, W.E., 1998, River longitudinal profiles and bedrock incision models: Stream power and the influence of sediment supply, in Tinkler, K., and Wohl, E.E., editors, Rivers over Rock, Fluvial Processes in Bedrock Channels: American Geophysical Union Geophysical Monograph 107, p. 237–260.

Sklar, L.S., and Dietrich, W.E., 2001, Sediment and rock strength controls on river incision into bedrock: Geology, v. 29, p. 1087–1090.

Slate, J.L., editor, 1999, Proceedings of Conference on Status of Geologic Research and Mapping, Death Valley National Park: U.S. Geological Survey Open-File Report 99-153, 189 p.

Slemmons, D.B., 1957, Geological effects of the Dixie Valley-Fairview Peak, Nevada, earthquakes of December 16, 1954: Seismological Society of America Bulletin, v. 47, p. 353–375.

Slemmons, D.B., 1967, Pliocene and Quaternary crustal movements of the Basin and Range Province, USA: Osaka City Journal of Geosciences, v. 10, p. 91–103.

Smart, J.S., and Surkan, A.J., 1967, The relation between mainstream length and area in drainage basins: Water Resources Research, v. 3, p. 963–974.

Smith, G.I., 1975, Holocene movement on the Garlock fault: U.S. Geological Survey Professional Paper 975, p. 202.

Smith, G.I., Troxel, B.W., Gray, C.H., Jr., and Von Huene, R., 1968, Geologic reconnaissance of the Slate Range, San Bernardino and Inyo counties, California: California Division of Mines and Geology Special Report 96, 33 p.

Smith, R.S.U., 1976, Late Quaternary pluvial and tectonic history of Panamint Valley, Inyo and San Bernardino counties, California: California Institute of Technology Ph.D. thesis, 295 p.

Smith, R.S.U., 1979, Holocene offset and seismicity along the Panamint Valley fault zone, western Basin and Range Province, California: Tectonophysics, v. 52, p. 411.

Snyder, N.P., Whipple, K.X., Tucker, G.E., and Merritts, D.J., 2000, Landscape response to tectonic forcing; digital elevation model analysis of stream profiles in the Mendocino triple junction region, northern California: Geological Society of America Bulletin, v. 112, p. 1250–1263.

Snyder N.P., Whipple, K.X., Tucker, G.E., and Merritts, D.M., 2003. Importance of a stochastic distribution of floods and erosion thresholds in the bedrock river incision problem: Journal of Geophysical Research, v. 108, doi:10.1029/2001JB001655

Soil Survey Staff, 1975, Soil taxonomy: A basic system of soil classification for making and interpreting soil survey: U.S. Department of Agriculture, Soil Conservation Service, Agriculture Handbook No. 436, U.S. Government Printing Office, Washington D.C., 754p.

Sólyom, P.B., and G.E. Tucker, 2004, Effect of limited storm duration on landscape evolution, drainage basin geometry, and hydrograph shapes: Journal of Geophysical Research, v. 109, F03012, doi:10.1029/2003JF000032.

Southern California Earthquake Center Group C, 2001, Active Faults in the Los Angeles Metropolitan Region: 47 p. http://www.scec.org/research/special/SCEC001activefaultsLA.pdf

Sowers, J.M., Unruh, J.R., Lettis, W.R., and Rubin, T.D., 1994, Relationship of the Kickapoo fault to the Johnson Valley and Homestead faults, San Bernardino

County, California: Bulletin of the Seismological Society of America, v. 84, p. 528-536.

Spencer, J.E., 1984, Role of tectonic denudation in warping and uplift of low-angle normal faults: Geology, v. 12, p. 95–98.

Spotila, J.A., and Anderson, K.B., 2004, Fault interaction at the junction of the Transverse Ranges and Eastern California shear zone: a case study of intersecting faults: Tectonophysics v. 379, p. 43–60.

Spotila, J.A, Bank, G.C. Reiners, P.W, Naeser, C.W., Naeser, N.D., and Henika, W. S., 2004, Origin of the Blue Ridge escarpment along the passive margin of Eastern North America: Basin Research, v.16, No.1 p. 41-63. doi:10.1111/j.1365-2117.2003.00219

Stanford S.D., Ashley, G. M., and Brenner G.J., 2001a, Late Cenozoic fluvial stratigraphy of the New Jersey piedmont; a record of glacioeustasy, planation, and incision on a low-relief passive margin: Journal of Geology, v. 109, p. 265–276.

Stanford, S.D., Ashley, G.M., Russell, Emily W.B., and Brenner, G. J., 2002, Rates and patterns of late Cenozoic denudation in the northernmost Atlantic Coastal Plain and Piedmont: Geological Society of America Bulletin, v. 114, p. 1422–1437

Starkel, L., 2003, Climatically controlled terraces in uplifting mountain areas: Quaternary Science Reviews, v. 22, p. 2189–2198.

Stein, R.S., Barka, A.A., and Dieterich, J.H., 1997, Progressive failure on the North Anatolian fault since 1939 by earthquake stress triggering: Geophysical Journal International, v. 128, p. 594–604.

Stein, R.S., and Barrientos, S.E., 1985, Planar high-angle faulting in the Basin and Range, geodetic analysis of the 1983 Borah peak, Idaho, Earthquake: Journal of Geophysical Research, v. 90, p. 11,355–11,366.

Stein, R.S., and Bucknam, R.C., 1986, Quake replay in the Great Basin, Natural History, 95, p. 28–35. http://quake.wr.usgs.gov/research/deformation/modeling/papers/naturalhistory/naturalhistory.html

Stein, R.S., and Ekstrom, G., 1992, Seismicity and geometry of a 110-km-long blind thrust fault: 2, Synthesis of the 1982-1985 (Coalinga) California earthquake sequence: Journal of Geophysical Research, v. 97, p. 4865-4883.

Stein, R.S., King, G.C.P., and Rundle, J.B., 1988, The growth of geological structure by repeated earthquakes; field examples of continental dip-slip faults: Journal of Geophysical Research, v. 93, p. 13,319–13,331.

Stein, R.S., and Yeats, R.S., 1989, Hidden earthquakes: Scientific American, v. 260, p. 48-57.http://quake.wr.usgs.gov/research/deformation/modeling/papers/

scientam/scientam.html

Stephenson, W.J., and Kirk, R.M., 1998, Rates and Patterns of Erosion on Intertidal Shore Platforms, Kaikoura Peninsula, South Island, New Zealand: Earth Surface Processes and Landforms, v. 23, p.1071–1085.

Stephenson, W.J., and Kirk, R.M., 2000a, Development of shore platforms on Kaikoura Peninsula, South Island, New Zealand, Part One, The role of waves: Geomorphology, v. 32, p. 21–41.

Stephenson, W.J., and Kirk, R.M., 2000b, Development of shore platforms on Kaikoura Peninsula, South Island, New Zealand, Part Two, The role of subaerial weathering: Geomorphology, v. 32, p. 43–56.

Stephenson, W.J., and Kirk, R.M., 2001, Surface swelling of coastal bedrock on inter-tidal shore platforms, Kaikoura Peninsula, South Island, New Zealand: Geomorphology, v. 41, p. 5–21.

Stewart, I.S., 1996, A rough guide to limestone fault scarps: Journal of Structural Geology, v. 18, p. 1259–1264.

Stewart, G.H., and Veblen, T.T., 1982, Regeneration patterns in southern rata (*Metrosideros umbellata*)–kamahi (*Weinmannia racemosa*) forest in central Westland, New Zealand: New Zealand Journal of Botany, v. 20, p. 55–72.

Stock, J.D., and Montgomery, D.R., 1999, Geologic constraints on bedrock river incision using the stream power law: Journal of Geophysical Research, v. 104, p. 4983–4993.

Strahler, A.N., 1952, Dynamic basis of geomorphology: Geological Society of America Bulletin, v. 63, p. 923–938.

Strahler, A.N., 1950, Equilibrium theory of erosional slopes approached by frequency distribution analysis: American Journal of Science, v. 248, p. 673-696; 800–814.

Strahler, A.N., 1957, Quantitative analysis of watershed geomorphology: American Geophysical Union Transactions, v. 38, p. 913–920.

Strahler, A.N., 1964, Quantitative geomorphology of drainage basins and channel networks: *in* Chow, V.T., editor, Handbook of Applied Hydrology, section 4-11, McGraw-Hill, New York.

Stuiver, M., Reimer, P.J., Bard, E., Beck, J.W., Burr, G.S., Hughen, K.A., McCormac, G., van der Plicht, J., and Spurk, M., 1998, INTCAL98 radiocarbon age calibration, 24,000-0 cal BP: Radiocarbon, v. 40, p. 1041–1083.

Sutherland, R., 1994, Displacement since the Pliocene along the southern section of the Alpine Fault, New Zealand: Geology, v. 22, p. 327–330.

Sutherland, R., and Norris, R.J., 1995, Late Quaternary displacement rate, paleoseismicity, and geomorphic evolution of the Alpine Fault, Evidence from Hokuri Creek, South Westland: New Zealand Journal of Geology and Geophysics, v. 38, p. 419–430.

Swan, F.H., III, Hanson, K.L., Schwartz, D.P., and Knuepfer, P.L.K., 1981, Study of earthquake recurrence intervals on the Wasatch fault at the Little Cottonwood site, Utah: U.S. Geological Survey Open-File Report 81-450, 30 p.

Sykes, L.R., and Nishenko, S.P., 1984, Probabilities of occurrence of large plate rupturing earthquakes for the San Andreas, San Jacinto, and Imperial fault, California, 1983–2003: Journal of Geophysical Research, v. 89, p. 5905–5927.

Sylvester, A.G., 1988, Strike-slip faults.: Geological Society of America Bulletin, v. 100, p. 1,666–1,703.

Sylvester, A.G., and Smith, R.R., 1976, Tectonic transpression and basement-controlled deformation in the San Andreas fault zone, Salton trough, California: American Association of Petroleum Geologists Bulletin, v. 60, p. 74–96.

Talling, P.J., 1998: How and where do incised valleys form if sea level remains above the shelf edge?: Geology: v. 26, p. 87–90.

Taylor, R., 1870, New Zealand and its inhabitants: William Macintosh, London, 713 p.

Thatcher, W., and Lisokowski, M., 1987, Long-term seismic potential of the San Andreas fault southeast of San Francisco, California: Journal of Geophysical Research, v. 92, p. 4771–4784.

Thornes, J.B., and Brunsden, D., 1977, Geomorphology and time: John Wiley, New York, 208 p.

Tibaldi, A., Ferrari L., and Pasquarè G., 1995, Landslides triggered by earthquakes and their relations with faults and mountain slope geometry: an example from Ecuador: Geomorphology, v. 11, p. 215–226.

Tinkler, K., and Wohl, E., 1998a, A primer on bedrock channels, in Tinkler, K., and Wohl, E., editors, Rivers over rock: Fluvial processes in bedrock channels: Geophysical Monograph Series, v. 107, American Geophysical Union, Washington, D.C., p. 1–18.

Tinkler, K., and Wohl, E., 1998b, Field studies of bedrock channels, in Tinkler, K., and Wohl, E.E., editors, Rivers over Rock, Fluvial Processes in Bedrock Channels: Geophysical Monograph Series, v. 107, American Geophysical Union, Washington, D.C., p. 261–277.

Tippett, J.M., and Kamp, P.J.J., 1993, Fission track analysis of the late Cenozoic vertical kinematics of continental Pacific crust, South Island, New Zealand: Journal of Geophysical Research, v. 98, p. 16,119–16,148.

Toksoz, M.N., Arpat, E., and Saroglu, F., 1977, East Anatolian earthquake of 24 November 1976: Nature, v. 270, p. 423–425.

Toppozada, T.R., and Borchardt, G., 1998, Re-evaluation of the 1836 "Hayward Fault" earthquake and the 1838 San Andreas Fault earthquake: Bulletin of the Seismological Society of America, v. 88, 140–159.

Treiman, J.A., Kendrick, K.J., Bryant, W.A., Rockwell, T.K., and McGill, S.F., 2002, Primary surface 23 rupture associated with the Mw 7.1 16 October 1999 Hector Mine earthquake, San Bernardino County, California: Bulletin of the Seismological Society of America, v. 92, p. 1171–1191.

Tsai, Y.B., and Aki, K., 1969, Simultaneous determination of the seismic moment and attenuation of seismic surface waves: Bulletin of the Seismological Society of America, v. 59, p. 275–287.

Tsutsumi, H., and Yeats, R.S., 1999, Tectonic setting of the 1971 Sylmar and 1994 Northridge earthquakes in the San Fernando Valley, California: Seismological Society of America Bulletin, v. 89, p. 1232–1249.

Tsutsumi, H., Yeats, R.S., and Huftile, G.J., 2001, Late Cenozoic tectonics of the northern Los Angeles fault system, California: Geological Society of America Bulletin, v. 113, p. 454–468.

Tucker, A.Z., and Dolan, J.F., 2001, Paleoseismologic evidence for a >8 ka age of the most recent surface rupture on the eastern Sierra Madre fault, northern Los Angeles metropolitan region, California: Seismological Society of America Bulletin, v. 91, p. 232–249.

Tucker, G.E., and Bras, R.L., 1998, Hillslope processes, drainage density, and landscape morphology: Water Resources Research, v. 34, p. 2751–2764.

Tucker, G.E., and Slingerland, R., 1994, Erosional dynamics, flexural isostacy, and long-lived escarpments: a numerical modeling study: Journal of Geophysical Research, v. 99, 12229–12244.

Turko, J. M., and Knuepfer, P.L.K., 1991, Late Quaternary fault segmentation from analysis of scarp morphology Geology: v. 19, p. 718–721

Unruh, J., 2001, Seismic hazards associated with blind thrusts in the San Francisco Bay area: California Geological Survey Bulletin, v. 210, p. 211–228.

Unruh, J.R., Hauksson, E., Monastero, F.C., Twiss, R.J., and Lewis, J.L., 2002, Seismotectonics of the Coso Range-Indian Wells Valley region, California: Transtensional deformation along the southeastern margin of the Sierran microplate, in Glazner, A.F., Walker, J.D., and Bartley, J.M., editors, Geological evolution of the Mojave Desert and southwestern Basin and Range: Geological Society of America Memoir 195, p.

277–294.

Unruh, J.R., and Moores, E.M., 1992, Quaternary blind thrusting in the southwestern Sacramento Valley, California: Tectonics, v. 11, p. 192-203.

Upton, P., Koons, P.O., and Chamberlain, C.P., 1995, Penetration of deformation-driven meteoric water into ductile rocks; Isotopic and model observations from the Southern Alps, New Zealand: New Zealand Journal of Geology and Geophysics, v. 38, p. 535–543.

U.S. Geological Survey, 1971, The San Fernando, California, earthquake of February 9, 1971: U.S. Geological Survey Professional Paper 733, 254 p.

U.S. Geological Survey, 1996, USGS Response to an Urban earthquake, Northridge 1994: U.S. Geological Survey Open-File Report 96-263, 68 p.

Vacher, H.L., 1999, Computational geology 8, the power function: Journal of Geological Education, v. 47, p. 473–481.

Vandenberghe, J., 2003, Climate forcing of fluvial system development, an evolution of ideas: Quaternary Science Reviews, v. 22, p. 2053–2060.

van der Beek, P., Summerfield M.A., Braun J., Brown R.W., and Fleming A., 2002, Modeling postbreakup landscape development and denudational history across the southeast African (Drakensberg Escarpment) margin: Journal of Geophysical Research, v. 107 (B12): 2350. doi:10.1029/ 2001JB000744

Van Dissen, R.J., 1989, Late Quaternary faulting in the Kaikoura region, southeastern Marlborough, New Zealand: Oregon State University, M.S. Geology thesis, 72 p.

Van Dissen, R.J., and Berryman, K.R., 1996, Surface rupture earthquakes over the last 100 years in the Wellington region, New Zealand, and implications for ground shaking hazard: Journal of Geophysical Research, v. 101, p. 5999–6019.

Van Dissen, R.J., and Yeats, R.S., 1991, Hope fault, Jordan thrust, and uplift of the Seaward Kaikoura Range: Geology, v. 19, p. 393–396.

Vassallo, R., Ritz, J.F., Braucher, R., and Carretier, S., 2005, Dating faulted alluvial fans with cosmogenic ^{10}Be in the Gurvan Bogd mountain range (Gobi-Altay, Mongolia), climatic and tectonic implications: Terra Nova, v.17, p. 278–285. doi:10.1111/ j.1365-3121.2005.00612.x

Vincent, K.R., 1995, Implications for models of fault behavior from earthquake surface-displacement along adjacent segments of the Lost River Fault, Idaho: University of Arizona Ph.D. Dissertation, Tucson, AZ, 152 p.

Vincent, K.R., and Bull, W.B., 1990, Patterns of Holocene surface-rupture displacement along two contiguous segments of the Lost river fault, Idaho: U.S. Geological Survey Contract Report 14-08-0001-G-1525, Office of Earthquakes, Volcanoes, and Engineering, Menlo Park, California, 36 p.

Vincent, K.R., Bull, W.B., and Chadwick, O.A., 1994, Construction of a soil chronosequence using the thickness of pedogenic carbonate coatings: Journal of Geological Education, v. 42, p. 316–324.

Vita-Finzi, C., and King, G.C.P., 1985, The seismicity, geomorphology, and structural evolution of the Corinth area of Greece: Philosophical Transactions of the Royal Society of London, v. A314, p. 379–407.

Vittoz, P., Stewart, G.H., And Duncan, R.P., 2001, Earthquake impacts on old-growth *Nothofagus* forests, north-west South Island, New Zealand: Journal of Vegetation Science, v. 12, p. 417–426.

von Blanckenburg, F., 2006, The control mechanisms of erosion and weathering at basin scale from cosmogenic nuclides in river sediment: Earth and Planetary Science Letters, v. 242, p. 224–239.

Vry, J.K., A.C., Storkey, and Harris, C., 2001, Role of fluids in the metamorphism of the Alpine Fault Zone: Journal of Metamorphic Geology, v. 19, p. 21– 31.

Wahrhaftig, C., 1965, Stepped topography of the southern Sierra Nevada: Geological Society of America Bulletin, v. 76, p. 1165–1190.

Wakabayashi, J., and Smith, D.L., 1994, Evaluation of recurrence intervals, characteristic earthquakes and slip rates associated with thrusting along the Coast Range-Central Valley geomorphic boundary, California: Bulletin of the Seismological Society of America, v. 84, p. 1960–1970.

Walker, J., Douglas, Kirby, E., and Andrew, J.E., 2005, Strain transfer and partitioning between the Panamint Valley, Searles Valley, and Ash Hill fault zones, California: Geosphere, v. 1: p. 111-118. doi:10.1130/ GES00014.1

Wallace, R.E., 1968, Notes on stream channels offset by the San Andreas fault, southern Coast Ranges, California, *in* Dickinson, W.R., and Grantz, Arthur, editors, Proceedings of conference on geologic problems of San Andreas fault system: Stanford University Publications in Geological Science, v. 11, p. 6–21.

Wallace, R.E., 1970, Earthquake recurrence intervals on the San Andreas fault: Bulletin of the Geological Society of America, v. 81, p. 2875–2889.

Wallace, R.E., 1977, Profiles and ages of young fault scarps, north-central Nevada: Bulletin of the Geological Society of America, v. 88, p. 1267–1281.

Wallace, R.E., 1978, Geometry and rates of change of

fault-generated range fronts, north-central Nevada: U.S. Geological Survey Journal of Research, v. 6, p. 637–649.

Wallace, R.E., 1980, Discussion--Nomograms for estimating components of fault displacement from measured height of fault scarp: Bulletin of the Association of Engineering Geologists, v. 17, p. 39–45.

Wallace, R.E., 1984, Patterns and timing of late Quaternary earthquakes in the Great Basin Province and relation to some regional tectonic features: Journal of Geophysical Research, v. 89, p. 5763–5769.

Wallace, R.E., 1987a, Grouping and migration of surface faulting and variations in slip rates on faults in the Great Basin Province: Seismological Society of America Bulletin, v. 77, p. 868–876.

Wallace, R.E., 1987b, A perspective of paleoseismology: U.S. Geological Survey Open-File Report 87–673, p. 7–16.

Wardle, P., 1980, Primary succession in Westland National Park and its vicinity, New Zealand: New Zealand Journal of Botany, v. 18, p. 221–232.

Warr, L.N., and Cox, S., 2001, Clay mineral transformations and weakening mechanisms along the Alpine Fault, New Zealand: in The Nature and Tectonic Significance of Fault Zone Weakening, R. E. Holdsworth et al., editors: Geological Society Special Publication 186, p. 85–101.

Watters, R.J., and Prokop, C., 1990, Fault-scarp dating utilizing soil strength behavior techniques: Bulletin of the Association of Engineering Geologists, v. 27, p. 291–301.

Weber, G.E., and Cotton, W.R., 1981, Geologic investigation of recurrence intervals and recency of faulting along the San Gregorio fault zone, San Mateo County, California: U.S. Geological Survey open-File Report 81-263, 99 p.

Webb, R.H., 1996, Grand Canyon, a century of change: University of Arizona Press, Tucson, 290 p.

Webb, R.H., Melis, T.S., Wise, T.W., and Elliott, J.G., 1996, "The Great Cataract," The effects of late Holocene debris flows on Lava Falls Rapid, Grand Canyon National Park, Arizona: U.S. Geological Survey Open-file Report 96-460, 96 p.

Webb, R.H, Melis, T.S., Griffiths, P.G., Elliott, J.G., Cerling, T.E, Poreda, R.J., Wise, T.W., and Pizzuto, J.E., 1999, Lava Falls Rapid in Grand Canyon; Effects of late Holocene debris flows on the Colorado River: U.S. Geological Survey Professional Paper 1591, 90 p.

Webb, R.H., Pringle, P.T., and Rink, G.R., 1989, Debris flows in tributaries of the Colorado River in Grand Canyon National Park, Arizona: U.S. Geological Survey Professional Paper 1492, 39 p.

Weber, G.E., and Cotton, W.R., 1981, Geologic investigation of recurrence intervals and recency of faulting along the San Gregorio fault zone, San Mateo County, California: U.S. Geological Survey Open-File Report 81–263.

Wegmann, K.W., and Pazzaglia, F.J., 2002, Holocene strath terraces, climate change, and active tectonics; The Clearwater River basin, Olympic Peninsula, Washington State: Geological Society of America Bulletin, v. 114, p. 731–744.

Weissel, J.K., and Seidl, M.A., 1997, Influence of rock strength properties on escarpment retreat across passive continental margins: Geology, v. 25, p. 631–634.

Weldon, R.J., and Sieh, K.E., 1985, Holocene rate of slip and tentative recurrence interval for large earthquakes on the San Andreas fault in the Cajon Pass, southern California: Geological Society of America Bulletin, v. 96, p. 793–812.

Wells, A. 1998. Landscape disturbance history in Westland, New Zealand: the importance of infrequent earthquakes along the Alpine fault: Lincoln University Ph.D thesis.

Wells, A., and Goff, J.R., 2006, Coastal dune ridge systems as chronological markers of paleoseismic activity: a 650 year record from southwest New Zealand: The Holocene, v. 16, no. 4.

Wells, A., Duncan, R.P., and Stewart, G.H., 2001, Forest dynamics in Westland, New Zealand, the importance of large, infrequent earthquake-generated disturbance: Journal of Ecology, v. 89, p. 1006–1018.

Wells, A., and Yetton, M., 2004, Earthquake tree-ring impacts in the middle and upper Buller River catchment: New Zealand Earthquake Commission Research Report 03/492, 39 p.

Wells, A., Yetton, M.D., Duncan, R.P., and Stewart, G.H., 1999, Prehistoric dates of the most recent Alpine fault earthquakes, New Zealand: Geology, v. 27, p. 995–998.

Wells, D.L., and Coppersmith, K.J., 1994, New empirical relationships among magnitude, rupture length, rupture width, rupture area, and surface displacement: Bulletin Seismological Society of America, v. 84, p. 974–1002.

Wentworth, C.M., and Zoback, M.D., 1989, The style of Late Cenozoic deformation at the eastern front of the California Coast Ranges: Tectonics, v. 8, p. 237–246.

Wentworth, C.M., Zoback, M.D., and Bartow, J.A., 1983, Thrust and reverse faults beneath the Kettleman Hills anticlinal trend, Coalinga earthquake region, California, inferred from deep seismic-reflection data:

EOS, Transactions, American Geophysical Union, v. 64, p. 747.

Wernicke, B., 1981, Low-angle normal faults in the basin and range province: nappe tectonics in an extending orogen: Nature, v. 291, p. 645–647.

Wernicke, B., 1992, Cenozoic extensional tectonics of the U.S. Cordillera, in Burchfiel, B.C., Lipman, P. W., and Zoback, M.L., editors, The Cordilleran Orogen: Conterminous U.S., The Geology of North America, Geological Society of America, Boulder, Colorado, v. G-3, p. 553–581.

Wernicke, B.P., 1995, Low-angle normal faults and seismicity: A review: Journal of Geophysical Research, v. 100, p. 20,159–20,174.

Wernicke, B.P., and Axen, G.J., 1988, On the role of isostasy in the evolution of normal fault systems: Geology, v. 16, p. 848–851.

Wernicke, B., Clayton, R., Ducea, M., Jones, C.H., Park, S., Ruppert, S., Saleeby, J., Snow, J.K., Squires, L., Fliedner, M., Jiracek, G., Keller, R., Klemperer, S., Luetgart, J., Malin, P., Miller, K., Mooney, W., Oliver, H., and Phinney, R., 1996, Origin of high mountains in the continents: The southern Sierra Nevada: Science, v. 271, p. 190–193.

Wernicke, B., and Snow, J.K., 1998, Cenozoic tectonism in the central Basin and Range; motion of the Sierran–Great Valley block: International Geology Review, v. 40, p. 403–410.

Wesnousky, S.G., 1986, Earthquakes, Quaternary faults, and seismic hazard in California: Journal of Geophysical Research, v. 91, p. 12,587–12,631.

Wetmore, C.M., 1986, Lichens and air quality in Sequoia National Park: National Park Service Contract Report CX0001-2-0034.

Wheeler, R., Crone, A., and Omdahl, E., 1987, Boundaries between segments of normal faults; criteria for recognition and interpretation: U.S. Geological Survey Open-File Report 87-673, p. 385–398.

Whipple K.X., 2004, Bedrock rivers and the geomorphology of active orogens: Annual Review of Earth and Planetary Sciences, v. 32, p. 151–185.

Whipple, K.X, and Tucker, G.E., 1999, Dynamics of the stream-power river incision model: implications for height limits of mountain ranges, landscape response time scales, and research needs: Journal of Geophysical Research, v. 104, p. 17,661–17,674.

Whitehouse, I.E., and McSaveney, M.J., 1983, Diachronous talus surfaces in the Southern Alps, New Zealand, and their implications to talus accumulations: Arctic and Alpine Research, v. 15, p. 53–64.

Whitehouse, I.E., McSaveney, M.J., Knuepfer, P.L.K., and Chinn, T.J., 1986, Growth of weathering rinds on torlesse sandstone, Southern Alps, New Zealand, in S.M. Colman and D.P. Dethier editors, Rates of chemical weathering of rocks and minerals: Academic Press, New York, p. 419–435.

Wieczorek, G.F., Snyder, J.B., Alger, C.S., and Isaacson, K.A., 1992, Rock falls in Yosemite Valley, California: U.S. Geological Survey Open-File Report 92-387, 38 p., 2 appendixes, 4 plates.

Wieczorek, G.F., and Jäger, S., 1996, Triggering mechanisms and depositional rates of postglacial slope-movement processes in the Yosemite Valley, California, Geomorphology, v. 15, p. 17–31.

Wieczorek, G.F., Morrissey, M.M., Iovine, G., and Godt, J., 1998, Rock-fall hazards in the Yosemite Valley: U.S. Geological Survey Open-file Report 98-467, 1:12,000, 7 p. http://pubs.usgs.gov/of/1998/ofr-98-0467/

Wieczorek G.F., and Snyder, J.B., 1999, Rockfalls from Glacier Point above Camp Curry, Yosemite National Park, California: U.S. Geological Survey Open-file Report 99-385, http://pubs.usgs.gov/of/1999/ofr-99-0385/.

Wieczorek, G.F., and Snyder, J.B., 2003, Historical rockfalls in Yosemite National Park, California: U.S. Geological Survey Open-File Report 03-491, Online Version 1.0, http://pubs.usgs.gov/of/2003/of03-491/.

Wieczorek, G.F., Snyder, J.B., Waitt, R.B., Morrissey, M.M., Uhrhammer, R., Harp, E.L., Norris, R.D., Bursik, M. I., and Finewood, L.G., 2000, The unusual air blast and dense sandy cloud triggered by the July 10, 1996, rockfall at Happy Isles, Yosemite National Park, California: Geological Society of America Bulletin, v. 112, n. 1, p. 75–85.

Willett, S.D., Slingerland, R., and Hovius, N., 2001, Uplift, shortening, and steady state topography in active mountain belts: American Journal of Science, v. 301, p. 455–485.

Wills, C.J., and Borchardt, G., 1993, Holocene slip rate and earthquake recurrence on the Honey Lake fault zone, northeastern California: Geology, v. 21, p. 853–856.

Wischmeier, W.H., and D.D. Smith, 1965, Predicting rainfall-erosion losses from cropland east of the Rocky Mountains: U.S. Department of Agriculture Handbook 282, Washington, D.C.

Wischmeir, W.H., and D.D. Smith, 1978, Predicting rainfall erosion losses–A guide to conservation planning: U.S. Department of Agriculture Handbook 537, Washington D.C.

Wohl, E., 1994, Energy Expenditure in deep, narrow bedrock canyons: Geological Society of America Abstract with Programs, v. 26, p. 233–234.

Wohl, E.E., and Merritt, D.M., 2001, Bedrock channel morphology: Geological Society of America Bulletin, v. 113, p. 1205–1212.

Wolman, M.G., and Gerson, R., 1978, Relative scales of time and effectiveness of climate in watershed geomorphology: Earth Surface Processes, v. 3, p. 189–208.

Wood, A., 1942, The development of hillside slopes: Geologists Association Proceedings, v. 53, p. 128–140.

Wright, C.A., 1998, Geology and paleoseismicity of the central Alpine Fault, New Zealand: M.Sc. thesis, University of Otago, Dunedin, New Zealand.

Wright, C.A., Norris, R.J., and Cooper, A.F., 1998, Paleoseismic history of the central Alpine Fault, Geological Society of New Zealand Annual Meeting Abstracts.

Wright, L.A., Otton, J. K., and Troxel, B.W., 1974, Turtleback surfaces of Death Valley viewed as phenomena of extensional tectonics: Geology, v. 2, p. 53–54.

Xiao, H., and Suppe, J., 1992, Origin of rollover: American Association of Petroleum Geologists Bulletin, v. 76, p. 509–529.

Xiong, L., 1995, A dendroclimatic study of *Libocedrus bidwillii* Hook. f. (kaikawaka): Lincoln University Ph.D. Dissertation, 295 p.

Xiong, L., and Palmer J.G., 2000, Reconstruction of New Zealand temperatures back to AD 1720 using *Libocedrus bidwillii* tree-rings: Climatic Change, v. 45, p. 339–359.

Yair, A., Sharon, D., and Lavee, H., 1978, An instrumented watershed for the study of partial area contribution of runoff in the arid zone: *in* Field Instrumentation and Geomorphological Problems, O. Slaymaker, A. Rapp, T. Dunne, editors; Zeitschrift für Geomorphologie, v. 29, 206 p.

Yeats, R.S., 2001, Reverse fault earthquake pairs: San Fernando Valley 1971 and 1994 and Nelson, New Zealand 1929 and 1968: Geological Society of America Cordilleran Section Abstracts, Session No. 32.

Yeats, R., Sieh, K., and Allen, C., 1997, The geology of earthquakes: Oxford University Press, New York, 568 p.

Yetton, M.D., 1998, Progress in understanding the paleoseismicity of the central and northern Alpine fault, Westland, New Zealand: New Zealand Journal of Geology and Geophysics, v. 41, p. 475–483.

Yetton, M.D., 2000, Probability and consequences of the next Alpine Fault earthquake. Ph.D thesis, Department of Geological Sciences, University of Canterbury, New Zealand.

Yetton, M.D., and Nobes, D.C., 1998, Recent vertical offset and near-surface structure of the Alpine Fault in Westland, New Zealand, from ground penetrating radar profiling: New Zealand Journal of Geology and Geophysics: v. 41 p. 485–492.

Yetton, M.D., Wells, A., and Traylen, N.J., 1998, The probability and consequences of the next Alpine fault earthquake: New Zealand Earthquake Commission Research Report 95/193, 193 p.

Yielding, G., Jackson, J.A., King, G.C.P., Sinvahl, H., Vita-Finzi, C., and Wood, R.M., 1981, Relations between surface deformation, fault geometry, seismicity, and rupture characteristics during the El Asnam earthquake of 10 October 1980: Earth and Planetary Science Letters, v. 56, p. 287–304.

Young, A., 1972, Slopes: Oliver and Boyd, Edinburgh/London, 288 p.

Zandt, G., 2003, The southern Sierra Nevada drip and the mantle wind direction beneath the southwestern United States: International Geological Review, v. 45, p. 213–223.

Zandt, G., Gilbert, H., Owens, T.J., Ducea, M., Saleeby, J., and Jones, C.H., 2004, Active Foundering of a Continental Arc Root Beneath the Southern Sierra Nevada, California: Nature, v. 432, p. 41–46.

Zaprowski, B.J., Evenson, E.B., Pazzaglia, F.J., and Epstein, J.B., 2001, Knickzone propagation in the Black Hills and northern High Plains: A different perspective on the late Cenozoic exhumation of the Laramide Rocky Mountains: Geology, v. 29, p. 547–550.

Zaprowski, B.J., Pazzaglia, F.J., and Evenson, E.B., 2005, Climatic influences on profile concavity and river incision: Journal of Geophysical Research, v. 110, F03004, doi:10.1029/2004JF000138.

Zreda, M., and J. Noller, 1998a, Ages of prehistoric earthquakes revealed by cosmogenic ^{36}Cl in a bedrock fault scarp at Hebgen Lake: Science, v. 282, p. 1097–1099.

Zreda, M.G., and J.S. Noller, 1998b, Cosmogenic ^{36}Cl dating of the Hebgen Lake, Montana, fault scarp, *in* Dating and Earthquakes: Review of Quaternary Geochronology and Its Application to Paleoseismology, J.M. Sowers, J.S. Noller and W.R. Lettis editors, 4-39-4-48, Nuclear Regulatory Commission Report NUREG/CR-5562.

Zreda, M.G., and F.M. Phillips, 1998, Quaternary dating by cosmogenic nuclide buildup in surficial materials, *in* Dating and Earthquakes: Review of Quaternary Geochronology and Its Application to Paleoseismology, J.M. Sowers, J.S. Noller and W.R. Lettis editors, 2-101-2-127, Nuclear Regulatory Commission Report NUREG/CR-5562.

Index

Where feasible, this index organizes multiple entries under topics that organize subjects and themes of this book. Examples include "base level", "concepts", "earthquakes", "modeling", "paleoseismic studies", "process-response models" and "reaches".

Semibold italics page numbers (*261*) refer to illustrations. **Bold** page numbers (**236**) refer to tables. Use of – (such as 105–11) flags a continuous discussion of a topic